3/97

Head-Mounted Displays

Head-Mounted Displays

Designing for the User

James E. Melzer

Kirk Moffitt

McGraw-Hill

New York San Francisco Washington, D.C. Auckland
Bogotá Caracas Lisbon London Madrid
Mexico City Milan Montreal New Delhi
San Juan Singapore Sydney
Tokyo Toronto

Library of Congress Cataloging-in-Publication Data

Melzer, James E.
    Head-mounted displays: designing for the user / James
E. Melzer, Kirk Moffitt.
        p.   cm. — (Optical and electro-optical engineering series)
    Includes bibliographical references and index.
    ISBN 0-07-041819-5 (alk. paper)
    1. Helmet-mounted displays—Design and construction. 2. Virtual
reality.   I. Moffitt, Kirk Wayne, 1951–   . II. Title.
III. Series.
TK7882.I6M45 1997   96-49041
621.39'9–dc21   CIP

McGraw-Hill

A Division of The McGraw·Hill Companies

1 2 3 4 5 6 7 8 9 0 DOC/DOC 9 0 1 0 9 8 7 6

ISBN 0-07-041819-5

*The sponsoring editor for this book was Steve Chapman, the editing
supervisor was Bernard Onken, and the production supervisor was
Pamela Pelton. It was set in Century Schoolbook by Publication
Services.*

Printed and bound by R. R. Donnelley & Sons Company

McGraw-Hill books are available at special quantity discounts to use
as premiums and sales promotions, or for use in corporate training
programs. For more information, please write to the Director of Special
Sales, McGraw-Hill, 11 West 19th Street, New York, NY 10011. Or
contact your local bookstore.

This book is printed on recycled, acid free paper
containing a minimum of 50% recycled de-inked fiber.

Contents

Contributors

Preface, Chapter 1: James E. Melzer
Kaiser Electro-Optics
2752 Loker Avenue West
Carlsbad, CA 92008
jmelzer@aol.com

Preface, Chapter 1, Chapter 5: Kirk Moffitt, Ph.D.
79245 Camino del Oro
La Quinta, CA 92253
619/360-0204
macpr@cyberg8t.com

Chapter 2: Robert G. Eggleston, Ph.D.
Fitts Human Engineering Division
Armstrong Laboratory (AL/CFH)
2255 H Street
Wright-Patterson AFB, OH 45433-7022
reggleston@al.wpafb.af.mil

Chapter 3: H. Lee Task, Ph.D.
Fitts Human Engineering Division
Armstrong Laboratory (AL/CFH)
2255 H Street
Wright-Patterson AFB, OH 45433-7022
ltask@al.wpafb.af.mil

Chapter 4: Robert E. Fischer
OPTICS 1, Inc.
3050 Hillcrest Drive, suite 100
Westlake Village, CA 91362
refischer@optics1.com

Chapter 6: Chris E. Perry
Escape and Impact Protection Branch
Armstrong Laboratory (AL/CFBE)
2800 Q Street, Bldg. 824
Wright-Patterson AFB, OH 45433-7022
cperry@al.wpafb.af.mil

Chapter 6: John R. Buhrman
Escape and Impact Protection Branch
Armstrong Laboratory (AL/CFBE)
2800 Q Street, Bldg. 824
Wright-Patterson AFB, OH 45433-7022
jbuhrman@al.wpafb.af.mil

Chapter 7: Jennifer J. Whitestone
Fitts Human Engineering Division
Armstrong Laboratory (AL/CFHD)
2255 H Street
Wright-Patterson AFB, OH 45433-7022
jwhitestone@al.wpafb.af.mil

Chapter 7: Kathleen M. Robinette
Fitts Human Engineering Division
Armstrong Laboratory (AL/CFHD)
2255 H Street
Wright-Patterson AFB, OH 45433-7022
krobinette@al.wpafb.af.mil

Chapter 8: Elizabeth Thorpe Davis, Ph.D.
School of Psychology
Georgia Institute of Technology
Atlanta, GA 30332-0170
ed15@prism.gatech.edu

Chapter 9: David A. Southard, Sc.D.
Charles Stark Draper Laboratory MS-3F
555 Technology Square
Cambridge, MA 02139-3563
southard@draper.com

Chapter 10: Victoria Tepe Nasman, Ph.D.
Logicon Technical Services, Inc.
P.O. Box 317258
Dayton, OH 45437-7258
vnasman@al.wpafb.af.mil

Chapter 10: Gloria Calhoun
Fitts Human Engineering Division
Armstrong Laboratory (AL/CFHP)
2255 H Street
Wright-Patterson AFB, OH 45433-7022
gcalhoun@al.wpafb.af.mil

Chapter 10: Grant R. McMillan, Ph.D.
Fitts Human Engineering Division
Armstrong Laboratory (AL/CFHP)
2255 H Street
Wright-Patterson AFB, OH 45433-7022
gmcmillan@al.wpafb.af.mil

Chapter 11: Jeff Gerth, Ph.D.
Georgia Institute of Technology
Georgia Tech Research Institute
ELSYS/SEV - Human Factors Branch
Atlanta, GA 30332-0840
jeff.gerth@GRTI.gatech.edu

Preface

One of our early experiences with a commercial head-mounted display (HMD) was at an evening reception following a virtual reality conference. Although we had been building military HMDs for several years, this was going to be a new experience. After fortifying ourselves with wine and cheese, it was our turn to view the HMD. We were disappointed. It was front-heavy and uncomfortable. It had a fuzzy appearance, and it did not allow for eyeglasses. When we mentioned this, the man demonstrating the device assured us that it was not the wine and that it would not have made any difference if we *had* worn eyeglasses.

At the same show we had the opportunity to try on a headset that was billed as *the VR HMD*. One of us has a rather large head, and this device did not fit over it. After a quick modification by the vendor, we managed a tight fit, but it was not worth the effort. The imagery was so badly misaligned that viewing it for more than a few moments was painful.

Another experience was a series of meetings we were having with a group of Army flight-safety officers. We wanted them to fly one of our company's HMDs in their helicopter, but first we needed approval from their safety committee. The process took quite some time, because of what we perceived as incessant questions about the most minute details of our design. After a particularly grueling face-to-face session, the meeting broke up and we left with our flight-safety approval. As we were walking out of the building, one of the flight surgeons took us aside. He told us not to take the interrogations personally, because the people we were talking with were the ones responsible for investigating accidents—an unpleasant task, considering how a helicopter crashes.

These three examples show why in designing or buying an HMD we need to understand who the user is, how the user will interact with the display, and what the environment will be like. The first two examples show the results of a lack of this understanding—poorly aligned displays that don't fit. The third shows the results of having that understanding. The flight surgeons learned about the delicate balance between the HMD *as display* and the HMD *as life-support* through their experiences.

An HMD is something you wear *and* something you view. It is a personal device that can provide you with information, train you to do a job by simulating what it would really be like, or entertain you by transporting you to a fantasy world. At the center of these experiences is the human who wears and views the HMD. Properly designed, an HMD can suspend belief sufficiently to train a pilot to fly an airplane or a surgeon to perform a new operation, or transport you to the surface of Mars. Improperly designed, the HMD can be uncomfortable to wear, difficult to use, and even painful to view.

This is not surprising, as it seems to be the fate of many new technologies when first introduced. One example is the early DOS-based computers. To perform a routine task like saving a file to disk, the user had to enter a string of seemingly unrelated and unintelligible characters. This turned off some people, confused others, and convinced still more that the personal computer was not a solution for everyday tasks.

Early HMDs took a similar path. It was thought that a display on the head was simply that—glass and electronics mounted in front of the eyes, with no serious regard given to what was really needed by the user. Early designers were rushing toward a vision of virtual and interactive imagery, and they placed their emphasis on the technology, not on the user. The result was displays that were uncomfortable to wear and difficult to use. HMDs have received a lot of publicity recently—some good as a result of excellent new applications, and some bad as a result of poor designs that were poorly implemented.

It is for all of these reasons that we decided to focus this book on the fundamental needs of the user. We know that the technology will improve over the next few years—we have seen it change just during the writing of this book—but the human who wears the HMD will not appreciably change over the next several millenia. If we understand what these fundamental needs are, we can take the developments in technology, implement them in our designs, and provide an HMD that will benefit the user. There will still be trade-offs to be made as technology improves, but understanding the user's essential needs will help us make intelligent decisions.

This book is a compilation of the many subjects that relate to the design of HMDs. It is by its nature a multidisciplinary discussion, because to adequately address the needs of the user, we must cross numerous behavioral, psychological, performance, and anthropometric boundaries. The authors of the chapters are experts in their fields with academic, commercial, and military backgrounds and we thank them for their fine work. We hope that this book will benefit both users and designers of HMDs.

We would like to extend our thanks to Kaiser Electronics for support during Kirk Moffitt's tenure with the company, and to Kaiser Electro-Optics for continued support of Jim Melzer. Finally, we would like to thank Warren Smith for his support and guidance during the preparation of this book.

HMD Design—
Putting the User First

James E. Melzer

Kirk Moffitt

The head-mounted display (HMD) is a critical link in virtual-environment and visually coupled systems. HMD users can experience immersion in computer-generated virtual environments, privately view a movie, perform a delicate endoscopic surgical procedure, or fly an attack helicopter nap-of-the-earth in darkness. The success of these tasks depends on the design of the HMD system. Given the intimate interface to the human, the user should be the central focus of the design process. An HMD will be successful only if full consideration is given to the characteristics and tasks of the user.

1.1   The Richness of an HMD

The head-mounted display (HMD) provides the user with a set of capabilities that conventional displays cannot duplicate. An HMD can be personal, interactive, expansive, *and* virtual. Handheld televisions and video games, personal computer monitors, panoramic theater screens, and head-up displays share one or two of these attributes at most. Only an HMD provides the user with an intimate display that can be reactive to head and body movement and surround him or her with a virtual environment that extends far beyond the confines of the miniature image source.

Unlike televisions, computer monitors, and movie screens, which usually vary only in size, HMDs come in many types that accommodate a wide range of uses. An HMD can be any of the following:

- A simple reticle projector that a pilot uses to designate an enemy aircraft

- A more thorough symbology display that gives the pilot orientation and status information

- A small offset display that a technician can glance at for reference data

- A private view of a selected movie by an airline passenger

- Stereo imagery relayed from head-steered cameras located on a remote vehicle

- A computer-generated, panoramic world that can be navigated with simple movements and gestures

This wealth of applications makes a book on HMD design worthwhile. It is not our intent to provide a formula for building each variation, but rather to engage the reader in a discussion of fundamental HMD design concepts that center on the characteristics and capabilities of the user. The chapters in this book cover topics as diverse as fitting HMDs to human heads, perceptual requirements of HMDs, and incorporating brain-actuated control into HMDs. The common thread is the need to put the user at the center of the design process.

1.2   What Is an HMD?

In its simplest form, an HMD consists of an image source and collimating optics in a head mount (see Fig. 1.1). The HMD can then become more elaborate in several ways. There may be one or two display channels. These channels may display graphics and symbology with or without video overlay. They may be viewed directly and occlude external vision for a fully immersive experience, or they may use a

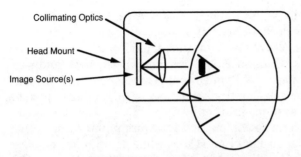

Figure 1.1 An HMD consists of an image source, collimating optics, and a head mount—providing a virtual image to the wearer.

semitransparent combiner with see-through to the outside world. In this augmented reality mode, the HMD may overlay symbology or other information on the world view.

The HMD image source may be a CRT or LCD mounted on the head, or the image may be brought up to the head through a fiberoptic bundle. An HMD may use a simple headband for mounting on the head, or the optics and the displays may be integrated into an aviator's flight helmet. This latter device is a specialized case of the head-mounted display—the *helmet-mounted display*.

The HMD is part of a larger system that can include an image generator, a head tracker, audio, and a manual input device (see Fig. 1.2). The image generator may be a sophisticated image rendering engine or a personal computer. A tracker, which communicates the orientation of the user's head to the image generator, immerses the user in a virtual environment. This immersion is often enhanced by 3D or directional audio. Input/output devices can include brain- or voice-actuated control, a joystick, and a 3D mouse or glove to manipulate virtual objects.

Properly designed, this deceptively simple arrangement of optics and electronics can fit comfortably and be worn for several hours. It can instruct you in new ideas, provide important information, or transport

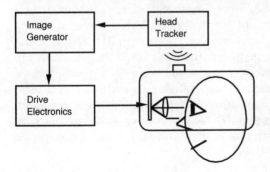

Figure 1.2 An HMD is part of a visually coupled display system consisting of an image generator, drive electronics, and a head tracker.

you to alternative realities. Improperly designed, an HMD can quickly strain your eyes, your neck, or your stomach with symptoms that can last for several hours. Why? Our evolutionary development occurred in the natural world, with the neck supporting only the head and with visual imagery being perfectly aligned and correspondent. HMDs can place a burden on the neck, can easily be misaligned, and may put the visual and vestibular systems into conflict.

These are many diverse, complex, and interdisciplinary issues that are associated with the design of an HMD. Some are subtle; some are obvious. All are centered on the needs of the users, who will be wearing the device, and the tasks they are performing. If we ignore these needs, we will produce an HMD that can have negative side effects that will not be accepted. Only by understanding the users and treating them as an integral part of the process can we assure a successful design.

1.3 Early HMDs

In the 1960s Ivan Sutherland married CRTs with focusing optics, head orientation hardware, and an early computer image generator to produce the first HMD system. Because of the weight, Sutherland suspended it from the ceiling, from which it got the name "Sword of Damocles" (Rheingold, 1991). Sutherland used this system to conduct early experiments in virtual environments and HMD-based stereo-vision with a wide-field-of-view binocular display.

The U.S. military briefly experimented with helmet-mounted sights in the 1970s. The Visual Target Acquisition System, or VTAS, was a simple HMD used to aim air-to-air missiles. It was a lightweight device that attached to a standard flight helmet, and it reflected light from a series of light-emitting diodes off a special visor. Unfortunately, the VTAS was abandoned because of the limitations in missile technology at that time (Dornheim, 1995b).

Improvements in HMD technology and their applications proceeded slowly until the mid 1980s, when there were four key HMD developments.

First, researchers at NASA Ames Research Center developed their Virtual Interactive Environment Work Station (VIEW), consisting of a wide-field-of-view (WFOV), head-tracked HMD with 3D virtual sound connected to a computer image generator. This HMD design, shown in Fig. 1.3, used two commercial LCD image sources with projector lenses, and it looked like an oversized scuba mask. It provided the wearer with visual and audio feedback in a WFOV immersive environment, but with low resolution. This HMD design became synonymous with the emerging technology of virtual reality, and variations of this design were sold through the mid 1990s.

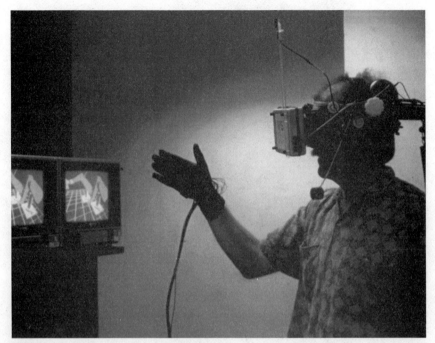

Figure 1.3 Early virtual reality HMD developed by NASA for the Virtual Interactive Environment Work Station (*photo courtesy of NASA Ames Research Laboratory*).

Second, the United States Air Force sponsored the Super Cockpit program under the direction of Tom Furness at Wright-Patterson Air Force Base. A key Super Cockpit development was the Visually Coupled Airborne Systems Simulator (VCASS), sometimes called "The Bug That Ate Dayton" for its mantis-like appearance. This device, shown in Fig. 1.4, provided an extremely large field of view and gave the Air Force its first insights into HMDs for flight simulation. Perhaps the most important outcome was to establish Wright-Patterson as a center of excellence in understanding the human factors of HMDs.

Third, the U.S. Army deployed the Honeywell Integrated Helmet and Display Sighting System (IHADSS) on the AH-64 Apache helicopter, the first operational HMD for military applications. The IHADSS monocular HMD, shown in Fig. 1.5, is mounted on the side of the pilot's helmet and is linked to the infrared sensors and weapons. It provides a monochrome image with user-accessible adjustments. More important, it was attached to a helicopter flight helmet and was certified by the Army as crash-worthy. Through its experience with the IHADSS, the Army has recognized the value of the HMD to the pilot for navigation and weapons delivery. Its new RAH-66 Comanche helicopter, due for

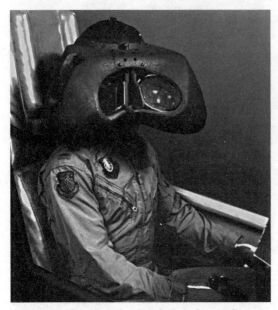

Figure 1.4 The Visually Coupled Airborne Systems Simulator (VCASS), developed by the Air Force (*photo courtesy of the Armstrong Laboratory Human Engineering Division, Wright-Patterson Air Force Base, Ohio*).

Figure 1.5 The Integrated Helmet and Display Sighting System (IHADSS) for the Army's AH-64 Apache helicopter (*photo courtesy of Honeywell Military Avionics, Minneapolis, Minnesota*).

Figure 1.6 The Agile-Eye™ HMD (*photo courtesy of Kaiser Electronics, San Jose, California*).

deployment after the turn of the century, will carry a new binocular HMD.

Fourth, the U.S. Air Force sponsored the VISTA SABRE program. This was a series of simulations that picked up where the VTAS program left off in the 1970s, linking an HMD (the Kaiser Electronics Agile-Eye™, shown in Fig. 1.6) with off-bore-sight missiles for air-to-air combat scenarios (Arbak, 1989). These simulations demonstrated the significant advantage a pilot would have over a non-HMD-equipped adversary. This technology has recently been produced by Russian and Israeli avionics companies (Dornheim, 1995a), and a monocular-symbology HMD will find its way onto a domestic high-performance fighter aircraft within the next few years.

These four HMD systems had an important impact on the development of HMD technology by demonstrating their usefulness in a variety of commercial and military applications. Whereas they were sophisticated by the standards of the day, the technology is advancing rapidly, so current systems are appearing more like visors and headbands, as shown in Fig. 1.7. Future developments in both commercial and military HMDs will benefit from similar developments in the following areas:

- Advances in optical design
- High-resolution matrix image sources with high-speed digital display electronics

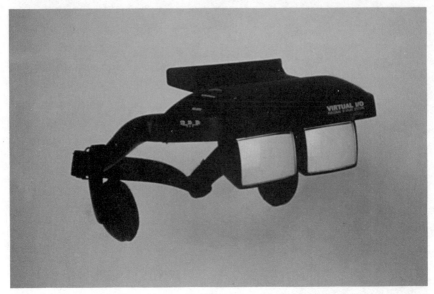

Figure 1.7 New, lightweight designs for HMDs resemble a pair of sunglasses (*"i-glasses"* *photo courtesy of Virtual i-O, ©1995 Virtual i-O, all rights reserved*).

- Head trackers that are fast, accurate, and sourceless
- Compact and inexpensive image generators

1.4   User Requirements

Designs for military and commercial HMDs are challenged by similar problems:

- The need for wide-field-of-view *and* high-resolution imagery
- The need to maintain image alignment of a complex electro-optical system
- The need to fit a wide range of head shapes and sizes
- The need to minimize head-supported weight for comfort and safety

These are primarily hardware requirements, most of which will improve as the technology grows. A more fundamental problem in designing HMDs is the lack of specifications—accepted numerical values— that bound the limits of human performance. Besides a small set of commonly held rules of thumb, the human factors database for HMD design is simply inadequate. Even military specifications, while suitable for human-machine interfaces in conventional displays, are insufficient to address head-supported weight, head-tracking update and lag, and FOV requirements for various tasks.

A variation of the "human-factored" approach is to design a "human-centered" HMD system, where the HMD has been designed from the perspective of the user to support the roles and tasks of the user (Riley, 1995; Rouse, 1991). This methodology differs from the purely engineering approach, which looks at solutions from a technology-first perspective—an approach that commonly results in interesting but unusable designs.

It is useful to consider the user as part of the total HMD system—what Gibson (1986) calls a *perceptual system,* as illustrated in Fig. 1.8. If we start at the bottom of the hierarchy in Fig. 1.8, we consider what the monocular eye needs in a display—sufficient resolution, contrast, luminance, and focus to see the information. We also know that this monocular eye has a limited range of motion (Bahill, Adler, and Stark, 1975). Moving up the hierarchy, we see that the human is more than just two monocular eyes. We know they function as a binocular pair, with the monocular and binocular functions linked together, driven by the visual environment and the observer's intent (Moffitt, 1989) and actions that impose requirements on the alignment and quality of the binocular imagery.

Further up on our hierarchy, we have the binocular eyes on a moving head. We know that the nose and the forehead impose limitations on the extent of the visual field, dividing it into a central binocular field with

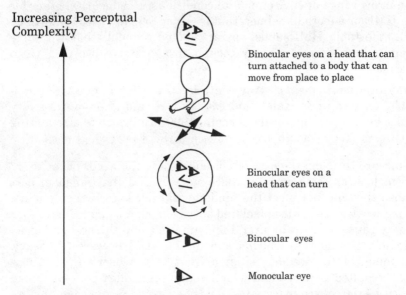

Increasing Perceptual Complexity

Binocular eyes on a head that can turn attached to a body that can move from place to place

Binocular eyes on a head that can turn

Binocular eyes

Monocular eye

Figure 1.8 "The eye is part of a dual organ, one of a pair of mobile eyes, and they are set in a head that can turn, attached to a body that can move from place to place. These organs make a hierarchy and constitute what I have called a perceptual system" (Gibson, 1986, p. 53).

flanking monocular fields (Melzer and Moffitt, 1991). We also know that in searching for objects, the head and eyes work together through coordinated gaze movements. (Robinson, Koth and Ringenbach, 1976).

At the top of our hierarchy, we have the binocular eyes that move together on a head that moves on a body that moves in space. This mobile head imposes system requirements for how smoothly and accurately the displayed imagery is updated. All these individual parts of the human as an observer within the world work together to provide information about the world around us.

With this understanding, we can identify four fundamental areas that must be satisfied in the HMD design or selection process:

Visual requirements.   These are threshold visual requirements of contrast, luminance, focus and alignment, color, and resolution. They are derived from a fundamental understanding of how the human visual system functions and from a knowledge of the ambient viewing conditions.

Physical requirements.   Humans come in a surprisingly large range of sizes with low correlations between head dimensions. Critical issues of anthropometry, head and neck center of gravity, and comfort must be addressed.

Environmental requirements.   Humans function comfortably within a narrow range of environmental conditions. Headgear, whether for an HMD or a football helmet, that is either too hot or too cold will be uncomfortable. HMDs must operate in environments that can range from a high-g-force military cockpit to aggressive use in a video arcade.

Interface requirements.   Any controls on the HMD must be accessible, easy to understand, and easy to use, and their functionality must be properly mapped (Norman, 1990). Interface to a rendering engine must be smooth and seamless, without lag or jerkiness.

What are the consequences of not addressing the needs of the user? We can look at some early virtual-reality HMDs that failed in part because they did not meet the fundamental needs for image quality. The picture was either too pixelated or too blurry, misaligned binocular imagery caused eyestrain after a few minutes, and the system lagged noticeably behind the head movement. The HMDs were too heavy, with much of the weight being in front of the user's eyes. These factors resulted in user dissatisfaction and discomfort plus much bad publicity—the consequence of technology-centered rather than human-centered design.

Negative side effects can result partially from poor HMD design and partially from an incomplete understanding of how humans and

HMDs interact (Peli, 1990; 1995). These range from cybersickness, or simulator sickness (Regan, 1993; Regan & Price, 1994; Kennedy et al., 1993), to visual stress (Miyashita and Uchida, 1990), to dissociation of the accommodation-vergence response (Mon-Williams, Wann, and Rushton, 1993; Wöpking, 1995).

Some early investigations on this latter effect were done with poorly designed systems, making the data suspect when the results were extrapolated to HMDs in general. More recent research has started to separate the hardware deficiencies from the visual effects (Rushton, Mon-Williams, and Wann, 1994; Wann, Rushton, and Mon-Williams, 1995).

The problems with HMDs have not been limited to commercial devices. During the late 1980s the Navy and the Air Force worked with three companies to produce an integrated night-vision HMD. Competing designs underwent stringent safety-of-flight testing to determine their crashworthiness. Even after thorough testing was completed, development was halted because of safety concerns (Fulghum, 1990). Earlier, studies of Marine Corps helicopter pilots flying with night-vision goggles found that they were incorrectly adjusting them, causing latent phorias (Sheehy and Wilkinson, 1989). Proper training in how to adjust and use the night-vision goggles remedied this problem, although illusions related to depth and distance perception remain (Crowley, Rash, and Stephens, 1992).

1.5 Task Requirements

If we have met the basic requirements for the user, we can concentrate on requirements that are task-driven. To do so, it is instructive to look at what the end users will be doing with the HMD. If we can understand the task they will be performing and the role they will be playing within the system, we can identify the critical design issues.

A good example is the fundamental division between military and commercial applications. Military HMDs carry the burden of the dual role as both display and life-support equipment. The military environment also demands that hardware be more durable than typical commercial equipment. This requirement must be balanced by the need for protection with minimum weight and optimized center of gravity.

Commercial HMDs do not have the same life-support needs. They do require the same fundamental display features, although the ambient luminance requirements are not as stringent and the environmental requirements such as temperature and humidity are not as harsh. They also require a design that is lightweight and balanced, since the headset may be used for long durations by people who are not as physically capable as pilots.

We can look at each of the following scenarios and consider the task-based design requirements.

> Returning from a reconnaissance mission, the young helicopter pilot is flying at treetop level to hide from threats. But these trees are also threats, their branches being sometimes higher than the rotor blades of his aircraft. This is a nap-of-the-earth mission, the most difficult flying possible—especially on a moonless night. His HMD provides WFOV imagery from the infrared sensor located in the nose of his helicopter. Global Positioning System information is combined with mission route information and displayed on the HMD as waypoints. The sensor imagery, though seen as shades of green against the black nighttime background, provides the pilot with a feeling of control over his aircraft and a sense of where he is to avoid the trees.

This is the most demanding HMD design environment possible. In the nighttime nap-of-the-earth scenario, the pilot is in an immersed environment. There is no ambient illumination, so he relies on the infrared sensor that converts the temperature differences in the surrounding terrain to video imagery. This imagery is displayed with critical aircraft and targeting data that changes according to the aircraft location and the pilot's head orientation. Here the critical requirements are field of view, resolution, video contrast, and low tracker lag.

During daylight operations this *immersed-environment HMD* becomes an *augmented-reality HMD*. Here the primary task involves overlaying symbology on the outside world, enabling the pilot to always maintain visual contact with his external environment and to efficiently switch his attention as dictated by task demands. The higher ambient illumination level places tougher demands for luminance and contrast on the HMD, and the need to balance these demands with crash protection, communications, and hearing protection further complicates the design.

> Sitting safely at a distance, the operator carefully maneuvers the robot around the suspicious-looking container. The HMD she wears is linked to a pair of video cameras on top of the robot that provide a high-resolution stereo image of the container. Remotely reaching into the container, she deftly disarms the explosive device using the robotic arms. She prefers this method of bomb disposal, because it is much faster and safer than the traditional methods that encumbered her in layers of armor and placed her in proximity to the explosives.

Some tasks require a view of the outside world, while others are best served by occluding all external visual cues. For this remote teleoperation task, total visual immersion is appropriate. The operator relies on high-quality stereo imagery from gimballed cameras to guide the remote-controlled hands in real time. She needs a system that provides the resolution, color, stereovision, and low system latency that give her the feeling of *presence* to accomplish the task.

The surgeon backs into the operating theater, keeping her hands held high. She moves to the patient, who is already under the control of the anesthesiologist. A circulating nurse places the HMD on the surgeon's head so that she can keep her gloved hands sterile. As the operation proceeds, the surgeon inserts the endoscope and instruments through the trocars into the patient's chest. She moves quickly to perform the coronary bypass operation, aided by the stereoscopic imagery that provides the much-needed display of depth. The video is transmitted to an adjacent viewing room, where her running commentary provides a valuable teaching tool to the attending medical students.

This minimally invasive surgical scenario would not be considered an immersive environment: the surgeon needs a stereoscopic, high-resolution, color display that is comfortable to wear for long periods. She does not need a large field of view, and she may want her peripheral vision to be unoccluded so that she can see the operating room around her.

Here is a situation in which we must concern ourselves not only with the user—the surgeon—but also with the patient she is operating on. The headset and endoscope are well aligned to prevent eye strain. To give her stable hand/eye coordination, so that she does not risk injury to the patient, the stereo imagery must be accurate and repeatable.

The therapist used to have difficulty explaining his young patient's vision problems to the parents, as they usually felt that providing eyeglasses was sufficient. Recently he has been using the HMD with a computer program that let him change the imagery from the two video cameras to show the parents what their child saw with poor binocular vision—a convincing way to help them visualize the problem.

This is another immersive scenario, with a twist over the previous ones. Here the vision therapist takes advantage of a different capability of the HMD, one that enables him to *purposely* distort the imagery to convey to the parents their child's visual problems. The HMD is a truly unique display device because it is easy to create a visual scene that does not really exist—poor alignment, ill-constructed stereo imagery, and the like—a feature of the virtual-world capability. While the therapist can use this feature to his benefit, it illustrates a potential drawback—one that has been inadvertently implemented by designers, with serious visual side effects.

Before committing any design to wood and stone, the architect preferred to virtually construct the home in his computer. This time it was particularly important for him to see how his client would view it from her wheelchair. It did not take more than a few minutes to see that he had placed the cabinets out of reach and that the preparation area in the kitchen would have to be changed to allow her to get her legs under the counter space. These were not expensive changes to make, especially before construction began. Turning his head toward the south-facing wall, he decided that perhaps the window sill could be lowered to take advantage of the view of the garden he would put in.

Here is another immersive application: walking through a virtual building before it is built. The architect does not need an extremely expensive system, just one that will provide the essentials—a full-color, head-tracked display. This system has some similarities to the pilot's in that it must be comfortable and it must meet basic viewing requirements, although it can do so without the stringent daytime contrast or life-support requirements we discussed in previous scenarios.

1.6 Summary

HMDs can enhance the performance of many tasks by providing information in a new, exciting, and highly personal way. The imagery is presented naturally—where the user is looking—in contrast with traditional monitors, which constrain visual scanning and exploring. HMDs can be efficient and intuitive task enablers for any user—such as a scientist, architect, or surgeon—in much the same way as for a military pilot, if the design process concentrates on the capabilities of the users and the tasks they will be performing. The key to achieving success with HMDs is to put the user at the center of the design process.

1.7 References

Arbak, C. J. (1989) Utility evaluation of a helmet-mounted display and sight. *Helmet-Mounted Displays, Proceedings of the SPIE, 1116,* 138–141.

Bahill, A. T., Adler, D., and Stark, L. (1975). Most naturally occurring human saccades have magnitudes of 15 degrees or less. *Investigative Opthalmology, 14,* 468–469.

Crowley, J. S., Rash, C. E., and Stephens, R. L. (1992). Visual illusions and other effects with night vision devices. *Helmet-Mounted Displays III, Proceedings of the SPIE, 1695,* 166–180.

Dornheim, M. (1995a, October 16). U.S. intensifies efforts to meet missile threat. *Aviation Week & Space Technology,* p. 36.

Dornheim, M. (1995b, October 23). VTAS sight fielded, shelved in 1970s. *Aviation Week & Space Technology,* p. 51.

Fulghum, D. A. (1990, December 3). Navy orders contractors to stop work on ejectable night vision helmets. *Aviation Week & Space Technology,* pp. 67–68.

Gibson, J. J. (1986). *The ecological approach to visual perception.* Boston: Houghton Mifflin.

Kennedy, R. S., Jones, M. B., Lilienthal, M. G., & Harm, D. L. (1993). Profile analysis of aftereffects experienced during exposure to several virtual reality environments. *AGARD Conference Proceedings—Virtual Interfaces: Research and Applications* (AGARD-CP-541) (pp. 2-1–2-9). Neuilly-Sur-Seine, France: Advisory Group for Aerospace Research & Development.

Melzer, J. E., & Moffitt, K. (1991). An ecological approach to partial binocular-overlap. *Large-Screen Projection, Avionic, and Helmet-Mounted Displays, Proceedings of the SPIE, 1456* 124–131.

Miyashita, T., & Uchida, T. (1990). Cause of fatigue and its improvement in stereoscopic displays. *Proceedings of the Society for Information Displays, 31,* 249–254.

Moffitt, K. (1989). Ocular responses to monocular and binocular helmet-mounted display configurations. *Helmet-Mounted Displays, Proceedings of the SPIE, 1116,* 142–148.

Mon-Williams, M., Wann, J. P., & Rushton, S. (1993). Binocular vision in a virtual world: Visual deficits following the wearing of a head-mounted display. *Opthalmic and Physiological Optics, 13,* 387–391.

Norman, D. (1990). *The design of everyday things.* New York: Doubleday.

Peli, E. (1995, July). Real vision & virtual reality. *Optics & Photonics News, 7,* pp. 28–34.

Peli, E. (1990). Visual issues in the use of head-mounted monocular display. *Optical Engineering, 29,* 883–892.

Regan, E. C. (1993) Some side effects of immersion virtual reality. *AGARD Conference Proceedings—Virtual Interfaces: Research and Applications* (AGARD-CP-541) (pp. 16-1–16-8). Neuilly Sur Seine, France: Advisory Group for Aerospace Research & Development.

Regan, E. C., & Price, K. R. (1994). The frequency of occurrence and severity of side-effects of immersion virtual reality. *Aviation, Space, and Environmental Medicine, 65,* 527–530.

Rheingold, H. (1991). *Virtual reality.* New York: Summit Books.

Riley, V. (1995, Fall). Toward a standard definition of the term "human centered" and a next generation flight deck. *The Flyer* (Newsletter of the Aerospace Systems Technical Group of the Human Factors & Ergonomics Society), pp. 5–7.

Robinson, G. H., Koth, B. W., & Ringenbach, J. P. (1976). Dynamics of the eye and head during an element of visual search. *Ergonomics, 19,* 691–709.

Rouse, W. B. (1991). *Design for success.* New York: Wiley.

Rushton, S., Mon-Williams, M., and Wann, J. P. (1994). Binocular vision in a bi-ocular world: New-generation head-mounted displays avoid causing visual deficit. *Displays, 5,* 255–260.

Sheehy, J. B., & Wilkinson, M. (1989). Depth perception after prolonged usage of night vision goggles. *Aviation, Space, and Environmental Medicine, 60,* 573–579.

Wann, J. P., Rushton, S., & Mon-Williams, M. (1995). Natural problems for stereoscopic depth perception in virtual environments. *Vision Research, 35,* 2731–2736.

Wöpking, M. (1995). Viewing comfort with stereoscopic pictures: An experimental study on the subjective effects of disparity magnitude and depth of focus. *Proceedings of the Society for Information Displays 3,* 101–103.

1.8   Annotated Bibliography

Kalawsky, R. S. (1993). *The science of virtual reality and virtual environments.* Reading, MA: Addison-Wesley. Kalawsky's text provides several good sections on HMDs. These sections include a look at the history of HMDs, a discussion of how HMDs are integrated into complex systems, and in-depth treatment of the electronics and optics of specific systems. Unfortunately, Kalawsky includes several graphics and concepts without giving due credit. Specifically, figures 6.3, 6.4, and 6.10 were developed by the editors of this book in support of various Kaiser Electronics HMD systems.

Kocian, D. F., & Task, L. (1995). Visually coupled systems hardware and the human interface. In W. Barfield & T. A. Furness III, (Eds.), *Virtual environments and advanced interface design.* New York: Oxford University Press. This chapter provides an extensive treatment of HMD optics, displays, and trackers. The appendix is a survey of 20 commercial and military HMDs.

Peli, E. (1995, July). Real vision & virtual reality. *Optics & Photonics News 7,* pp. 28–34. A recent spate of articles in the popular press took a very uncritical look at the perceptual and physical problems associated with HMDs and virtual reality applications. Peli's article looks at the HMD-related vision issues in a very concise and insightful manner and should be read by those who wish to clear the air.

Rash, C. E., & Verona, R. W. (1992). The human factor considerations of image intensi-fication and thermal imaging systems. In M. A. Karim (Ed.), *Electro-optical displays*. New York: Marcel Dekker. This chapter looks at the specific issues of night-vision gog-gles and FLIR systems from the perspective of U.S. Army aviation programs. A large part of the chapter deals with the technology of these two night-vision approaches, and there are also discussions of crash biodynamics, field of view, and display resolution.

Rheingold, H. (1991). *Virtual reality*. New York: Summit Books. This book presents a historical perspective on virtual reality and HMDs through the early 1990s.

Wells, M. J., & Haas, M. (1992). The human factors of helmet-mounted displays and sights. In M. A. Karim (Ed.), *Electro-optical displays*. New York: Marcel Dekker. This chapter examines the human factors of HMDs, divided into four areas: visibility, comfort, fidelity, and intuitiveness. The treatment is a useful introduction to the display issues of HMDs.

Monthly publications, such as *RealTime Graphics, CyberEdge Journal,* and *VR News*, provide timely information on virtual-reality technology and provide annual updates of HMD and related technology.

User-Centered Design in the Trenches: Head-Mounted Display System Design and User Performance

Robert G. Eggleston

This chapter describes the user-centered design philosophy and attempts to reduce it to a practical decision-making process that a designer can execute within prevailing engineering design practices. It presents the position that all designers are responsible for including human performance factors in their design decisions, even when these decisions address detailed implementation issues. A three-step, user-centered decision process is described as a means to assist the designer in achieving this goal. The essential idea is to have the designer transform a design problem from a machine-centered form to a psychophysical form that inherently considers user performance.

Once a solution is found for the psychophysical problem, an inverse transform (conceptually speaking) is used to express the solution in machine terms. Examples are used to illustrate the user-centered decision process, explain psychophysical variables, and introduce techniques to facilitate application of the user-centered approach.

2.1 Introduction

The ultimate goal for any head-mounted display (HMD) system is to enable the user to achieve task objectives to an acceptable level and with a reasonable expenditure of effort. This goal has important implications for the design and development of an HMD system. It indicates that to produce a successful HMD product, the design team must first be able to recognize the relation between system properties and specific aspects of user performance, and then make good design decisions based on this knowledge.

The prevailing approach to human-machine system design can be characterized as machine-centered. From this orientation, design problems are, in general, posed in terms of machine functions and features that the designer tries to realize in the final medium of the artifact. The search for solutions to design problems typically concentrates on characteristics of component technologies and finding clever ways to mold them into the desired functional features. The interaction between the design problem and user performance may be considered early in the design process. Typically, human and task interaction issues are considered mainly in relation to the establishment of machine functions (e.g., allocation of functions between the machine and the user) and with respect to the design of displays and controls that the human will use. Throughout the majority of design activities, however, relatively little attention is given to user performance factors during design problem-solving and implementation activities.

It is often not easy to recognize the relation between a detailed design issue and its impact on human performance. As a result, the designer tends to concentrate on technology factors during design problem solving. This is one reason why important human performance factors fail to receive attention during detailed design decision-making activities. More broadly, the machine-centered paradigm by nature discourages the designer from even looking for human performance connections during detail design. Features are instantiated in hardware and software, not in interactions with the user. Unfortunately, it is precisely at the point of user-machine interactions where many human-machine systems fall short of achieving development goals. This suggests that there is a need to approach design from a new perspective, one that actively considers user performance as an integral factor in design decision making throughout the entire development process.

User-centered design is an alternative design approach that attempts to place the human at the center of design decision making. This view embraces the notion that a human-machine system, such as an HMD, is intended to *support the user* in accomplishing a job. Because the system is explicitly cast in a support role, the user-centered approach attempts to ensure that the design team will focus on the user when making design decisions. From the user-centered orientation, design issues are cast in terms of user and task properties, and hardware and software solutions are formed on the basis of integration with the user to satisfy joint user and task performance criteria. It is in this way that support of the user becomes a central feature in the design decision-making process.

This chapter presents a specific view of user-centered design in relation to the design decision-making process. It tries to make clear how user-centered problem solving differs from machine-centered problem solving all the way down to and including detailed design decisions. It offers a framework and an approach for implementing user-centered design, and it provides examples to illustrate various features of the framework.

The user-centered design concept is presented at two levels. Initially, a conceptual overview of the approach and how it contrasts with machine-centered design is presented. This is followed by a more down-to-earth, practical discussion of the user-centered orientation as a design decision-making process that can be incorporated within current system engineering practices. The process description itself is presented in two ways. One focuses on providing an experiential understanding and feel for user-centered design decision making. The process is also described at a meta-level that attempts to reveal and reflect on the rationale behind aspects of the process. The hope is that by seeing both views of the process, readers will be able to modify it, as appropriate, to fit their own situations while preserving the spirit of the user-centered philosophy.

2.2   User-Centered Design

The term *user-centered design* is used loosely in many quarters. Some developers treat it as a process to incorporate input about the task and system features from current members of the target user community.

Other designers go further and regard it as the full process of mission and user task analysis, including a workload and cognitive task analysis, that is used to set design requirements and establish display/control formats, symbology, layout, and operational procedures. While these are all important aspects of design and indeed do concentrate on user issues, by themselves they do not capture the essence of the user-centered design concept.

If these views do not define user-centered design, then what does? In short, user-centered design is a philosophy intended to pervade the entire design and development process. Rouse (1991) provides the clearest and most complete treatment of user-centered design, which he refers to as *human-centered design*. From his view, user-centered design establishes a focus and a design objective, which in turn lead to design issues. The design focus is on the role of the human, not on the machine or hardware system. The design objectives for a project need to be expressed in terms of human roles, not in terms of system functions. From this orientation, Rouse identifies four design issues that need to be solved: (1) formulating the right problem, (2) designing an appropriate solution, (3) implementing the solution so that it performs well, and (4) ensuring user satisfaction with system use.

Formulating the right problem means viewing the design project from the viewpoint of how the product will support the user in achieving the goals of a job. This is different from looking at the problem as developing a tool that the human can use to accomplish the job. The former problem statement includes the user *in* the design problem. The latter statement places the user *outside* the design problem. The user is merely the recipient or benefactor of a tool that was developed from the latter perspective. The design activities mentioned in the earlier paragraph, while focused on the user, can and often are accomplished from the function-oriented position. In this case they do not embrace user-centered concepts. The focus is on the tool or machine, once the user's problem has been analyzed. In user-centered design, the focus remains on the user throughout design and development.

Rouse has suggested that user-centered design needs to occur at three levels: an interface level, a job level, and an organization level. From the conventional view of human-machine systems, emphasis is given to the human during the design of displays and controls that allow the user to (1) operate the system, (2) gain knowledge of system states and behaviors, and (3) gain information about the task situation and task state. Because considerable attention is given to the user in making these design decisions, many designers probably believe they are engaged in user-centered design. According to the view expressed by Rouse and endorsed here, this is not necessarily true. To be faithful to the spirit of the user-centered design philosophy, all design decisions must be approached in such a way that they actively address support for the user.

Human-machine systems are intended to support users in accomplishing job tasks. Invariably, jobs are accomplished inside an organizational structure. It follows, therefore, that the designer must address issues at this higher level as well to achieve the goal of the user-centered approach.

Another level of user-centered design is emphasized in this chapter. The perspective presented here espouses that user-centered design must reach all the way down to the detailed process and structural level of design decision making. By this I mean that *the designer of specific processes and structures shares responsibility for how the system will support the user.* In other words, in a very real sense, the designer is responsible for user performance with the system. This view on user-centered design places it squarely in the middle of the design decision-making process for *all aspects of the system,* not just for the design of the user interface. This is meant to imply that the designer cannot turn over responsibility for human design issues to the human factors specialist and press ahead with the "real" engineering problems without further regard for the human. The designer must come to understand the design problem in terms related to user support, and form solutions from this perspective. The human factors specialist is a resource that the designer can draw upon in the process of preparing to make a design decision, but a "user-centered" oriented decision is in the hands of every designer. The following discussion and illustrations elaborate on this theme for HMD system design.

2.3 Applying User-Centered Design Philosophy to HMD System Design

When applied to HMD system design, the user-centered design philosophy may be expressed in terms of a guiding framework for design decision making. As shown in Fig. 2.1, the framework asserts that the role of an HMD system is to assist the user in the performance of a job or task. Since assistance cannot be demonstrated without including user performance, user performance must actively be considered in all design decisions. Because of this emphasis on performance, the framework is referred to as the User Performance Framework for HMD System Design.

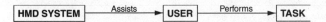

Figure 2.1 User performance framework for design decision making.

The user-centered perspective may seem like a subtle shift, since the machine-centered view, with its emphasis on the design of functions and features, can be interpreted as providing assistance to a user. An HMD feature such as head tracking, for example, can assist the user in directing the movement of a tool in a work space. From the User Performance Framework, however, the satisfactory design of a feature, even if it can be used to assist the user, is not enough. The user-centered paradigm requires the designer to take responsibility for user performance in the feature design. From the machine-oriented perspective, a design is successful if useful features have been included in the system. From the user-centered view, the design is not successful until the user demonstrates good performance with the feature. This is an important shift in emphasis.

The user-centered paradigm places the user in the center of the design decision-making process, as depicted in Fig. 2.2. This is in contrast to the peripheral location of the user in the machine-centered approach (see Fig. 2.3). User performance, in part, is contingent on processes and properties intrinsic to the user; hence, since the designer takes responsibility for user performance in the user-centered approach, the designer must actively consider these properties when making design decisions. It is in this sense that the user is given a central role in the design decision-making process.

Another way to contrast the alternative design paradigms is in terms of the definition of a feature or property. In the machine framework, features are defined by physical (hardware or software) variables. For example, a button or a menu would be a user interface feature. In the

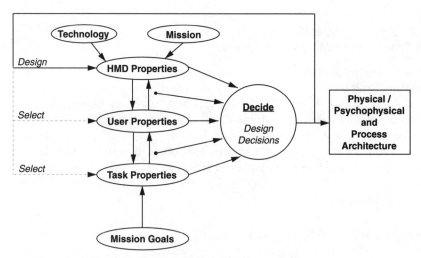

Figure 2.2 Conceptual model of user-centered design.

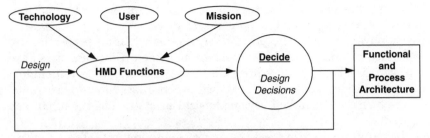

Figure 2.3 Conceptual model of machine-centered design.

User Performance Framework, features are defined by psychophysical variables. A psychophysical variable links one or more physical properties with one or more human properties. In other words, a psychophysical variable represents the interaction between variables from these two domains. For example, when display resolution, a physical variable, is thought of in terms of visibility, it has been transformed into a psychophysical variable. By emphasizing psychophysical variables as properties to be designed into the HMD system, the issue of user performance is implicitly taken into consideration during design decision making.

The shift in emphasis expressed by the User Performance Framework is represented, in an abstract way, in Fig. 2.2, which depicts a general user-oriented process for making detailed design decisions. As the figure suggests, relevant properties (functions and features) from the HMD system and its component technologies are considered interactively with properties from the user and task. Task properties are also included in the main decision-making pathway, since they provide the extrinsic criteria against which user performance is measured. Design decision making includes the design of functions and features, which is represented by the arrow from the HMD properties oval to the decision activity. Psychophysical expression of these properties is represented by the arrow with the solid dot on the end that conveys the interactive links between HMD physical and user properties to the decision activity. Similarly, the arrow with the solid dot from the links between the user and task properties conveys the required success criteria to the decision activity.

The design decision-making process starts when a design problem has been posed. Typically, problems are initially expressed in abstract terms. The problem statement leads to a set of analysis and synthesis activities that result in the definition of HMD system properties. The initial property descriptions also tend to be stated in abstract terms. The process of defining properties continues in cyclical fashion, with each cycle yielding progressively more detailed and concrete expressions of the properties, including division of some properties into separate entities. Further, as stated above, a most important part of the decision-making process is to reexpress properties in a psychophysical form. The process terminates when all properties are expressed physically in the final medium of the artifact and also have a linked psychophysical expression of the property as a companion. Physical design decisions (i.e., refined property statements) are made on the basis of the linked psychophysical property, not the other way around.

The requirement to express design properties in terms of psychophysical variables is different, and many designers may not be sure how to satisfy this requirement derived from the detailed user-centered approach. One method that can be used to implement the User Performance Framework is to follow a three-step design decision-making procedure that has been structured to facilitate the formation of psychophysical property definitions.

The procedure begins with the statement of a design problem. This statement may express a need to achieve something or identify a property that needs to be refined in some way, such as establishing the image resolution requirement for an HMD display. The first step in the procedure involves a search for important ways the hardware or software property may interact with intrinsic properties of the user to impact user performance. This is called the *Discovery Step.* It is an initial attempt to understand the problem in terms of one or more psychophysical variables. The *Analysis Step* is next. This step involves the acquisition of information about technology, user, and task properties that may interact. One goal of the analysis is to reduce the number of factors that are to be considered in the final *Decision Step.* A second goal is to increase the designer's understanding of all potentially relevant factors, with the hope that the designer will gain an appreciation for technology, user, and mission constraints on design solutions, as well as gain greater clarity about design degrees of freedom. Finally, by immersing designers into the problem in a way that invites a psychophysical perspective, it is hoped that the designers will be inspired and enabled to recognize user-centered design solutions.

The Decision Step focuses on the integration of knowledge acquired from the analyses performed earlier. Further, it also encourages the designer to perform, formally or informally, "what if" testing of candidate solution ideas. Of course, the final activity in this step is to select a solution path. The solution statement includes a refined description of a system property, its psychophysical analog, and a rationale or justification for the solution. A conceptual overview of the procedure is depicted in Fig. 2.4.

In the figure, a technology property ($P_{i,j}$) is indexed by the notation $HMD_{1,0}$. This indicates that, in the present case, the technology

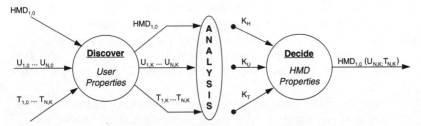

Figure 2.4 Conceptual model of three-step user-centered design decision-making procedure.

property that is the basis for a design problem refers to an HMD system or subsystem. The first subscript is a sequence number assigned to each problem, and the second subscript identifies the level of decomposition of the problem/property that is currently under review. This numbering system provides a convenient way to keep track of property descriptions and design decisions made throughout development activities. $U_{i,j}$ and $T_{i,j}$ stand for user and task properties or features, respectively. In theory, all properties $U_{i,j}$ and $T_{i,j}$ (the i represents properties running from 1 to N, and the j represents key elements of a property running from 1 to K) are available as inputs for the Discovery Step. The outcome of this step consists of the selected U and T properties believed to be relevant to the design of the $HMD_{i,j}$ property. This reduced list is submitted to a more detailed analysis in the Analysis Step. As indicated earlier, the Analysis Step consists of accumulating, analyzing, and synthesizing information from all available sources to achieve a deeper understanding of the properties of each domain, how they interrelate, and their implications for design. The outcome of this step may be characterized as knowledge about the HMD, U, and T properties. Since the knowledge and perspective gained from the analysis may be achieved in part by way of analogy and other forms of creative insight, this step is not restricted to a formal analytic method. To represent this possibility and make clear that the process is not a simple information flow process, Fig. 2.4 shows a gap on the output side of the Analysis Step.

The final step in the decision-making procedure involves integrating knowledge in such a manner that the next level of design detail for the property in question can be expressed in terms that are sensitive to user and task performance. That is, the new property expression either implicitly or explicitly addresses user performance and sets one or more user performance criterion for the property. In the latter stages of design, the output from this final step should also contain a statement of boundary conditions that specify the region in which the performance criteria can be met. In short, the Decision Step reflects an integration of knowledge from a user-centered perspective and includes a weighing of evidence regarding cost, schedule, and other factors. To reflect the integrative nature of the decision, the output of the process is a new, refined property description that denotes relevant user and task factors as parameters to the refined property expression: $HMD_{1,0}(U_{N,K}; T_{N,K})$.

The complete user-centered design decision-making procedure does not enforce a sharp transition between analysis and decision-making activities. Certainly, some decisions are made during analysis. A decision happens when the designer has acquired enough insight to "see" the solution or is forced to converge on one because of schedule pressure. It should also be clear by now that the resultant HMD structure/process decision description provides an audit trail of design decisions in a form that is explicitly connected to user performance. Finally, each new property description serves as input to the next round of design decision making. This process continues until all property descriptions are expressed in a way that can be directly realized in the final medium of the artifact.

This introduction to the User Performance Framework has been presented largely in an abstract manner. Although it provides a useful conceptual model of how the user-centered design philosophy flows down to detailed design decision-making activities, the reader may have several questions about how such activities would actually be performed. The remainder of the chapter attempts to address these practical concerns through illustrations that demonstrate the three-step procedure in action.

2.4   The User Performance Framework in Action

We shall use a fictitious design project as a means of establishing specific HMD design problems and to provide detailed information for use in a user-centered design decision-making process. For practical reasons, many details of the complete systems design and development process are not covered. As a result, illustrations of user-centered decision making will also be incomplete. This should not be cause for concern, because there is more than enough information provided to guide the reader through an implementation of the three-step design decision process.

Imagine that the U.S. military is planning to develop a new helicopter system to be used primarily in a scout mission. The main objective of the scout mission is to acquire information about enemy activity in a potential or actual war zone. Based on analysis of expected scout missions, it was decided to include an HMD system as a subsystem in the scout development project. The HMD is expected to assist the user in many different tasks. One of these tasks is known as a *zone reconnaissance task*. In this task, a geographical zone of interest is identified for visual inspection. The scout must fly to the zone, which may be deep inside enemy territory; maneuver to a safe vantage point in the zone, possibly moving from one vantage point to another; make observations about terrain features and human activities of military importance in the zone; and finally, report findings of military significance in a timely manner to a higher command. A second use of the HMD system that is under consideration deals with mission training and rehearsal. To support these activities, the idea is to use the HMD system to produce an interactive synthetic or virtual environment. By simulating the planned mission in this dynamic synthetic environment, the crew can acquire important training in system operations under mission conditions and at the same time become familiar with the geographical, procedural, and cooperative aspects of the assigned mission. That is, it provides an opportunity to rehearse the mission.

2.4.1 Identification of HMD, task, and user properties

As the designer approaches the development of an HMD for this application, it is important to keep in mind that the HMD's role is to assist the user in achieving mission goals. To this end, the designer needs to link properties of the HMD to properties of the user, which are in turn linked to properties of the task or mission as a whole. Within the User Performance Framework, mission properties are called *tasks*. Tasks specify a goal and describe user activities needed to achieve the goal. Initial task descriptions usually contain several terms that communicate the nature of the activity, but they often leave open the physical actions required to complete it. The same task, therefore, may be completed in different ways.

The mission description just presented is sufficient to allow the designer to form a top-level characterization of the mission in terms of task properties. Clearly, the crew will have to navigate to the area to be reconnoitered, maneuver the helicopter to a suitable observation point, make observations, and then report findings. Both for reasons of being covert and for avoiding the dangers created by enemy weapon systems, the crew will most likely have to use terrain features as screens. This will necessitate flying extremely low and, in some instances, close to protruding objects. With this knowledge, additional task properties may be deduced. Some of these will be coded as new tasks, while others will be coded as refinements of one or more tasks that have been identified previously. For example, the crew will have to increase altitude to a level above terrain features to gain an unobstructed view of the zone and then remask for protection. The activity of masking and unmasking defines a new task. Flying low adds the hazard of aerial wires that must be avoided when maneuvering the helicopter. Avoiding wire strikes, therefore, is an element of maneuvering that is important in the local environment. Since safe maneuvering has already been identified as a task, the added feature of avoiding wire strikes reflects a decomposition of the task into more specific terms. An example of two levels of some task properties for the zone reconnaissance mission are shown in Fig. 2.5.

The figure also contains property lists for an HMD system and the user, since a top-level mission and systems analysis indicated that an HMD is a candidate technology to assist the user with the zone reconnaissance task. From the user-centered perspective, the analysis needs to be extended to include an identification of HMD and user properties because the interaction between properties in these two domains impacts user performance, which in turn impacts task performance. The linkage between one or more user properties and one or more HMD properties defines a psychophysical issue that requires attention in the design process. The representation of linked property lists shown in this manner invites attention to human performance issues. For example, it is easy to see that a decision about the resolution

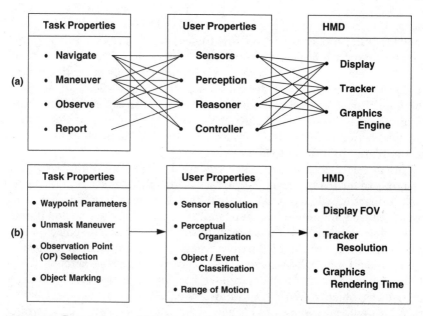

Figure 2.5 Descriptive model that maps task factors, expressed as properties, to user and HMD properties. Part (*a*) shows mapping at a top level. Part (*b*) shows mapping at a lower level of task decomposition. (Individual property connections have been suppressed.)

of the HMD display should be based at least in part on knowledge of the resolution of human vision as it relates to aspects of the mission task(s) (see Fig. 2.5b).

The HMD properties shown in Fig. 2.5a and b can be easily derived from general knowledge of previous HMD systems and from a cursory examination of the design problem, as initially stated. Some knowledge of how humans perceive objects and scenes, make controlled movements, and reason about actions is needed to identify property factors for the user like those shown in Fig. 2.5b. Each round of design decision making produces a more detailed listing of properties and solutions to previous ones. The solution to a design problem may be a refined property or a process for achieving an HMD property. The process of identifying properties and resolving problems continues in all three domains (task, user, and HMD) until the design is finalized in the physical medium of the artifact.

The linkage between HMD properties and user properties is critically important to how well the user can discharge his or her abilities to accomplish a task. HMD properties need to potentiate user abilities in a way that does not activate limitations on these abilities. This is the heart of user-centered design. When mission tasks are inspected at a top level, it is clear that the linkages between the HMD, the user, and task features are connected in a many-to-many manner (see Fig. 2.5). This indicates that multiple aspects of the user and the HMD are used to achieve each desired aspect of task performance. Ideally, one would like to reduce the many-to-many mappings to simple one-to-one mappings, once properties have been decomposed to their lowest form. In reality this is not a reasonable goal, especially when a system needs to support a range of user activities. It is efficient to have a single HMD property that can support many user activities. Therefore, a more appropriate goal is to discover how to cluster HMD properties into ability classes and then link HMD and user ability classes to a mission task. An ability class consists of a set of properties and processes that can produce a complete behavior. Although ability classes still represent a many-to-many mapping function, they provide relief from a dependency between a task property and the full capabilities of the user and/or HMD. Thus, only limited aspects of the HMD design need to be addressed to support any specific user and task performance issue. This allows the design to be treated in a modular fashion, just as it is when approached in a conventional functional manner.

2.4.2 Example of the three-step decision-making procedure

The partial mission analysis and accompanying linked property description for the zone reconnaissance problem do not indicate how design decisions are made in the user-centered paradigm. A linker property description, like the one in Fig. 2.5, will show the outcome of design decisions, but it does not reveal anything about the decision-making process itself. To see user-centered design decision making in action, initially we will limit the decision to a single HMD property as a design problem. The immediate goal is to provide a quick "walk through" of the entire decision-making procedure. We have deliberately chosen a relatively simple and straightforward design problem for inspection so that the discussion can move easily between comments on the design problem content, the decision-making process, and documentation methods. A more detailed look into each step of the process is presented later.

Assume that a design engineer has been assigned the task of determining what image resolution is needed for the HMD to be used in the helicopter system. This design problem may be regarded as a request to specify a subproperty of an existing property of an HMD system, namely the visual display (see Fig. 2.5a and 2.5b). In other words, the designer is being asked to produce a property, display resolution, for a previously designed or selected property of the system. Note that both the design problem and its solution are expressed in terms of system properties.

Before digging into details of the display resolution problem, it is helpful to introduce a general form that can be used both to guide and to record decision-making activities. Figure 2.6 contains a form that can be used to support the three-step decision-making process. It has already been partially filled in for the display resolution problem. It is structured around the user-centered decision-making process. At the top, the design problem is coded as an HMD property, using the notation introduced in Fig. 2.4 (see previous discussion). The goal, of course, is to specify a value for this property. The form is partitioned into four sections, one for each step in the decision-making process and a preliminary section used to record the initial list of user and task properties that bear on the design problem. Given this structure, the form can serve as a prompt for the steps in the decision-making process and also for the type of information that needs to be generated. The notation scheme provides a concise and systematic way to record information and decisions. In general, it is worth the time to produce a good format for recording information and supporting the decision process in other ways. Besides its value during the process, it will be useful for (1) tracking and updating, (2) communicating among individuals and other design teams, and (3) several system management activities.

Once a design problem is given to a staff engineer, the user-centered design process starts by inspecting the job or mission analysis to identify areas of task performance that appear to depend in some way

Design Problem $H_{1,1}$ (Select Display Resolution)

Initial Property List:

HMD: $H_{1,1}$ (Display Resolution) : [Design problem expressed in physical terms.]
User: $U_{1,0}$ (Vision)
Task: $T_{1,0}$: Observe Recon Area

$T_{1,1}$:	Passageways		$E_{1,0}$:	Environmental conditions
$T_{1,1-1}$:	Troop movement routes		$E_{1,1}$:	Weather
$T_{1,1-2}$:	Vehicle Routes		$E_{1,2}$:	Day/night
$T_{1,2}$:	Terrain features analysis		$E_{1,3}$:	Smoke
$T_{1,3}$:	Vehicle Detection			
$T_{1,3-1}$:	ID : wheel, track, towed			
$T_{1,3-2}$:	Number by type			
$T_{1,4}$:	Identify vehicle choke points			
$T_{1,5}$:	Identify trafficability level			

Discover User Properties:

User: $U_{1,0}$ (Vision)
$U_{1,1}$: Static acuity
$U_{1,2}$: Dynamic acuity
$U_{1,3}$: Interaction of $U_{1,1}$ & $U_{1,2}$ with lighting conditions
($E_{1,1}$; E_{1-2}; E_{1-3}) by viewing distance
$U_{1,4}$: Interaction with transmissivity of optics ($H_{5,1}$)

Analysis:

K_U: • Types of acuity and measurement methods
        • Luminance effects

K_H: • CRT phosphor characteristics
        • Relay Optics MTF

K_T: • Detection, recognition, identification, priorities by mission type
        • Environmental effects on object/event detection, recognition, ID

K_U-K_H: • Optics transmissivity interaction
            • Field-of-view interaction

Design Decision:

DD: $H_{1,1}$ (Display Resolution/Visibility) : [Note psychophysical expression attached to design problem]

Side Effects:
Information coding concept (object x,y,z; viewing condition 1,2,3)
Zoom concept (increase pixels/objects a,b,c)
Transmissivity contingency
Field-of-view contingency
Task performance boundary states

Figure 2.6 Working form for use with the three-step user-centered design decision-making process.

on the design issue. Similarly, the designer tries to identify properties of the user that are involved with task performance in these areas. Based on the information given about the zone reconnaissance task and other analyses, the designer has produced the items listed in the initial property section shown in Fig. 2.6. These task factors can be used to derive task-level evaluation criteria for use in assessing alternative design options.

It is obvious that vision is the user property that interacts with display resolution. Therefore, vision is identified on the property list as the top-level user factor. Because user-centered design requires the designer to take responsibility for user performance, visual performance is the relevant area for this design problem.

Broadly speaking, the HMD assists user performance and thus indirectly assists task performance. Both levels of performance are represented in the initial property list. This section also includes entries for environmental conditions. These are included because these factors may interact with vision to make it more or less difficult to achieve task goals. In other words, they impact visual performance and therefore need to be considered by the designer.

The first step in the user-centered decision-making process focuses on the discovery of user properties that are relevant to any interactions with the HMD property under design. For the display resolution problem, discovery amounts to finding out what aspects of vision the designer needs to know more about. In this instance it probably will be obvious to the designer that knowledge about human visual acuity is needed. However, a review of the task factors and environmental conditions derived from prior analyses may stimulate more detailed questions about visual acuity. For example, the discovery process could lead the designer to ask if visual acuity is the same for stationary and moving objects. Since some objects of military interest will at times be moving, this is a relevant question. Thus, the designer makes a note that information on dynamic visual acuity neds to be investigated. Once questioning of this nature gets started, the designer might also note that acuity may change as a function of viewing conditions such as reduced light due to haze or smoke. This, in turn, may stimulate a question about a possible interaction between image resolution and the transmissivity of the relay optic system of the HMD, because the designer knows there is considerable light loss with many relay optical systems for HMDs. The discovery activity continues in this manner until new suggestions begin to seem marginal.

In effect, the discovery process is a focused brainstorming activity that results in one observation flowing from another, stimulated by the initial seed and emerging offshoots generated by the process itself, with occasional back tracks and referrals to the property list and other documentation. It is in this way that the Discovery Step helps the designer focus on user performance and begin to recognize in a more specific and detailed manner possible interactions among the HMD, user, and task properties. A partial list of the discovered user interaction issues is summarized in Fig. 2.6.

Once the designer is satisfied that all major issues relevant to user performance have been identified, attention is turned to analysis. On the basis of information gained from a variety of sources, the designer may learn that there are several types of visual acuity and ways to measure them; that human dynamic acuity is worse than static acuity and depends on object velocity; that acuity varies with retinal location; and that acuity also varies with luminance level. It is then clear that display resolution will interact jointly with the transmissivity of the optical relay assembly and several scene and viewing condition factors. Although overlooked in the Discovery Step, an interaction between resolution and display field of view was also noted when the designer was going through different data sources. Thus, this factor was added to the list of items to be analyzed. New discoveries like this can and usually will take place as part of the analysis phase, even though the analysis process is not geared to this end.

It is important to note that the analysis step is not limited to an examination of user properties deemed to be germane to the design problem. Rather, when needed, it also includes a detailed survey of properties of component technologies and subsystems that may be incorporated in the design solution. In addition, some analysis may also be devoted to an inspection of how properties from the two property domains may interact. In the present case the analysis activities leave the designer with a clearer understanding of the capabilities and limitations of human visual acuity and what task attributes should be detectable under different viewing conditions. In addition, the analysis may also reveal the relative importance of task attributes under different mission and environmental conditions. Armed with this knowledge and knowledge about available options for display resolution, field-of-view options, transmissivity options, etc., the designer is now in a position to make a design decision that effectively takes user performance into consideration (see Fig. 2.6).

As part of the final decision step, the designer is expected to concentrate on assessing interactions between user and HMD system properties. Various trade-off and what-if assessments will probably be performed. Eventually this leads to a display resolution decision that includes hardware and/or software options that can achieve the requirement. The decision, its assumptions and rationale, and its potential side effects are all recorded for further use in the systems engineering process.

With a little reflection, it should be clear that all aspects of the design decision that result from this user-centered process stem fundamentally from viewing the design problem in terms of *user performance*. In effect, the design problem is redefined from a physical form (display resolution) to a psychophysical form (object/scene visibility) that, by definition, includes user performance. The design goal, then, is to achieve a certain level of visibility under specified conditions. The psychophysical property directly represents the interaction between the HMD property and a user property (e.g., visual information processing). Thus, in a very real sense the designer has included visibility as a property designed into the HMD system. One manifestation of this property is the resolution made available by the system display. There may be times when the designer may not be able to crisply define the psychophysical property, but it will be implied through active consideration of how system and user properties interact. This may tend to occur when several interactions are involved between user properties and machine ones. When the psychophysical variable is clearly recognized and easily labeled, it is desirable to include it in the final design solution expression, as has been done in Fig. 2.6.

It is expected that the user performance–oriented decision process may have to be iterated several times for each HMD design issue (system property) until a design concept is formed that properly addresses all major interactions. The full process requires integrating, updating, and exchanging information about HMD property designs throughout the development process. Clearly, the slice of the process shown here only begins to show the complexities involved in design decision making, how user-centered design operates for detailed design decisions, and the value of systematic record keeping. The user-centered perspective does not make it any easier to reach a design decision, but it does increase the odds that final integration problems will be reduced and task performance will meet expectations.

2.4.3 Facilitating discovery

The Discovery Step is a vital component of the user-centered decision-making process. This is the time when the designer is first asked to identify human factors that interact with the design problem. If an adequate list of human properties is not generated, the entire decision-making process is weakened. Unfortunately, knowledge about human properties happens to be the area where the designer's technical knowledge is apt to be weakest. For this and other reasons, it may also be the place where the least amount of effort is expended. As a result, the designer may need encouragement even to use the user-centered decision-making process. Moreover, once the designer is prepared to start, further assistance may be needed to facilitate activation of the process. We begin this section with a brief look at an impediment to using a user-centered design approach during detailed design, and then offer a few simple methods to facilitate the discovery activities, once the designer is committed to its use.

Junior staff engineers are especially prone to believe that a user-centered approach is not applicable to the design problems handed to them, even if user-centered design is formally embraced by the senior staff. The problem stems from the fact that low-level design problems are usually assigned to junior engineers as implementation and not design issues. As a result, the engineer is only expected to "implement" the design in either the current or the final medium of the artifact. Since the task is couched in terms of implementation, the design engineer is encouraged to focus exclusively on characteristics and constraints offered by the medium of the artifact. Aspects of human performance seem irrelevant to the implementation problem.

Unfortunately, there is no way to guarantee that a designer at any level will use a user-centered approach to decision making. The odds of following a user-centered design approach and employing the three-step procedure are increased, however, (1) if it is a formal part of the organization's system engineering practices, (2) if it is actively endorsed by technical managers, and (3) if there are adequate tools to support its use. When it comes to low-level, detailed design decision making, it also helps if the junior staff members, who usually work at this level, are encouraged to view themselves as designers who are equally responsible for user performance. Perhaps the simple fact of knowing that there may be a tendency to think of "implementation" problems as lying outside the bounds where user-centered design is needed may be enough to prevent this from happening.

If it is assumed that a user-centered decision process will be used during detail design, then there are at least two simple techniques a designer can use to facilitate the Discovery Step. One method is to ask user-centered questions about the design problem. A second method involves the use of one or more user models to stimulate and focus questions.

The type of user-centered questions that are useful for the discovery process focuses on user performance. Although at times designers may not be able to generate user-oriented questions from the perspective of the hardware design problem, it should be easier for them to do so by starting with previously identified task factors, like the list shown in Fig. 2.6 for the display resolution problem. These task factors relate directly to human activities, which should be readily apparent. Then all the designer needs to do is ask what specific aspects of the human are used when these activities are being accomplished. Since the task factors are connected, a priori, to the design problem, the human factors will be too. By initiating the discovery process in this way, the designer will be able to break free of thinking about the design only in terms of detailed aspects of the hardware and software factors and thus be able to shift to the user or psychophysical perspective.

Another way to jump-start the discovery process with questions is for the designer to ask psychophysical questions about the design problem. To form such questions, a physical property must be connected to a psychological aspect of the user. For example, consider the design problem of determining when contact is made between graphically rendered objects. From the traditional machine design stance, the designer would tend to focus on physical questions such as, What range of shapes must the detection algorithm handle? Do material properties of objects need to be taken into account? What computer processing resources are available? What is the processing time budget? In contrast to these questions, psychophysical ones are stated from the perspective of the user. For this design problem some reasonable questions to ask are, How will the algorithm affect the *appearance* of the collision and postcollision behavior of the objects? How will the appearance change as a function of object shape, trajectory approach angle, or velocity or acceleration profile? From there it is easy to see that aspects of human perceptual processing are relevant to the design problem. Once this is established, the designer is positioned to search for and uncover (i.e., discover) other relevant psychophysical connections with the design problem.

Another tool that can be used in all three steps of the decision-making process is a model of the user or user-machine system. In the discovery phase, one might simply consult a standard human information processing model, for example, as a way to stimulate more concrete thinking about areas and ways the design problem may impact information processing. In some situations a manual control model might be more useful. Many different types of human performance models are available that could be useful. In general, it is desirable to use several different model types during discovery to help ensure that important items are not missed. The use of models will be illustrated in the design problem to be discussed in the next section.

2.5 Analysis Step

This section attempts to provide a flavor of the Analysis Step in the user-centered design decision-making procedure. The main goal is to show the types of analyses and level of detail that may be necessary before a high-quality design decision can be made. The discussion centers around the need to make design decisions regarding the temporal properties for an HMD system. Several interacting problems must be considered, as is typical of many design problems. The presentation begins at a point where the designer is starting to think about potential interactions between temporal characteristics and user performance. Hence, a portion of the discovery process is included in the example.

2.5.1 Use of models in the design decision-making process

Assume that a design engineer, who is a member of an HMD system development team, has been given responsibility for selecting the graphics computer to be included in the system, and the task of designing the process control architecture for the system. The designer must make several interrelated decisions. Should the control structure be fully synchronous, fully asynchronous, or hybrid? When can parallel processing be used? When is serialization required? What schedule prioritization is needed? What image update is needed? The designer is aware that, historically, processing time delays in HMD systems have produced image update lags that have resulted in user complaints and, in some instances, poor performance. The designer is determined to avoid such problems with the new system and has therefore decided to approach the design from a user performance perspective.

The first task, as previously discussed, is to gain a better understanding of how human properties might interact with the temporal properties of the HMD system. To assist in the discovery process, the designer has elected to employ a model of user behavior. This helps to keep properties of the user in front of the design team and also provides a point of departure for asking user-oriented questions. The selected process model is shown in Fig. 2.7. The model represents a human as a general data processor with distributed processing elements identified as labeled boxes. The boxes may be regarded as human properties, and the arrows as data-flow pathways.

As shown in the figure, the overall model forms a perception-action loop and indicates that perception may be influenced by reasoning and other cognitive factors. The human perceptual processor is set into action by stimulation from the external world. A form of energy is captured by one or more sensors, which, generally after local processing,

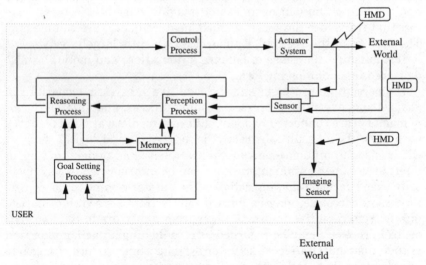

Figure 2.7 Process model of user performance. An HMD has been inserted based on planned use in the zone reconnaissance design problem.

transduce the energy into a data signal that serves as input for the perceptual processor. This unit, which may actually be a set of processors, is believed by some scientists to be amodal. This means that it does not preserve the sensory source of data after it is integrated into perceptual information (the output), which is called a *percept.*

A percept may be thought of as a phenomenal picture, extensive in space and time, that is projected outward and represents one's momentary awareness or understanding of the surrounding external world (including the perceiver as an entity in the world). Any particular understanding may be based in part on stored information retrieved from memory or resulting from cognitive processes that operate on the basis of symbolic information. The impressed sensed data may dominate the perceptual process (i.e., data driven) or it may be dominated by symbolic information (i.e., conceptually driven). The notion of a perceptual moment has been proposed as a cyclic process that takes approximately 50–200 milliseconds to form a percept (Card, Moran, and Newell, 1983). Other evidence indicates that some energy pattern changes that take on the order of 5 milliseconds to occur can stimulate different percepts (Westheimer and McKee, 1977). Scientists still have much to learn about the perceptual process. The important point here is to recognize that an HMD user experiences the external world largely in terms of integrated percepts consisting of objects, surfaces, and events defined by their interactions.

Two classes of human sensors are identified in the model: imaging and nonimaging. A sensor is considered to be of the imaging type if it creates a spatially and/or temporally extensive icon through local processing, prior to data delivery to the perceptual processor. Vision and audition are prominent examples of imaging sensors. Mechanoreceptors that transduce the sense of touch, heat, cold, etc., are nonimaging. The data from the vestibular apparatus, which is largely responsible for establishing the sense of balance, orientation, and motion, is also delivered in a nonimaging form.

The perceptual processor and the cognitive subsystem (consisting of goal setting, executive process, reasoning process, and different forms of memory) are connected by many different feedforward and feedback pathways. As a result, abstractly defined goals and logical analysis can be used to influence perceptual awareness and understanding, or alternatively, perceptual information can be incorporated in a symbolically based decision-making activity. The data presented to the control processor, therefore, may be formed purely from sensed information, purely from cognitive information, or from a mixture of the two. The control processor may be regarded as establishing an action plan that is then distributed to local actuator systems that produce changes in muscle state that lead to physical movements or actions by the limbs, head, eyes, vocal cords, etc.

As shown in Fig. 2.7, this basic process model of a human is expressed at a high level of symbolic abstraction. This level of presentation will probably be adequate for use during the discovery process early in a development project, but later the designer may wish to use a more detailed process model. (Elkind et al. (1990) provide a good review of different models of human performance that may be useful to designers.) Since the designer here is using the model to stimulate initial thinking about how temporal properties of the HMD system might interact with properties of the user to impact user performance, the level of detail shown will suffice.

A good way to begin the user-centered discovery process is to determine the different ways the HMD system attaches to or enters into user processing. Let us assume that the HMD system to support the zone reconnaissance mission consists of a single tracker attached to an HMD that supports both see-through to the outside world and sensor-based imagery and contains a visual cursor whose position is coupled to head look angle. It should be easy to see that this system will be inserted into the human processing streams in three ways. As shown in Fig. 2.7, the most obvious point of insertion is between the external world and the user's eye (represented generically as an imaging sensor). Because the HMD generates visual images, as well as allows visual access to the external world, there are two different data paths going into the eye, as shown. The remaining pathway involves the tracker portion of the HMD system. The tracker is attached to the head (via a helmet), which acts as a control actuator used to move a visual cursor. But head movement also stimulates mechanoreceptors, such as stretch receptors located in the neck and the vestibular apparatus, both of which provide data to the perceptual processor through nonimaging pathways. This internal information feedback path into the perceptual processor is easy to overlook without the aid of the model.

After the human process model has been used to visualize the areas where the HMD will interact with the user, the designer is ready to ask questions related to the design problem at hand. It should be obvious that there will be a temporal delay between the time the user's head is moved and the time when the cursor coupled to this movement changes position in the HMD image. Such a delay must occur, because a finite amount of time is needed to process head movement data and update the image. The visual image of the cursor depends on HMD processing time, but proprioceptive sensing of head movement does not go through the HMD; hence head movement data is made available to the perceptual processor in advance of that from the cursor, which comes from the imaging sensor pathway. This may create a problem since the intent is for the cursor to move with the head, not after it. Hence there will be a sensory cue mismatch when the HMD user tracks an object with head movements. This invites the question, How much time lag can be tolerated without degrading tracking performance? And this leads naturally to, What is the functional relation between tracking performance and time lag in updating the visual image of the object? Notice that these are psychophysical questions. The user-centered adjustment is beginning to take hold.

In general, the dynamics of any graphical object imaged on the HMD whose behavior is dependent on sensed data will be delayed and perhaps altered in other ways as well. This implies that the motion of a computer-generated image of a real object relative to the motion of other real objects contained in a see-through scene with graphical overlays will be delayed. These observations invite framing another general user performance question: How will lag affect the appearance of an interaction across virtual and real objects?

Usually, more performance-related insights will be gained if the designer uses multiple models that capture user and system aspects. Each model should highlight different aspects or treat the same ones from a different perspective. During discovery the goal is to identify as many human factors relevant to the design problem as possible. These will be reduced later, on the basis of analysis.

A second model is shown in Fig. 2.8 and 2.9. This model is somewhat unique in that it includes both hardware and image properties of an HMD along with various aspects of human perception and motor control. All of the listed system properties have an effect on the

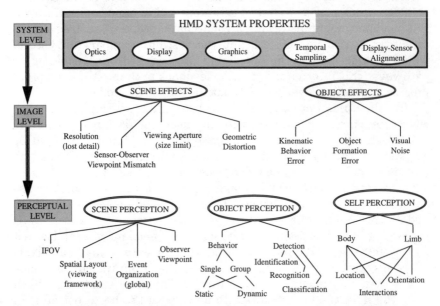

Figure 2.8 Multilevel, joint HMD-user performance model. It points out how hardware and software design decisions propagate to impact HMD images, which in turn impact user perception. Selected specific areas of effect are identified by line connections.

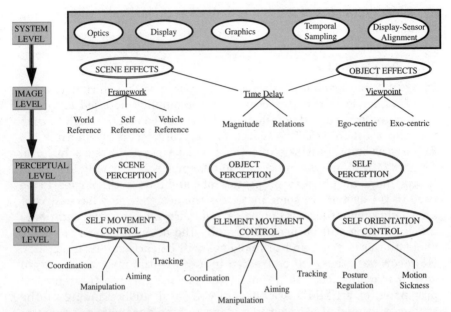

Figure 2.9 Multilevel, joint HMD-user performance model extension. The model is extended to include user motor control.

graphically rendered image, both at the global scene and individual object levels. (Lines showing these connections have been omitted to reduce clutter.) Lower-level properties of the HMD, which may have surfaced initially as design problems, are depicted under the image effect bubbles. Note that different properties are featured in the two figures. Those in Fig. 2.8 are believed to be most directly related to human perceptual performance, whereas those shown in Fig. 2.9 may tend to manifest their impact more on motor performance. Obviously, a human focus appears in the model at the perception and motor control levels. The factors attached to the bubbles at these levels address human performance. (The perceptual performance factors have been omitted from Fig. 2.9 to simplify the presentation.)

With respect to the temporal properties of an HMD, this descriptive model suggests that timing factors may impact the entire scene, as well as the kinematic accuracy of object movement and location within the scene. With respect to the user, these properties may disturb the perception of object interactions and the perception of self-orientation or movement, and cause problems for self-movement, self-orientation, and object movement control. These factors were not as strongly suggested by the process model shown in Fig. 2.7. Further, the model invites the designer to question whether timing factors may contribute to the onset of motion-induced sickness. These newly observed issues are added to the list of items for analysis previously identified during the discovery phase.

2.5.2 Detailed analysis

Once user performance issues are expressed with sufficient clarity, the designer is in a better position to accomplish detailed analyses, since more focused and detailed questions can be posed, which in turn will allow a more effective search for relevant technical information. Equally as important, the questions will be psychophysical in form. As part of the analysis process, the designer will normally also have to dig more deeply into properties of candidate technologies to be used in the system. In some instances the designer may have current knowledge about these technologies, but more often than not additional understanding will be needed. Usually the acquisition of knowledge about the user, task, and technologies will be carried out in parallel. However, here they will be treated in a sequential manner. We begin with a machine focus in an effort to gain insight into the temporal properties of an HMD system. Once a good understanding of the system is formed, it is easier to understand the dimensions of a design problem, and there is more opportunity for insights to occur as attention is turned to user factors.

Technology-centered analysis.   To support this analysis, we will employ an HMD system composed of a single magnetic-type tracker subsystem, a graphics processor with a hardware geometry pipeline to facilitate coordinate transformation calculations, and a raster-type display. Digital signal processing is assumed to be used in each subsystem. Hence, all signals are sampled in time. The tracker sensing system is assumed to include an active transmitter that broadcasts a signal for pickup by the sensing unit. This type of arrangement is found in many tracking

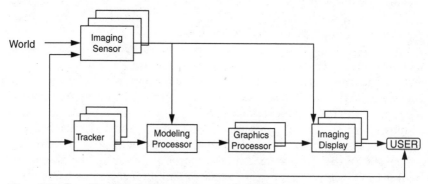

Figure 2.10 Functional block diagram of a generic HMD system.

systems used in HMD systems. The analysis begins with a separate inspection of the major HMD components. A functional block diagram for the system is shown in Fig. 2.10.

Magnetic tracker. The basic structure of a magnetic tracker system is shown schematically in Fig. 2.11. The system is activated by the reception of an electromagnetic signal by the sensor/receiver unit. The transmitter creates a pulsating energy field that serves to induce a current flow in each of three orthogonally wound coils contained in the receiver (sensor). A computational processor extracts (calculates) position and orientation information (referenced to the transmitter) sequentially from each of the coils, and is able to provide x, y, z, roll, pitch, and yaw information for the receiver state. The processor usually contains a set of temporal filters that can be used to minimize the effects of magnetic field disturbances introduced into the signal either from external sources or from unintended micromovements (e.g., tremor) made by the object to which the sensor is attached (e.g., the HMD system user).

Because trackers are made as separate products by vendors for use in different systems, the temporal characteristics of the tracker are generally included in product specifications. Unfortunately, what features are specified and the methods used for defining their values are not standardized within the industry. Obviously, this makes it difficult to make informed comparisons of competing products. For the following analysis we will use a BIRD™ tracker system, made by Ascension Technology Corporation.

Receiver. The receiver unit (sensor) for this system samples the broadcast signal once every 8.33 milliseconds, which is equivalent to a 120-Hz sampling rate. The sampled data is delivered to a processing unit that filters the data and then computes position and orientation values for the sensor location/orientation relative to the transmitter. The sampling process is controlled by a sync pulse. A cycle begins with a wait period for stabilization (about 5 milliseconds when the

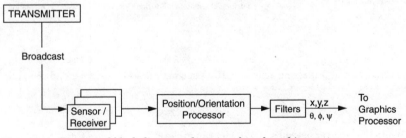

Figure 2.11 Functional block diagram of a sensor-based tracking system.

complete BIRD system is operating in 60-Hz update mode),[1] and then samples in series the x, y, and z axes from the corresponding loop antennae of the transmitter. Because of this sequential processing scheme, the system is not able to provide complete spatial positioning and orientation information for any specific time, t. Rather, the position coordinates assigned to tx represent the average position over the sampling interval (8.33 milliseconds). Thus, the system has a built-in spatial error whose magnitude is unspecified and varies as a function of receiver velocity. Under steady-state conditions, static spatial accuracy is advertised to be 0.1 inch (0.25 cm) for position and 0.5 degree visual angle for orientation.

Computational unit. Loop currents are transformed into position and orientation information by a set of algorithms contained in this unit. The algorithms complete processing in 8.33 milliseconds. Thus, with proper synchronization, the computational unit is operating on sample packet $n + 1$ when the receiver is acquiring sample packet n. This means that it takes 16.67 milliseconds for any one sample packet to be made available by the tracker to other subsystems in the HMD system (See Fig. 2.12).

Filtering. The BIRD system has several filters that may be used to minimize the effect of electromagnetic disturbances in the ambient environment. These include a wideband and a narrowband notch filter (primarily for canceling 60-Hz AC signals), and a bank of tuned

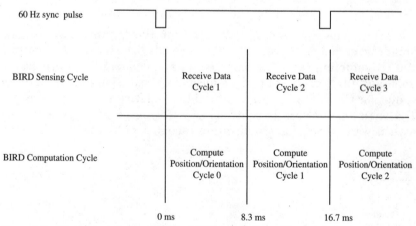

Figure 2.12 Signal processing chart for BIRD magnetic tracker, receiver, and computational units.

---

[1]While the receiver operates at 120 Hz, two cycles of data are needed to complete a position/orientation computation. Thus, minimum transport delay for the tracker as a whole is 16.67 msec, or 60 Hz.

DC filters, which can be selected adaptively. At a nominal 12-inch transmitter-to-receiver (sensor) separation distance, the manufacturer indicates that the DC filters, coupled with either the wide or narrow notch filter, add one 16.67 millisecond cycle to the processing time to detect a step input to a constantly changing sensor position. More processing time is added as the transmitter-to-receiver distance is increased. Thus, time delays introduced by the tracker may be as little as 16.67 milliseconds (no filters) or substantially longer, depending on filter combinations used and target velocity. According to the manufacturer, delays on the order of 33.4 to 66.67 milliseconds or even longer may be expected (S.S. Work, 1993, personal communication).[2]

Eggleston and his colleagues have measured the steady-state static and dynamic performance for both the original BIRD system and the newer FLOCK OF BIRDS™ system. The FLOCK OF BIRDS supports multiple receiver units. Eggleston and Janson (1993) used a three-dimensional Peg-Board arrangement to position a BIRD sensor at several calibrated positions (2-inch separation) in a 24-inch cubic volume. In general, they found steady-state position accuracies for static $x, y,$ and z inputs to be within the 0.1-inch specification. Performance for the FLOCK OF BIRDS was assessed in the same manner, except that the operating volume was increased in size to 8 ft (2.44 m) in length by 8 ft (2.44 m) in width by 6 ft (1.83 m) in height (Eggleston and Waltensperger, 1994). The results were comparable to those of the original study, with some fall-off due to the increased distances involved. While these findings confirm the advertised claims for the tracker system, it should be noted that performance may drop off significantly in a less than ideal working environment. For example, Eggleston, Janson, and Aldrich (1994) found position errors of over an inch (2.54 cm) 6 feet from the transmitter for the FLOCK OF BIRDS system when it was used in a room with standard computer flooring. Measurement accuracy will generally vary throughout the operating area. While it is possible to compensate for nonlinearities in magnetic field distortions via postprocessing, the added computation time will increase system lag (Bryson and Fisher, 1990). Static analysis of trackers has also been performed by Bryson (1992) and Williams (1994).

In an effort to determine the transport delay for a BIRD tracker under dynamic conditions, Eggleston, Janson, and Aldrich (1994) mounted a receiver on a sled that was moved at a constant velocity along a 6- or 18-inch track. End-to-end movement time was measured both with the BIRD system and with an independent digital counter arrangement. By comparing BIRD output with the output of this other timing system, time delays in the BIRD system could be determined to an accuracy

---

[2]The timing data provided by Ascension Technology reflects the time it takes the BIRD to respond to a step input.

of 1 millisecond. For this analysis, the BIRD was operated at 120 Hz. (Note: this rate is often not usable when the BIRD is placed in a system context.)

Three movement rates were used in the study (inches/second): 7 (17.78 cm/second), 12 (30.48 cm/second), and 30 (76.20 cm/second). All movement times were recorded with the notch filters turned off and with the DC filter fixed at an alpha level of 0.50. (With this setting, equal weight is given to old and new data in calculating position. That is, the current position is taken to be the average value of the position calculated at time $t - 1$ and that calculated at time t.) Data was collected in both point (polling) and stream (interrupt) modes.

The results were comparable across all track orientation conditions investigated. As expected, point mode added a constant 8.33 milliseconds of delay due to polling from the host computer. Results for the x-axis movement condition are representative of the findings. They are summarized for the 18-inch (45.72 cm) travel length in Fig. 2.13.

Mean distance from the stopping point (18 inches) is plotted as a function of mean time delay (i.e., lag in BIRD output relative to other timing circuit). All means are based on 20 samples. The dashed vertical line indicates the time the sled reached the stop position, as indicated by the alternative timing circuit. Thus, if the BIRD signaled the same position at that time, this would signify a zero time delay

Mean Distance by Delay and Response Speed
(Point Mode)

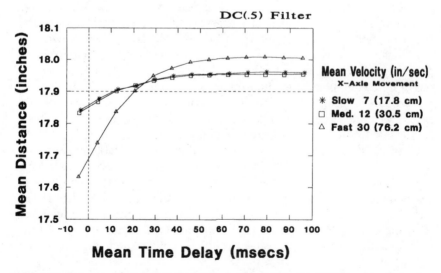

Figure 2.13 Dynamic assessment of position tracking with the BIRD tracker. Data shows time delay between actual object position and tracker-reported position. Actual terminal position is 18 inches.

for the BIRD system. The dashed horizontal line has been set at 17.9 inches (45.47 cm), since this value represents the nominal steady-state position (translation) accuracy for the BIRD system. With the 0.1 accuracy criterion, time delay for the BIRD was 13 milliseconds for the slow and medium speed conditions and 23 milliseconds for the fast one. However, if an asymptotic spatial performance criterion is used, then time delay values increase to 46 milliseconds and 56 milliseconds for the different speed conditions.[3] Clearly, any assessment of time delay is dependent on the level of spatial error that is deemed to be acceptable, as well as how error is defined.

Figure 2.14 shows the impact of different combinations of tracker filter settings on time delay. It indicates that time delay increases with movement velocity and depends on filter setting. Additional information on the dynamic performance of tracking systems can be found in a study by Adelstein, Johnston, and Ellis (1992). They assessed the performance of three different tracking systems in response to damped sinusoidal oscillations in the 1–3.8 Hz range—a frequency range that corresponds with continuous manual control behavior.

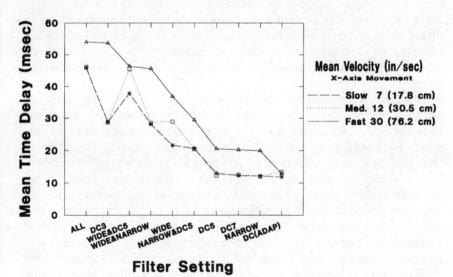

Figure 2.14 Dynamic assessment of position tracking with the BIRD tracker. Data shows time delay as a function of filter setting used.

---

[3] All times are reduced, on average, by 8.33 msec in stream mode.

As the designer reviews technical data on component technologies, it is important to attempt to relate the information found in the literature to the uses of interest for the current design problem. This may reveal important areas where adjustments are needed before the information can be properly applied to the present design problem. For example, it was noted in the previous data that BIRD time delay was measured to a precision of 1 millisecond. With a little effort the designer should be able to see that under normal operating conditions the tracker actually only updates at 16.67 milliseconds. Thus, when the tracker is used in a system, its output will impose a minimum lag of this amount, and not those indicated in the Eggleston et al. data (see Figs. 2.13 and 2.14). In a word, the designer must stay active and alert when processing technical information, always mindful of how it relates to the design situation.

This limited discussion of a particular tracker system was included to illustrate the level of detail the designer may need to form an adequate understanding of the factors that relate to the design problem. A simple review of vendor-supplied specifications will generally not be adequate. But, as noted, there can also be pitfalls when using published data, even if it is from an independent source or is contained in a reputable data bank or handbook.

The aim of the Analysis Step is for the designer to gain a deep understanding of technology and user properties that bear on the design problem. The hope is that this knowledge will produce insights about the psychophysical connection(s) between these two domains. Clearly, the fact that the decision-making approach is user-centered does not imply that details of the technology are unimportant and can be ignored.

Of course, a similar detailed analysis aimed at elucidating temporal behavior and constraints imposed by the component design would have to be performed for all candidate technologies being assessed to accomplish the tracking, image rendering, and visual display functions of the HMD system. In addition, an analysis that addresses the problem from a systems perspective is also needed. This analysis would bring out interactions across components and reveal how factors such as interface protocols and data buffering schemes might contribute to the delay in updating position information at the visual display. As part of a system-level analysis, it is useful to include a "best case" assessment of end-to-end transport delay. For example, if a fully synchronized HMD system is a hard constraint, and the best available component technologies process data at, say, 60 Hz, then the minimum time between when an event occurs in the world and when it is depicted on the visual display will be, on average, 58.34 milliseconds. This assumes that the system is composed of a separate tracker, a graphics processor (with video buffer), and a raster-based visual display that refreshes at

60 Hz (16.67 tracker processing + 16.67 graphics processing + 16.67 video buffer delay + 8.33 (average) time the raster paints the image on the display). The time required to satisfy interface protocols and other integration factors will increase delay in 16.67-millisecond increments.

Human-centered analysis.   Another very important aspect of the analysis activity addresses potential user performance consequences of the magnitudes and patterns of time delays associated with different design options that can solve the design problem. Time delays can affect sensory pickup, perception, decision making, and motor control. Time delay may cause images and information to be represented incorrectly (misrepresentation) or invite misinterpretation by internal processes of the human user. As indicated earlier, the designer needs to acquire as much detail as possible to understand how the HMD system properties under consideration interact with the user to impact user performance.

By consulting human performance data from many sources, such as human factors handbooks (Boff, Kaufman, and Thomas, 1986; Woodson, Tillman and Tillman, 1992; Van Cott and Kincaid, 1972), engineering data compendia (Boff and Lincoln, 1988), and recent findings broadcast on the World Wide Web, the designer can acquire a deeper understanding of the design problem from a psychophysical perspective. For example, a review of the various sources would indicate several different ways perceptual processing and motor control are adversely affected by the magnitude and pattern of time delays in an information transfer system such as an HMD. Based on a penetrating analysis of the user-centered items posted in the Discovery Step, as well as those added during analysis activities, the designer would acquire detailed information about time delay and its user consequences in terms of (1) apparent scene and/or object flicker, (2) erroneous appearance of simultaneously occurring objects or objects separated over time, (3) erroneous dynamic behavior of an object, including object formation, (4) multiple images of a single object, and (5) disruptions of user control and tracking performance.

Only a small portion of the knowledge gained in these areas is presented here. A great deal of information on human manual control is available, and some studies have addressed the effect of time delay on tracking and control performance. A study that is particularly relevant to the design problem has been performed by the author. The study assessed compensatory tracking and control performance when a glove mounted with a magnetic sensor served as the control actuator. Results from the study are summarized in Fig. 2.15. It is clear from inspection of the graph that human performance indexed by system gain at the point when control is lost (see Jex, McDonnell and Phatak, 1966) is degraded in a manner that is linear with time delay. Performance scores over 5.0 represent an ability to exercise good control over the

Virtual Hand Control: Time Delay Study

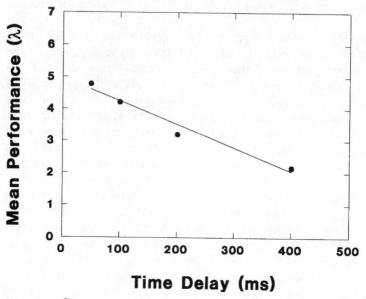

Time Delay (ms)

Figure 2.15 Compensatory tracking performance as a function of system time delay with the BIRD tracker.

nonlinearly varying control system, whereas scores around 2.0 indicate almost no control over the system.

Analogous to the situation with published data about hardware and software, the designer must also take care when relating human-centered data to the current design problem. A designer may have to work hard to properly relate such data parameters of the design problem (i.e., psychophysical variables) to user performance. A case in point is the variable of image flicker. It is well known that time-sample images can appear to flicker. The main source of the problem stems from an interaction between the integration time of visual processing and display update rate. The update rate at which flicker is observed or is bad enough to distract a viewer depends on a number of scene and user factors (see section 1.5 in Boff and Lincoln, 1988).

Design engineers have been aware of the image flicker problem for many years. The scheme to interlace alternating fields when forming a television image was largely motivated to reduce image flicker without increasing display bandwidth. The designer for the current design problem may be tempted to simply consult the standard data and stop there, since the data is readily available and the problem is "well understood." Further, the data may appear adequate to answer the question. Unfortunately, this would be a mistake. The conditions under which flicker appears vary as a function of image size. Flicker also interacts with object motion. Thus, the designer needs to look at published data on flicker in terms of the object size or sizes and motion profiles germane to the zone reconnaissance task. The designer will probably have to pull together information from different studies to acquire the neede level of understanding. There are no easy paths to solutions for complex problems.

The process control architecture for an HMD system offers a formidable design problem. It is multidimensional and complex. The brief treatment of the problem here is clearly not adequate to converge on a design solution. But this was not the intent. The objective of the illustration was to demostrate, in brief form, that a user-centered approach can be used with complex problems just as well as with simple ones. The result of problem formulation, user-centered discovery, and analysis for the problems that have been discussed, at least minimally, are summarized in Fig. 2.16. The integrated HMD-user knowledge shown in the analysis section represents a psychophysical expression of the design problem. The fact that such statements can be made is evidence that complex problems can be attacked in a user-centered design manner.

2.6 Final Comments

Few would argue with the goals of user-centered design. Any human-machine system that has useful features and is easy to use will be highly valued. In the standard machine-centered approach to design, human factors specialists are charged with making the system usable, while other design engineers are largely responsible for developing useful features. The user-centered approach to design decision making attempts to combine concerns for a useful and usable product into a common framework. This framework has been expressed here in terms of user performance. Within it, usefulness and usability are brought together by approaching design decision making psychophysically.

Considerable work is required to turn the user-centered design philosophy into an implementable design practice. I have tried to sketch out a route to this problem that focused on the design decision-making process. To go down this path, the HMD design engineer must do several things, all of which have been touched on in the chapter. In condensed form, they are the following:

1. Take responsibility for user performance.
2. Treat all design problems from a user orientation.
3. Learn to think about design problems psychophysically.

Design Problem $H_{2,0}$ (Select Process Control Architecture: Temporal Characteristics)

Initial Property List:

$HMD_{2,0}$ (Process Control Architecture: Temporal characteristics)
$\quad\quad\quad\quad$ $HMD_{2,1}$ $\quad$ Synchronous path(s)?
$\quad\quad\quad\quad$ $HMD_{2,2}$ $\quad$ Asynchronous path(s)?
$\quad\quad\quad\quad$ $HMD_{2,3}$ $\quad$ Serial structure?
$\quad\quad\quad\quad$ $HMD_{2,4}$ $\quad$ Parallel structure?
User: $\quad$ $U_{2,0}$ $\quad\quad\quad\quad\quad\quad$ (Perception)
$\quad\quad\quad$ $U_{3,0}$ $\quad\quad\quad\quad\quad\quad$ (Motor Control)
Task: $\quad$ $T_{2,0}$ $\quad\quad\quad\quad\quad\quad$ Observe and report activities and locations
$\quad\quad\quad\quad$ $T_{2,1}$ $\quad$ Mark vehicle(s)/location(s) on map
$\quad\quad\quad\quad$ $T_{2,2}$ $\quad$ Mark unit locations
$\quad\quad\quad\quad$ $T_{2,3}$ $\quad$ Track routes

Discover User Properties:

User: $\quad$ $U_{2,0}$ $\quad$ Perception
$\quad\quad\quad\quad$ $U_{2,1}$ $\quad$ Scene perception
$\quad\quad\quad\quad\quad\quad$ $U_{2,1\text{-}1}$ $\quad$ Perceptual organization
$\quad\quad\quad\quad$ $U_{2,2}$ $\quad$ Object perception
$\quad\quad\quad\quad\quad\quad$ $U_{2,2\text{-}1}$ $\quad$ Motion perception
$\quad\quad\quad\quad\quad\quad$ $U_{2,2\text{-}2}$ $\quad$ Perceptual organization
$\quad\quad\quad$ $U_{3,0}$ $\quad$ Motor Control
$\quad\quad\quad\quad$ $U_{3,1}$ $\quad$ Visual feedback path
$\quad\quad\quad\quad$ $U_{3,2}$ $\quad$ Proprioceptual feedback path
$\quad\quad\quad\quad$ $U_{3,3}$ $\quad$ Coordination between feedback channels

Analysis:

K_H:
- Tracker unit signal processing (each candidate)
- Graphics unit image processing (each candidate)
- Display unit processing
- Architecture effects on image update

K_U:
- Visual integration time constant (Bloch's Law)
- Apparent motion processing
- Motor control performance
  - aimed movement
  - continuous tracking

K_T:
- Time available to detect, mark, track/follow object(s)

K_U-K_H:
- Scene flicker (display update)
- Object flicker (tracker/image update)
- Spatio-temporal object location/behavior (temp. arch.)
- Multiple (ghost) images (temp. arch. w/ obj. vel./acc.)

Design Decision:

Figure 2.16 Summary of results for the process control architecture problem.

4. Actively collect, analyze, and synthesize published data and information.

5. Critically assess relations between conditions underlying published data to ensure proper connection with the conditions relevant to the design problem, and transform the data, if necessary.

6. Use systematic collection, recording, and reporting procedures.

7. Make good decisions!

2.7 Acknowledgments

Preparation of this manuscript occurred during a very busy and hectic time. The assistance provided by Bill Janson, Ken Aldrich, and Laura Mulford was instrumental in bringing the project to completion. I am grateful for their support in preparing the graphs contained in the manuscript and in handling all administrative details.

2.8 References

Adelstein, B. D., Johnston, E. R., and Ellis, S. R. (1992). A testbed for characterizing dynamic response of virtual environment spatial sensors. *The 5th Annual ACM Symposium on User Interface Software and Technology, Monterey, California,* 15–22.

Boff, K. R., and Lincoln, J. E. (1988). *Engineering data compendium: Human perception and performance.* Wright-Patterson AFB, OH: Armstrong Aerospace Medical Research Laboratory.

Boff, K. R., Kaufman, L. and Thomas, J. P. (1986). *Handbook of perception and human performance.* Wright-Patterson AFB, OH: Armstrong Aerospace Medical Research Laboratory.

Bryson, S. (1992). Measurement and calibration of static distortion of position data from 3D trackers. *Proceedings of the SPIE Conference on Stereoscopic Displays and Applications III, 1669,* 244–255.

Bryson, S., & Fisher, S. S. (1990). Defining, modeling, and measuring systems lag in virtual environments. *Proceedings of the SPIE Conference on Stereoscopic Displays and Applications, 1256,* 98–109.

Card, S. K., Moran, T. P., & Newell, A. (1983). *The psychology of human-computer interactions.* Hillsdale, NJ: Lawrence Earlbaum Associates.

Eggleston, R. G., & Janson, W. P. (1993). *Assessment of static accuracy and linearity of a magnetic tracker system.* Unpublished manuscript.

Eggleston, R. G., Janson, W. P., & Aldrich, K. A. (1994). *Assessment of the dynamic performance of a magnetic tracking system.* Unpublished manuscript.

Eggleston, R. G. & Waltensperger, G. M. (1994). *Assessment of static accuracy and linearity of a magnetic tracker system in an eight-foot cubic volume.* Unpublished manuscript.

Elkind, J. I., Card, S. K., Hochberg, J., & Huey, B. M. (eds.) (1990). *Human models for computer-aided engineering.* New York: Academic Press.

Jex, H. R., McDonnell, J. D., & Phatak, A. V. (1966). A "critical" tracking task for man-machine research related to operator's effective time delay. Part I: Theory and experiments with a first-order divergent controlled element. (NASA-CR-616). Hawthorne, CA: Systems Technology, Inc..

Rouse, W. B. (1991). *Design for success.* New York: John Wiley & Sons.

Van Cott, H. P., & Kinkade, R. G. (eds.) (1972). *Human engineering guide to equipment design.* Washington, DC: American Institutes for Research.

Westheimer, G., & McKee, S. P. (1977). The perception of temporal order in adjacent visual stimuli. *Vision Research, 17,* 887–892.

Williams, M. (1994). A comparison of two examples of magnetic tracker systems. In *AGARD Conference Proceedings—Virtual Interfaces: Research and Applications,* (AGARD-CP-541) (pp. 14-1–14-19). Neuilly-Sur-Seine, France: Advisory Group for Aerospace Research & Development.

Woodson, W. E., Tillman, B., and Tillman, P. (1992). *Human factors design handbook: Information and guidelines for the design of systems, facilities, equipment, and products for human use.* 2d ed. New York: McGraw-Hill.

HMD Image Source, Optics, and the Visual Interface

H. L. Task

This chapter presents many of the key parameters associated with head-mounted display image sources and optics and how these interface with human vision. After completing this chapter the reader should have a good understanding of most of the major parameters used to characterize head-mounted displays, including field of view, image quality (resolution), eye relief, exit pupil, luminance, and focus. In addition, the reader should have a better understanding of the interaction and trade-offs of these main parameters, such as field of view versus resolution, field of view versus eye relief, and exit pupil versus eye relief. The objective of this chapter is to provide the reader with the tools and background to be able to evaluate, select, and/or specify a head-mounted display.

3.1 Introduction

A head-mounted display (HMD, an abbreviation also used for *helmet-mounted display*) is a device that delivers a virtual image, such as from a cathode ray tube or liquid crystal display, visible to one or both eyes. A visually coupled system (VCS) is composed of three major parts: a head-mounted display, a means of tracking head or eye position (look angle), and a source of visual information that can be updated and changed to correspond with the head/eye look angle. Figure 3.1 is a block diagram of part of a visually coupled system with the HMD section further divided into the three major (numerically marked) elements: (1) the display or image source, (2) the optics through which the display is viewed, and (3) the means by which the display and optics are mounted on the head or helmet. The purpose of this chapter is to discuss the important characteristics of the image source and optics and how they interact to affect the overall HMD image.

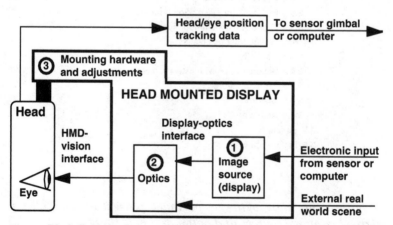

Fig 3.1. Block diagram of some of the key elements of a visually coupled system, including the three elements of an HMD.

3.2 Basic Optical System Approaches

Despite the fact that many HMDs have been designed and built over the years, there are really only two basic approaches to the optical design. These two can be called the simple magnifier and the compound microscope approaches (see Kingslake, 1972 or Smith, 1966 for detailed information). Each of these approaches is discussed in this section, although the majority of this chapter will deal primarily with the simple magnifier approach. Subdivisions of these two categories are possible and are discussed in the last section of this chapter.

There are two major differences between the simple magnifier and the compound microscope approaches. The first is that the compound microscope approach creates a real exit pupil, which may significantly limit the tolerance in eye location in order to see the HMD image. The second difference is that, in general, the compound microscope approach results in a longer physical distance between the display and the eye. This has the advantage of allowing room to fold the optics to improve the center of gravity of the HMD or to insert a beam splitter to allow superposition of the HMD image with a real-world scene or alternate image source. Table 3.1 is a listing of the differences between the two approaches. Some of them could be considered either advantages or disadvantages depending on the requirements of the application.

3.2.1 Simple magnifier

As its name implies, the simple magnifier is a fairly simple optical approach to the head-mounted display, as depicted in Fig. 3.2. It consists of a single positive lens (which may be composed of several elements to correct for aberrations and improve the image quality) that

TABLE 3.1. Differences between a Simple Magnifier HMD and a Compound Microscope HMD

Characteristic	Simple magnifier	Compound microscope
Eye-positioning tolerance	Large	Small
Physical path length	Shorter	Longer
Cost	Less expensive	More expensive
Size	Smaller	Larger
Weight	Lighter	Heavier
Center of gravity	Farther from head	Closer to head

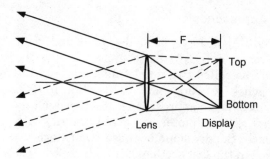

Fig 3.2. Ray trace for a simple magnifier showing rays from the top (dashed lines) and bottom (solid lines) of the display.

is typically located one focal length (F) away from the display surface. If the lens is located exactly one focal length away from the display surface, then the light rays emanating from any point on the display surface will be captured by the lens and bent so that they exit the lens parallel to one another. Figure 3.2 is a drawing of the light rays coming from a point at the top of the display (dashed lines) and coming from a point at the bottom of the display (solid lines) for a simple magnifier optical system. With the display located at the focal plane of the lens, the virtual image created is located at infinity. This is also referred to as a *collimated image*.

The simple magnifier approach has the advantages of simplicity, small size, low weight, and low cost. It is also relatively "forgiving" in terms of eye position with respect to the optical system, unlike the compound microscope approach. The major disadvantage of the simple magnifier approach is that the distance from the display to the lens and the lens to the eye are usually relatively short and make it difficult to fold the optical system or to insert a beam splitter or combiner.

3.2.2 Compound microscope

The compound microscope approach to HMD optical design uses two lenses instead of one, as drawn in Fig. 3.3. The lens closest to the display surface is called the *objective* lens (since it is near the object), and the second lens is referred to as the *eyepiece* lens (since it is closest to the eye). The objective lens (which may also be called the *relay* lens) produces an inverted real image (labeled as intermediate image in Fig. 3.3) of the display at a location between the eyepiece and objective lenses. The eyepiece is then used to produce a virtual image that can be viewed by the observer. The important difference between this approach and the simple magnifier is that this approach produces a real exit pupil. The exit pupil (for this example) is located where the parallel rays from the bottom of the display (solid lines) intersect the

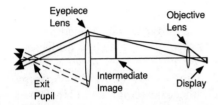

Fig 3.3. Compound microscope approach.

parallel rays from the top of the display (dashed lines; only the last two ray segments are shown to declutter the figure). This is also the real image of the objective lens as produced by the eyepiece lens. The significance of the exit pupil is that as long as the eye is within the exit pupil then the entire field of view (FOV) of the display is visible. Conversely, if the eye is totally outside of the exit pupil, then none of the FOV is visible. This is a significant disadvantage of this type of optical approach in addition to higher cost, more weight, larger size, and more complexity compared to the simple magnifier. The main advantage of this approach is that it is possible to use the extra path length to fold the optics to better conform to the head or helmet. Although both Figs. 3.2 and 3.3 show traditional lenses as the optical power components, it is possible to execute these designs using spherical, torroidal, or parabolic mirrors; diffractive or holographic optics; or gradient lenses.

3.3 HMD Optical Characteristics

The parameters used to describe an HMD are the same no matter what optical design approach is used, with the exception of eye relief and exit pupil. These two parameters are treated differently because the compound microscope approach forms a real exit pupil. For simplicity, the equations presented in this section describing the interrelationships between optical parameters and display parameters are for the simple magnifier approach only. The HMD characteristics presented in this section are listed here:

- Field of view
- Eye relief
- Image quality
- Exit pupil
- Luminance
- Focus

3.3.1 Field of view

Field of view (FOV) is probably one of the two most often used parameters to describe an HMD (the other is resolution, which is addressed in the following section). In simplest terms, the field of view is the angle subtended by the HMD virtual image as viewed by the observer, typically measured in degrees. For a monocular (single-eye) HMD, the optical field of view can be calculated using Eq. (3.1):

$$\text{FOV (degrees)} = 2\arctan\left(\frac{S}{2F}\right) \tag{3.1}$$

where FOV = field of view
S = size of display (inches or millimeters)
F = focal length of lens (inches or millimeters)

Figure 3.4a is a drawing of a simple magnifier lens with focal length F situated one focal length away from a display of size S. The two rays through the center of the lens (one from the top of the display and one from the bottom of the display) intersect at an angle that corresponds to the optical field of view. If the display is rectangular then the FOV can be expressed as a horizontal and vertical angle or it can be expressed as the diagonal of the rectangle (diagonal FOV). From Eq. (3.1) it is apparent that the FOV of the HMD can be increased by either increasing the size of the display (S) or by decreasing the focal length of the lens (F). These are shown in Figs. 3.4b and 3.4c, respectively.

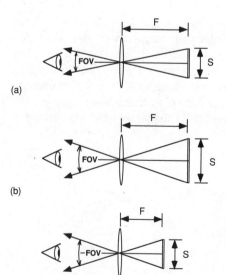

(a)

(b)

(c)

Fig 3.4. (a) Simple magnifier geometry for calculating optical field of view, (b) increasing field of view by increasing display size S, (c) increasing field of view by decreasing lens focal length F.

Note that the optical FOV calculated in Eq. (3.1) does not depend on the diameter of the simple magnifier lens, which may be counterintuitive. From a practical standpoint, the diameter of the simple magnifier lens (or eyepiece lens in the case of a compound microscope design) does play a role in the actual, or practical, field of view of the HMD. If the eye is too far back from the lens then the entire optical FOV may not be visible owing to the limited diameter of the magnifier lens. Figure 3.5 is a drawing of light rays coming from a simple magnifier HMD with the eye located as far away from the lens as possible while still close enough to see the very top and bottom of the display (with 50 percent vignetting at these extremes). If the eye is back further than this point (Fig. 3.6), there are no light rays from the extreme top or bottom of the display that get into the eye, which means that the actual (practical) FOV from this eye relief distance is less than the optical FOV calculated using Eq. (3.1). This makes the determination of (practical or actual) FOV more complicated than simply applying Eq. (3.1). If the angle subtended by the magnifier lens as measured from the eye is smaller than the FOV calculated using Eq. (3.1), the actual FOV available to the eye must be calculated using Eq. (3.2):

$$FOV = 2\arctan\left(\frac{D}{2L_e}\right) \qquad (3.2)$$

where FOV = field of view (actual)
D = diameter of magnifier lens
L_e = eye distance from lens

Note: this equation is valid only when $D < L_e(S/F)$.

The FOV calculated by Eq. (3.2) has also been referred to as the *instantaneous field of view* (for aircraft head-up displays) and is always smaller than the optical FOV calculated using Eq. (3.1). Figure 3.7 shows the geometry from which Eq. (3.2) is derived. For a well-designed HMD, the typical eye relief distance and lens diameter should be such that the optical FOV and the instantaneous FOV are the same (i.e., the full display is visible).

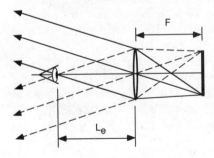

Fig 3.5. Eye positioned as far back as possible (maximum eye relief distance, L_e) and still be able to see the entire field of view with no more than 50% vignetting at the edges of the field of view (also assumes no pupil movement due to eye rotation).

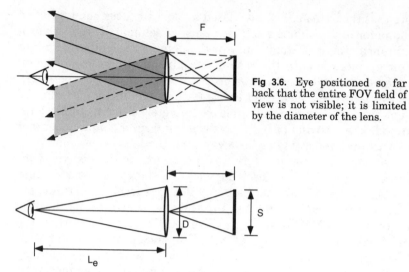

Fig 3.6. Eye positioned so far back that the entire FOV field of view is not visible; it is limited by the diameter of the lens.

Fig 3.7. Geometry for calculating FOV (actual) when lens diameter (D) is too small for the eye position distance from the lens [i.e., $D < L_e(S/F)$].

There is one more complicating factor that can occur with respect to characterizing the HMD field of view if the HMD is biocular or binocular. A binocular HMD is produced by using two independent monoculars, one for each eye, where a monocular consists of an eyepiece lens and a display. If the two monoculars are aligned with their optical axes parallel, then there is full binocular overlap and the total FOV is the same as the binocular FOV (that part of the FOV that can be seen by both eyes). For example, each monocular of a binocular HMD may present an optical FOV of 30° vertical by 40° horizontal, but the optical axes of the two monoculars may be rotated horizontally (about the eye position) to provide a different image view to the two eyes (this assumes the image on the display is also changed to correspond with the amount of monocular rotation). Usually this is done such that there is a central overlap region where both eyes can see some of the image, but the rest of the scene is observed by either the right eye or the left eye only. In this case the total FOV is the total angular coverage that can be seen by either eye, and the binocular field of view is that portion of the total angular coverage that both eyes can see. Figure 3.8 shows the total and binocular FOVs for a binocular HMD. Although Fig. 3.8 shows the right eye display on the right side and the left eye display on the left side, it is possible to rotate the optical axes of the two monoculars inward (instead of outward), resulting in the left eye seeing the scene to the right and the right eye seeing the

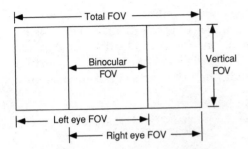

Fig 3.8. Schematic drawing of the different FOV that occur in a partially overlapped, divergent, binocular display. If the right and left eye FOVs were exchanged, this would illustrate a convergent, partially overlapped, binocular display.

scene to the left. There is some evidence that this latter approach to producing a partial overlap display is preferable (Melzer and Moffitt, 1991).

In the case of a biocular display that is produced using a single lens in front of both eyes, the right eye would see the left side of the display (and center) and the left eye would see the right side of the display (and center) as shown in Fig. 3.9. So if an HMD is biocular or binocular, the total field of view as well as the biocular or binocular FOV should be stated to clearly define the capability of the HMD.

3.3.2 Image quality

Image quality is one of the most important parameters for any display because it indicates how well the observer (or HMD user) will be able to perform with the display. The image quality of an HMD is typically summed up with a simple statement of its resolution. However, resolution is a relatively ill-defined term and HMD resolution can be stated in many ways. With the proliferation of solid-state displays, such as the liquid crystal display (LCD), a common way to specify resolution is to use the total number of picture elements (pixels) or the angular subtense of a single pixel as a measure of resolution. This latter measure becomes complicated if the pixel is rectangular instead of square because the angular subtense will be different in the vertical direction than in the horizontal direction. Even with this complication, the angular subtense of a single pixel is probably the best measure of resolution (for monochrome displays) because it is a simple means of comparing the display quality to a visual capability limit, which is usually assumed to

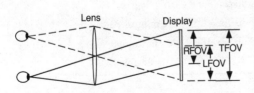

Fig 3.9. Right eye FOV (RFOV), left eye FOV (LFOV), and total FOV (TFOV) for a binocular HMD. Note that right and left refer to which eye see a particular part of the display, not which side of the display that particular part is on.

be one minute of arc. The average angular subtense of a single pixel is calculated by dividing the total number of pixels along one dimension by the angular FOV of that dimension. For example, if an LCD image source is 300 pixels high and the optical system produces a virtual image of this LCD that is 30° high, then the vertical resolution would be

$$\text{Res} = \frac{30 \text{ degrees}}{300 \text{ pixels}} = 0.1 \text{ degree per pixel}$$

Since there are 60 minutes of arc in a degree, this would correspond to 6 arc minutes, which is about 6 times more coarse than what the human eye can resolve. For this same example, the resolution could be expressed in terms of pixels per degree by dividing the number of pixels in the vertical dimension by the vertical FOV, which would yield

$$\text{Res} = \frac{300 \text{ pixels}}{30 \text{ degrees}} = 10 \text{ pixels/degree}$$

Resolution can also be expressed in terms of cycles per degree (2 pixels corresponds to 1 cycle), which for the example presented would yield

$$\text{Res} = \frac{300 \text{ pixels}}{30 \text{ degrees}} \frac{1 \text{ cycle}}{2 \text{ pixels}} = 5 \text{ cycles/degree}$$

Table 3.2 is a summary of the several ways that resolution can be stated or calculated and the corresponding visual limit (if applicable). The measure of visual acuity that is familiar to most people is Snellen acuity, used by optometrists and ophthalmologists. Snellen acuity is stated as a ratio of two numbers, such as 20/20 or 20/30. The top number refers to the distance in feet at which an observer can read a high-contrast letter of a specified size, and the lower number refers

TABLE 3.2.   Summary of Measures of Resolution

	Equation to calculate	Units	Visual limit
1.	$\text{Res} = (N/\text{FOV})$	pixels/degree	60 pixels/degree
2.	$\text{Res} = (N/2\text{FOV})$	cycles/degree	30 cycles/degree
3.	$\text{Res} = (8.74 \cdot N/\text{FOV})$	cycles/milliradian	1.72 cycles/milliradian
4.	$\text{Res} = (\text{FOV}/N)$	degrees/pixel	0.0167 degree
5.	$\text{Res} = (60 \cdot \text{FOV}/N)$	arc minutes/pixel	1 arc minute
6.	$\text{Res} = (17.5 \cdot \text{FOV}/N)$	milliradian/pixel	0.291 milliradian
7.	$\text{Res} = \text{TP}$	pixels	not applicable

to the distance at which a "normal" person can read the same size of letter. A Snellen letter that corresponds to 20/20 vision (normal vision) subtends an angle of 5 arc minutes high by 5 arc minutes wide. Since the Snellen letters are based on a 5-by-5 cell matrix, this means that 20/20 vision corresponds to a resolution limit of one arc minute. All of the visual limits provided in Table 3.2 are based on the one arc minute resolution assumption. In reality, many individuals can see better than one arc minute for high-contrast patterns and significantly worse than one arc minute for low-contrast or low-light-level patterns.

For the first three measures of resolution presented in Table 3.2, a higher number means better resolution, and for the second three measures, a lower number means better resolution (finer detail). For the last measure of resolution in Table 3.2, it is not possible to provide a corresponding visual limit since there is insufficient information to relate the total pixel count to what the eye would see. Of the seven measures of resolution listed, the fifth one is probably the best to use since its corresponding visual limit of one arc minute makes it easy to compare the display resolution to visual acuity. No matter which method is used to characterize resolution, the quality (or "goodness") of the display depends on how close the resolution is to the corresponding visual limit.

The preceding description of methods to assess HMD resolution is fairly simple, provided that the image source is composed of discrete picture elements, such as found in liquid crystal displays (LCDs). However, if the image source is a cathode-ray tube (CRT) or is a color display that achieves color through spatial integration, then it may be more difficult to determine what constitutes a "pixel" for purposes of resolution assessment. In the case of an LCD that has red, green, and blue pixels, the pixel count along any dimension should (for purposes of calculating resolution) correspond to the number of repeating color triads along that dimension. For example, if a color LCD is composed of elongated red, green, and blue pixels positioned in rows of triads as shown in Fig. 3.10, then the color pixel count in the horizontal

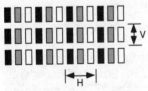

Fig 3.10. A color solid-state display with red, green, and blue pixels forming horizontal color triads. The horizontal (*H*) and vertical (*V*) dimensions of a color triad are marked. These are the dimensions that would be used to calculate the color resolution of this display.

dimension would be $\frac{1}{3}$ of the total pixel count in that dimension (one red, green, and blue triad constitutes one color pixel). However, the vertical color pixel count would be the same as the total vertical pixel count since the color triads are laid out only in the horizontal dimension. Other geometries of the color pixels could result in less clear determinations of horizontal and vertical color resolution. Sometimes the total pixel count (sum total of red, green, and blue elements) is used as a measure of the resolution of a color display. This, of course, results in a larger, more impressive number than stating the number of color triads. When evaluating an HMD it is important to determine which pixel count is provided in order to make an intelligent comparison between systems. It is this author's opinion that using the total pixel count instead of the triad count is misleading and should not be used.

In the case of a CRT, the image is usually formed in a raster structure with the raster lines running in the horizontal direction. In this case the vertical resolution can still be easily determined by counting the number of scan lines, but the horizontal resolution is difficult to determine since the scan line is continuous in this dimension. Probably the most popular method of determining the number of horizontal pixels on a CRT display in the computer/digital age is to simply quote the number of addressable cells in the horizontal dimension. This is not an unreasonable approach, provided that the spot size produced by the scanning electron beam in the CRT is sufficiently small that these addressable cells can indeed produce visible pixels. Another method, used in the past, is to determine the highest visible spatial frequency (vertical light and dark bars with one pair of bars corresponding to one cycle) that can be presented on the display. As spatial frequency is increased (a larger number of thinner light and dark bars) the display modulation contrast decreases. The limiting resolution is the spatial frequency at which the output modulation contrast has dropped to a value of about 3 to 5 percent. The width of the alternating light and dark bars can be used as the pixel width in calculating the resolutions presented in Table 3.2.

Although the complex topic of image quality is often reduced to a simple statement of the resolution of the HMD, that statement only partially addresses the question of image quality. Another useful concept in quantifying image quality is the modulation transfer function (MTF). A full discussion of MTF is beyond the scope of this chapter, but it is worthwhile to address a few of the features and limitations of this approach to image quality assessment. In short, the MTF of a display or optical system is the ratio of the output modulation to the input modulation as a function of spatial frequency. For an HMD it basically describes how much contrast (modulation) is available on the display as a function of spatial frequency (detail). Figure 3.11 is a graph of a typical MTF for a display such as a miniature CRT, used as an image source for an HMD. An extensive discussion of MTF and alternative image quality metrics can be found in Gaskill (1978, p. 497) and Task (1979).

Using the simple concept of resolution as a measure of image quality, it is important to note that there is always a difficult design trade-off between resolution and field of view for HMDs (or any virtual image display). Since the display has a fixed number of pixels, the number of pixels per degree becomes smaller when the field of view is made larger. This is readily apparent from the first equation in Table 3.2, which is repeated here for convenience:

$$ \text{Res} \;=\; \frac{N}{\text{FOV}} \tag{3.3}$$

where Res = average visual resolution in pixels/degree
       N = number of pixels along dimension of interest
       FOV = field of view along dimension of interest

If the FOV is made larger, then the number of pixels per degree becomes smaller, and vice versa. This fundamental equation is what makes it painfully obvious that the only way to have a large field of view and a good resolution (lots of pixels per degree) is to increase the number of pixels in the display. If technology (or cost) is limiting the number of pixels available, then one must make a trade-off decision between field of view and resolution.

Contrast is another key factor in assessing image quality, and there are several different equations that are used to calculate contrast. In

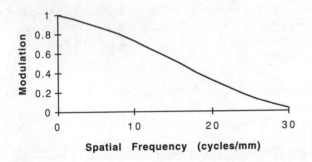

Fig 3.11. Typical modulation transfer function (MTF) for a miniature cathode ray tube (CRT) image source.

the preceding discussion it was noted that the MTF of a display describes how the modulation contrast available on the display decreases as spatial frequency increases. Modulation contrast is calculated using Eq. (3.4):

$$C = \frac{\text{Max} - \text{Min}}{\text{Max} + \text{Min}} \tag{3.4}$$

where C = modulation contrast
  Max = maximum luminance (luminance of light bars)
  Min = minimum luminance (luminance of dark bars)

Mathematically the value of modulation contrast can be as low as zero or as high as one. Quite often it is converted to a percentage (by multiplying by 100) with a range of zero to 100 percent. For an HMD, there may be other effects (besides display MTF) that can cause a loss of contrast in the virtual image seen by the observer. Contrast loss may be due to scattered light within the optical system, reflections from various surfaces, or ambient scene luminance if the HMD uses a combiner to superimpose the HMD image on the real world.

Another factor that relates to image quality is the presence or intensity of ghost images. These are unwanted images that can occur in an optical system due to reflections from surfaces of various optical elements. For HMDs that use a visor as the combiner or beam splitter element to superimpose HMD symbology on the real world, the visor is usually the source of secondary, or ghost, images. This arises because the visor has two surfaces (inner and outer), of which only one (the inner surface) is used as an optical element of the HMD system. The outer surface, however, can produce a slightly out-of-focus ghost image, the intensity of which depends on the surface treatments and absorption characteristics of the visor. Figure 3.12 shows how a ghost image is created from a visor HMD. A basic measure of this effect is the image-to-ghost ratio. It is very desirable to have a high image-to-ghost ratio, meaning the HMD image is much brighter than the unwanted ghost image. The image-to-ghost ratio for the visor case of Fig. 3.12 is calculated using Eq. (3.5):

$$G_r = \frac{R_1}{T_1 \cdot R_2 \cdot T_1} \tag{3.5}$$

where G_r = image-to-ghost ratio
  R_1 = reflection coefficient of first (inner) surface
  T_1 = transmission coefficient of visor material and inner surface
  R_2 = reflection coefficient of second (outer) surface

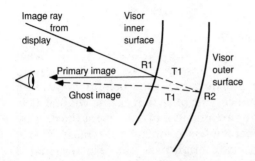

Fig 3.12. Ghost image from the outside surface of HMD visor helmet (visor thickness is greatly exaggerated.)

As an example of Eq. (3.5), suppose an HMD uses a tinted visor with a material transmission coefficient of 0.15 (15 percent) and a coating on the inner surface that is 50 percent reflective and 50 percent transmissive. For this case,

$$T_1 = (0.15)(0.50)$$
$$= 0.075 \text{ (for both material and coating transmission effects)}$$

$$R_1 = 0.50$$

$$R_2 = 0.04 \text{ (4 percent for uncoated plastic outer surface)}$$

Substituting these values into Eq. (3.5),

$$G_r, \text{tinted} = \frac{0.5}{0.075 \cdot 0.04 \cdot 0.075} = 2222$$

which means the ghost image would be only 1/2222 times as bright as the primary image. However, if the visor is clear plastic (not tinted) the material transmission would be essentially 1, and T_1 would be 0.50 (from the coating). The image-to-ghost ratio for the clear visor would then be

$$G_r, \text{clear} = \frac{0.5}{0.5 \cdot 0.04 \cdot 0.5} = 50$$

which is a much poorer ratio. Under low-light-level external conditions (which is when one would expect a clear visor to be used), a symbology presentation on an HMD with an image-to-ghost ratio of 50 would produce a very visible, distracting ghost image. Other elements in the optical system may also produce a secondary or ghost image, which can be very distracting for symbology-only displays. If the HMD is to be used with full imagery and there is no see-through capability, the amount of ghost imaging that can be tolerated is greater. Image quality includes much more than just resolution!

3.3.3 Luminance

Luminance is a measure of the "luminous intensity of any surface in a given direction per unit area of the surface as viewed from that direction" (Kingslake, 1972). Unless you have had a course in photometry, this definition is probably not very helpful in gaining an understanding of the luminance. For purposes of this chapter it is sufficient to think of luminance as an indicator of how bright the display is. Since luminance is the photometric quantity that most closely approximates the visual sensation of brightness, this is a reasonable approach.

Luminance is measured using a photometer or luminance meter that has approximately the same spectral sensitivity (sensitivity as a function of light wavelength) as the human eye. The most typical units for measuring luminance are the nit (equal to 1 candela per square meter) and the footlambert (equal to $1/\pi$ lumen per square foot per steradian). It is possible to convert from one unit to the other since 1 footlambert is equivalent to 3.43 candelas/square meter (cd/m^2). As a matter of reference, the luminance of white paper in good reading light is about 170 cd/m^2 (50 ftL), peak luminance of a good rear-projection home television is about 1,700 cd/m^2 (500 ftL), and the luminance of clean snow in bright sunlight is about 34,300 cd/m^2 (10,000 ftL). Human vernier acuity is close to its best when the luminance level is equal to or greater than about 17 cd/m^2 (5 ftL; Berry et al., 1950). This implies that it is desirable to ensure that the apparent luminance of the HMD is at least 17 cd/m^2 (5 ftL) assuming the HMD is in a controlled lighting environment. If the HMD is to be used in sunlight to present symbology overlaid on real-world scenes, then the luminance will need to be much greater than 17 cd/m^2 to obtain the contrast necessary to see the symbology.

In determining the luminance of an HMD, it is probably simplest to measure the luminance of the display (the CRT or LCD) that is used as the image source for the HMD and then calculate the luminance available to the observer using the transmission efficiency of the optical system. For simple magnifier optical systems using only a few optical elements with antireflection coated surfaces, the transmission efficiency may be quite high. If such a system has a transmission efficiency of 90 percent, then the observed luminance of the HMD would simply be 90 percent of the luminance measured for the display image source. Since photometric measurement devices exist that can easily measure the luminance of the display image source, one would think the measurement and specification of HMD luminance should be simple. However, a few ambiguities need to be addressed. These include (1) measurement angle to the surface, (2) space-average luminance versus pixel luminance, and (3) peak luminance versus average scene luminance. Referring back to the beginning of this section, the definition of luminance included a viewing (or measurement) direction with respect to the surface. This is necessary because some display types may have a higher luminance when viewed on-axis than when viewed off-axis. In an HMD, the effect of this is that the edges of the field of view (FOV) may be significantly less bright than the center of the FOV. This can be particularly true of back-lit LCDs if the illuminating source does not do a good job of diffusing (scattering) the light. The best way to characterize the HMD luminance with regard to angle is to specify both the luminance at the center of the FOV (on-axis; measurement angle is zero with respect to a line perpendicular to the surface of the display) and at the edge of the FOV (off-axis; measurement angle equal to one-half of the FOV size, i.e., angle to the edge of the FOV).

Devices used to measure luminance are designed to measure the average luminance of the display surface that is within the measurement aperture of the device. If the measurement aperture is relatively large and thus includes many pixels, then the resultant reading is the space-average luminance of the display. However, if the aperture is small enough to include only part of one pixel, then the reading is the luminance of a single pixel. This pixel luminance reading will typically be significantly higher than the space-average luminance since it does not include the non-light-emitting dead area between pixels. The space-average luminance is a better indicator of what brightness will be perceived by a human observer. It is therefore recommended that when stating or measuring the on-axis and off-axis HMD luminance a photometer be used that makes a space-average reading.

The third ambiguity with respect to luminance deals with how the display is driven. The peak luminance is the maximum luminance that can be achieved on the display (either on-axis or off-axis, and either space-averaged or single pixel) for the maximum input signal. If the display is to be used for symbology only (no imagery), then peak luminance is of primary importance because that is the level that an observer will see. However, if the display is to be used for imagery, then the more appropriate parameter would be average scene luminance. The average scene luminance will be less than the peak luminance but is somewhat ill-defined because it depends on the content of the scene used to make the measurement. Even with this deficiency, it still provides a better indicator of what the observer will see in the way of brightness when viewing the display.

In general, it is recommended that for imagery applications the space-average scene luminance be stated for on-axis and at the edge of the FOV. For symbology applications it is more appropriate to state the peak luminance (which may be either space-average or single pixel depending on whether or not the symbology stroke width is more than a single pixel wide) for the on-axis and edge-of-FOV positions.

3.3.4 Eye relief distance

The eye relief distance is simply the distance from the design eye position to the nearest element of the HMD optics or supporting structure. Its purpose is to provide some indication as to the compatibility of the HMD with the use of corrective eyeglasses. Specifications for military helmet-mounted systems have used the value of 20 mm or 25 mm for eye relief distance to ensure compatibility with eyeglasses. For simple magnifier systems, the eye relief distance and the size of the eye motion box available are inversely related, as described in the following section.

3.3.5 Exit pupil (or eye motion box) size

If the HMD optical system is designed as a compound microscope, then it will form a real exit pupil. A real exit pupil is relatively easy to see and measure by placing a white sheet of paper at the plane of the exit pupil and observing the size of the sharply defined circle of light (assuming the lenses used in the system are circular). The simple magnifier optical system, however, does not have a real exit pupil that can be observed in the same way. Instead there is a volume of space (usually cone-shaped) within which the observer's eye pupil must be positioned in order for the observer to see the entire FOV of the display. If the observer's eye is outside of this volume, then some of the FOV will be clipped (vignetted). The size of this volume is determined by the lens diameter, the lens focal length, and the size of the display. Figure 3.13 is a drawing of a simple magnifier with this observation volume designated by shading. This volume is what determines the relative trade-off between eye relief distance and eye motion box size. Referring to Fig. 3.13, if the eye is positioned a distance L_e back from the lens (the eye relief distance), then the eye pupil must be positioned within dimension E to see the whole image. If the eye relief distance is increased, then the amount of eye movement that can be allowed decreases. The relationship between optical eye relief, eye motion box, lens focal length, lens diameter, and display size is presented in Eq. (3.6):

$$E = D - \frac{L_e S}{F} \tag{3.6}$$

where E = eye motion box size
  D = lens diameter
  S = display size
  F = lens focal length
  L_e = optical eye relief distance

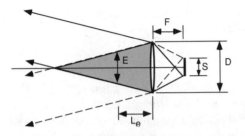

Fig 3.13. Eye relief distance (L_e) versus eye motion box size (E) for simple magnifier HMD optics.

The actual, physical eye relief distance may be somewhat less than that determined by Eq. (3.6) if other optical elements (such as a beam splitter) are positioned between the eye and the lens.

3.3.6 Focus and accommodation

All of the examples of simple magnifier HMD optics presented in this section have shown the display located one focal length away from the lens (e.g., Figs. 3.2, 3.4, and 3.13). Under this condition, the virtual image created by the lens is located at infinity, requiring the eye lens to accommodate (focus) to infinity in order to obtain a sharp image on the retina. If the eye relief distance on the HMD is sufficient to accommodate the use of eyeglasses, then it may be appropriate to have the HMD optics set for a fixed focus at infinity. The advantage of this is reduced size and weight since no moving parts are required. It may also be advantageous to set the fixed focus distance considerably less than infinity to reduce eye fatigue.

If the eye relief distance is too short to accommodate eyeglasses, then having adjustable focus provides a means of permitting nearsighted and farsighted individuals to adjust the HMD image to a distance where they can obtain a sharp image on their retina. Focusing is accomplished by moving the lens closer to or further from the display. Moving the lens closer than one focal length away from the display will compensate nearsighted individuals and moving the lens further than one focal length away will compensate farsighted individuals. Focusing range is typically stated in diopters (a unit of optical power); negative numbers correspond to nearsighted vision and positive numbers correspond to farsighted vision. For purposes of adjusting night vision goggles (which can be considered a special kind of HMD), the U.S. Army requires that the eyepiece be adjustable from -6 diopters to $+2$ diopters. This focus range is probably more than satisfactory to accommodate the vast majority of HMD users for most applications.

Equation (3.7) defines the relationship between the lens position (with respect to the display) and the location of the virtual image:

$$D_i = \frac{dF}{F - d} + l_e \tag{3.7}$$

where D_i = distance from eye to virtual image
$\quad d$ = distance from display to lens
$\quad F$ = focal length of lens
$\quad l_e$ = distance from eye to lens

If the display is at the focal point of the lens, then $d = F$, the denominator of Eq. (3.7) is zero, and the distance from the eye to the virtual image is infinity. If d is smaller than F, then the image distance will be positive; conversely, if d is larger than F, then the image distance will be negative. Note that a negative image distance actually means the image produced is real and on the same side of the lens as the eye (a positive distance corresponds to a virtual image located on the opposite side of the lens from the eye). This image distance can be converted to optical power (discussed previously) using Eq. (3.8):

$$\text{OP} = -\frac{1}{D_i} \tag{3.8}$$

where OP = optical power in diopters
$\quad D_i$ = image distance in meters calculated using (3.7)

The optical power values in diopters correspond to the eyeglass prescription values for which the focusing adjustment range can compensate. If the lens movement distance corresponds to a focusing range of $+2$ to -4 diopters (as calculated using Eqs. (3.7) and (3.8)), then the HMD optics will be suitable to compensate for observers' eyeglass prescriptions that range from $+2$ to -4 diopters. Note that this focusing adjustment does not compensate for astigmatism.

3.4 HMD-Vision Interface Issues

The previous section presented several optical HMD parameters that are a result of the interaction of the display characteristics and the optical system characteristics. This section addresses some of the human visual interface issues, which are a result of the interaction between the HMD parameters and human vision.

3.4.1 Ocularity

One of the first design or selection decisions that must be made when designing or selecting an HMD is whether it should be monocular, biocular, or binocular. The primary factor in this decision is the application. Other factors, such as cost, size, weight, ease and rapidity of fitting, comfort, etc., are also considered. In general, monocular HMDs (image to one eye only) are less costly, lighter weight, smaller, and easier to adjust and fit than binocular HMDs. Binocular HMDs (each eye viewing a separate, independent image) require two separate optical and display channels, which tends to make them more costly, heavier, larger, and more difficult to adjust because of the need to ensure the two images are properly aligned with the observer's eyes. Biocular HMDs use a single display and a single large lens through which both eyes can view the virtual image of the display. The biocular display has the advantage of providing an image to each eye (compared to the monocular HMD) but is incapable of providing a stereo pair of images to the two eyes (like a binocular display). If the HMD optical system is binocular but the electronic image source is capable of only providing a single channel of information, then the overall system might be considered to be pseudo-biocular since it cannot produce a stereo display. In this case the HMD has all the disadvantages of a binocular system (binocular image alignment, size, weight, etc.) without the advantage of producing a stereo image. Another potential advantage of a binocular HMD is the possibility of increasing the total horizontal field of view by adjusting the optical axes of each optical channel to produce a partially overlapping display (see previous section on field of view).

3.4.2 Superposition with external scene

In addition to ocularity, HMDs can also be subdivided according to whether or not they have "see-through" capability. If an HMD uses an optical combiner to superimpose the HMD image on a real world scene, then it has "see-through" capability. Conversely, if the HMD does not have an optical combiner for this purpose, it is a "non-see-through" system. The night-vision goggle (NVG) community also has both "see-through" and "non-see-through" devices, which have been designated Type II and Type I NVGs respectively (Kocian and Task, 1995). A mnemonic to aid in remembering the difference between Type I and Type II: "Type two has see-through, Type one has none."

A Type II HMD, used to overlay the HMD image on some real-world scene, must have sufficient luminance to be compatible with the scene luminance. If the real-world scene is outside daylight, such as in military helmet-mounted cueing systems, then the HMD luminance must be sufficient to be visible against several thousand footlamberts (sunlit white clouds).

However, for indoor use, with controlled room lighting, then the competing scene luminance may be only several tens of footlamberts. If the Type II HMD is binocular, then the image vergence and focus (apparent image distance) must also be compatible with the type of real-world objects that will be combined with the HMD image. For example, if the HMD image is to be combined with features of a relatively close (a few feet away) real-world object, then the HMD images must cause the eyes to converge to the same distance as the real object; otherwise double vision will occur—of either the real-world object or the overlay feature. Likewise the apparent image distance or focus of the HMD must be at about the same distance as the real object since the eyes can only focus on one distance at a time.

By contrast, Type I systems are relatively benign in that there is no competing real-world scene that dictates focus, vergence, or luminance.

3.4.3 Field curvature

The relationship between display location, lens position, and the location of the image formed by the lens described by Eqs. (3.6) and (3.7) assumes that the lens is a flat field lens. The term *flat field lens* means that all points on the display (which is usually a flat surface) are imaged to the same flat plane in image space. However, quite often a simple magnifier will not produce a flat field image; that is, the image will be formed as a curved surface. This means the observer's eyes will have to change focus when viewing the center of the display versus the edge of the display since they are at different optical distances. For younger eyes this is no problem, but for older eyes that do not have much accommodative (focusing) range, it may mean that part of the image will be out of focus because of the field (image) curvature.

Field curvature can be a significant problem for partial overlap binocular HMDs (see Sec. 3.3.1), as illustrated in Fig. 3.14. This arises because the image distance (focus) for the right and left eyes differs as a function of position on the display, requiring the two eyes to focus independently when shifting gaze from one point on the image

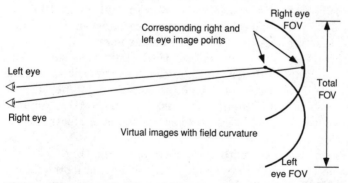

Fig 3.14. Illustration of right- and left-eye focus mismatch in a partial overlap binocular display as a result of field overture.

to another. For example, consider the case in Fig. 3.14 of a binocular HMD, composed of right and left oculars that have the equivalent of one diopter of field curvature between the center and the edge of the display. When the observer looks at the point indicated on the figure, the left eye would be focusing a diopter closer than the right eye. As the observer shifts gaze from the left to the center and then to the right of the total FOV, the focus mismatch goes to zero (at the center) and then to one diopter in the opposite direction (i.e., the right eye is now focusing a diopter closer than the left eye). This constant, independent right-eye and left-eye shifting of focus when scanning the display is not comfortable and should be avoided by minimizing the amount of field curvature, especially for partially overlapped HMDs.

3.4.4 Distortion

For a perfect lens, the angle at which the image of a point on the display is observed is related to the position of that point on the display by the following equation:

$$R = k \tan \phi \tag{3.9}$$

where R = radial distance on the display from the center to the point
k = proportionality constant (usually the focal length of the lens)
ϕ = observed angle to the image of the point

If the relationship between points on the display and observed images of those points departs from Eq. (3.9), then distortion is present. Distortion in monocular HMDs is normally not much of a problem if the distortion is reasonably small. Just as in the case of field curvature, distortion is of most concern for partially overlapping binocular displays due to the mismatch in image location for the right and left eyes. Distortion is typically worst at the edges of the field of view; it is the area where the mismatch between right-eye viewing and left-eye viewing of the same image point is the worst, just as in the field curvature case. Self (1986) reviewed the literature regarding tolerances for binocular instruments and found considerable differences in the recommended maximum allowed mismatch between right- and left-eye images. Tolerances were presented for vertical mismatch (dipvergence) and for horizontal misalignments (divergence and convergence). The most stringent tolerances were for dipvergence. Recommended tolerances for dipvergence ranged from a high of one prism diopter (equivalent to about 10 milliradians, or 34.5 arc minutes) to a low of 1 milliradian (3.4 arc minutes). For long-term wear of an HMD, it is probably desirable to have a binocular tolerance for dipvergence on the order of the stringent requirement, 1 milliradian. This means that the distortion tolerance in a partially overlapped binocular HMD must not be exceeded anywhere on the display. Divergence and convergence are discussed further in the following section.

3.4.5 Adjustments

Providing adjustments for different parts of an HMD has both advantages and disadvantages. In general, the more adjustments, the more moving parts are required, which usually means more expense and greater chance of component failure. On the other hand, more adjustments should provide for better fit to a greater population of potential users. This section presents descriptions with comments for several common types of HMD adjustments.

The basic characteristics of focus adjustment and visual accommodation are discussed in Sec. 3.3.6. However, a few other issues regarding focus adjustment and the human interface should be noted. Although the focus adjustment allows a user who normally wears simple prescription (spherical power only) glasses to bring the HMD image into sharp focus, it does not necessarily provide a means to adjust the focus to the user's prescription. If the user is relatively young and has considerable accommodative range, then he or she will be able to obtain a sharp image on the HMD for a wide range of focus settings. Therefore it is possible for the user to set the focus at a point that may cause visual discomfort with extended use. For a binocular HMD it is also possible to set the focus (image location) differently for the two eyes, forcing the eyes to accommodate to different distances and potentially resulting in visual discomfort during extended use. Probably the most desirable situation is for the HMD to have sufficient eye relief to permit the use of prescription eyeglasses and fix the HMD focus distance to between $\frac{1}{2}$ and 2 meters to minimize eye lens focus effort. The $\frac{1}{2}$ to 2 meters corresponds to 2 and $\frac{1}{2}$ diopters, respectively; this is close to the dark focus or "relaxed" focus condition of the human eye of about 1.7 diopters (Leibowitz and Owens, 1975). If older individuals (those who normally wear bifocals or trifocals) are going to be using the HMD, then they need to wear single prescription glasses that will allow them to focus at the distance the HMD is set to; they should not wear their bifocals or trifocals with the HMD!

Another parameter closely related to focus that may or may not be considered a user accessible adjustment is vergence. Tolerance on dipvergence (vertical mismatch between the oculars of a binocular HMD) is discussed in Sec. 3.4.4 and should not be user accessible. However, divergence or convergence (the horizontal mismatch between oculars of a binocular HMD) may be considered for user adjustability. For normal unaided vision, eye convergence and focus (accommodation) distance are naturally and involuntarily adjusted together in a coordinated manner. In other words, as an individual looks at objects at various distances the eyes automatically focus at the distance of the object and, at the same time, rotate inwards so that the line of sight from the two eyes converge at the same distance as the object. This natural, coordinated focus and convergence activity can be and usually is disconnected by a binocular HMD. Focus is controlled by the location of the image produced by the HMD optics and, once set by the user, remains fixed. However, if the binocular HMD is used to present a stereoscopic pair of images, the amount of eye convergence required to attend to a particular object in the display will vary depending on the stereoscopic distance of the object. This means the eye convergence will be changing as different objects on the display are viewed but the eye focus will remain the same, thus disconnecting the long-learned, involuntary coordination of the two. This disconnect may cause more severe problems for some individuals than others and may become more of a problem for extended viewing situations than for short-term viewing. The same disconnect occurs for night-vision goggle (NVG) users because NVGs produce a stereo pair of images but focus is fixed at a single distance. Individual differences in acceptance of this disconnect may explain why some NVG users experience visual discomfort, fatigue, and nausea after extended (more than 3 or 4 hours) NVG use and others do not. It is an area suitable for further research. An advantage of a biocular display is that the vergence and focus remain coordinated automatically (just as in natural viewing) when the single, simple magnifier lens is adjusted to change focus. Of course, monocular HMDs have no vergence adjustment since there is only one optical channel.

Another important adjustment for binocular HMDs is interpupillary distance (IPD), which is the distance between the centers of the pupils of the two eyes when the eyes are converged at infinity. IPD adjustment is especially important for binocular HMDs that use exit-pupil-forming optics and are used in environments in which the mounting apparatus may slip on the head. With this type of HMD and in a nonfriendly environment, slippage can cause total loss of the HMD image, so it is important to ensure that both eye pupils are centered in their respective HMD exit pupils by adjusting the IPD (and with other adjustments discussed next). The range of IPD adjustment recommended by the Military Standardization Handbook, MIL-HDBK-141, *Optical Design*, and repeated in several other human factors equipment design references, is 50 mm to 76 mm—which should be adequate to accommodate the vast majority of potential HMD users.

Other adjustments that may be appropriate for an HMD (depending on the application) are up-down, fore-aft, and tilt. The up-down

**TABLE 3.3. Adjustment
Ranges for ANVIS-6 NVGs**

1.	Focus: +2 to −6 diopters
2.	Fore-aft: 16 mm minimum
3.	Tilt: 8 degrees minimum
4.	IPD: 52 to 72 mm
5.	Up-down: 16 mm minimum

adjustment permits one to adjust the position of the exit pupil with respect to the eyes but does not change the angular position of the HMD image. The tilt adjustment permits the adjustment of the apparent location of the HMD image in the pitch direction (up-down angle). Finally, the fore-aft movement allows for the adjustment of the eye with respect to the eyepiece lens. This affects the eye relief distance and allows eyeglass users to position the HMD optics far enough away from their eyes to let them wear their eyeglasses while permitting those that do not use eyeglasses to position the optics closer to the eye and increase the allowed eye motion or HMD slippage.

The ANVIS-6 NVGs that are widely used in the U.S. military employ all of the adjustments discussed in this section (except vergence) to fit the wide range of characteristics of NVG users. Table 3.3 is a summary of the ranges specified for each of the adjustment dimensions.

One cautionary note with respect to adjustments: the probability that the HMD optical system will be misadjusted increases with the number of adjustments that are available! If the system has a considerable number of adjustments, then appropriate training or very good instructions may be necessary to minimize the potential for misadjustment.

3.5 Summary

This chapter has presented many of the key issues associated with HMD image sources and optics and how they interface with human vision. After completing this chapter the reader should have a good understanding of most of the major parameters used to characterize HMDs including field of view, image quality (resolution), eye relief, exit pupil, luminance, and focus. In addition, the reader should have a better understanding of the interaction and trade-offs of these main parameters, such as field of view versus resolution, field of view versus eye relief, and exit pupil versus eye relief. It has been the intent of this chapter to provide the reader with the tools and background to be in a better position to evaluate, select, and/or specify an HMD.

3.6 Bibliography

Berry, R. N., Riggs, L. A., & Duncan, C. P. (1950). The relation of vernier and depth discriminations to field brightness. *Journal of Experimental Psychology, 40*, 349.

Birt, J. A., & Task, H. L. (Eds.). (1972). *A symposium on visually coupled systems: development and application* (Aerospace Medical Division technical report AMD-TR-73-1). TX: Brooks Air Force Base.

Carollo, J. T. (Ed.). (1989). *Helmet-Mounted Displays, Proceedings of the SPIE, 1116.*

Farrell, R. J., & Booth, J. M. (1984, February). *Design Handbook for Imagery Interpretation Equipment* (Publication D180-19063-1). Seattle WA: Boeing Aerospace Company.

Ferrin, F. J. (1973, September). F-4 visual target acquisition system. In J. A. Birt & H. L. Task (Eds.), *Proceedings of a symposium on visually coupled systems: Development and application* (AMD TR-73-1). TX: Brooks Air Force Base.

Ferrin, F. J. (1991). Survey of helmet tracking technologies. *Large Screen Projection, Avionic, and Helmet Mounted Displays, Proceedings of the SPIE, 1456*, 86–94.

Gaskill, J. D. (1978). *Linear systems, fourier transforms, & optics.* New York: John Wiley & Sons.

Hertzberg, H. T. E., Daniels, G. S., & Churchill, E. (1954). *Anthropometry of flying personnel–1950* (WADC Technical Report 52-321). Wright-Patterson Air Force Base, OH: Wright Air Development Center.

Kingslake, R. (Ed.), (1972). *Applied optics and optical engineering* (Vol. 1, pp. 1–9, 232–236). New York Academic Press.

Klein, M. V. (1970). *Optics.* New York: John Wiley & Sons.

Kocian, D. F. (1987). Design considerations for virtual panoramic display (VPD) helmet systems. *AGARD Conference Proceedings 425: The Man-Machine Interface in Tactical Aircraft Design and Combat Automation* (pp. 22-1–22-32). Neuilly Sur Seine, France: NATO Advisory Group for Aerospace Research & Development. (NTIS No. AGARD-CP-425).

Kocian, D. F. (1990). Visually coupled systems (VCS): Preparing the engineering research framework. *Eleventh Annual IEEE/AESS Dayton Chapter Symposium, 28*–38.

Kocian, D. F. & Task, H. L. (1995). Visually-coupled systems: Hardware and human interface. In W. Barfield and T. Furness (Eds.), *Virtual Environments and Advanced Interface Design.* New York and Oxford, England: Oxford University Press.

Landau, F. (1990). The effect on visual recognition performance of misregistration and overlap for a biocular helmet mounted display. *Helmet-mounted displays II, Proceedings of the SPIE, 1290.*

Leibowitz, H. W. & Owens, D. A. (1975). Anomalous myopia and the intermediate dark focus of accommodation. *Science, 189*, 646–648.

Lewandowski, R. J. (Ed.). (1990). *Helmet-Mounted Displays II, Proceedings of the SPIE, 1290.*

Melzer, J. E., & Moffitt, K. W. (1991). Ecological approach to partial binocular overlap. *Large Screen Projection, Avionic, and Helmet-Mounted Displays, Proceedings of the SPIE, 1456*, 124.

Military Standardization Handbook, MIL-HDBK-141, *Optics.* (1962). U.S. Department of Defense.

RCA Corporation. (1974). *Electro-Optics Handbook* (RCA Technical Series EOH-11). Harrison, NJ: Author.

Ross, J. A., & Kocian, D. F. (1993). Hybrid video amplifier chip set for helmet-mounted visually coupled systems. *1993 Society for Information Display International Symposium Digest of Technical Papers, 24*, 437–440.

Self, H. C. (1972, November). The construction and optics problems of helmet-mounted displays. In J. A. Birt and H. L. Tast (Eds.), *Proceedings of a symposium on visually coupled systems: Development and application* (AMD-TR-73-1). TX: Brooks Air Force Base.

Self, H. C. (1986, May). Optical tolerances for alignment and image differences for binocular helmet-mounted displays. AAMRL-TR-86-019, Wright-Patterson Air Force Base, Ohio.

Sherr, S. (1979). *Electronic Displays.* New York: John Wiley & Sons.

Shmulovich, J., & Kocian, D. F. (1989). Thin-film phosphors for miniature CRTs used in helmet-mounted displays. *Proceedings of the Society for Information Display, 30* (4), 297–302.

Smith, W. J. (1966). *Modern Optical Engineering.* New York: McGraw-Hill.

Task, H. L. (1979). An evaluation and comparison of several measures of image quality of television displays. AAMRL-TR-79-7, Wright-Patterson Air Force Base, Ohio.

Task, H. L., & Kocian, D. F. (1992). *Design and integration issues of visually coupled systems* (SPIE SC54). Short course presented at SPIE's OE/Aerospace Sensing 1992 International Symposium, Orlando, FL.

Task, H. L., Kocian, D. F., & Brindle, J. H. (1980, October). Helmet mounted displays: design considerations. In W. M. Hollister (Ed.). *Advancement on visualization techniques* (AGARDograph No. 255). London: Harford House.

Widdel, H., & Post, D. L. (Eds.). (1992). *Color in electronic displays.* New York: Plenum Press.

Wells, M. J., Venturino, M., & Osgood, R. K. (1989, March). Effect of field of view size on performance at a simple simulated air-to-air mission. *Helmet-Mounted Displays, Proceedings of the SPIE, 1116.*

Winner, R. N. (1972). A color helmet mounted display system. In J. A. Birt, & H. L. Task (Eds.), *Proceedings of a symposium on visually-coupled systems* (AMD-TR-73-1). TX: Brooks Air Force Base.

Fundamentals of HMD Optics

Robert E. Fischer

The optical design of HMDs is an important ingredient in a successful, high-performance HMD. The image quality as seen by the user depends on both the display and its associated electronics as well as the viewing optics. With the ever increasing demands of wider fields of view, longer eye relief, and larger exit pupils, the viewing optics has become one of the more difficult and critical technologies in HMD design. This chapter will present the fundamentals of HMD optics. After reviewing the basic parameters and specifications for HMDs, I will review basic first-order optics of HMD systems and then extend this to lens aberrations as encountered in HMD designs. I will then present a parametric analysis of various HMD viewing optics designs.

4.1 Introduction

Head-mounted displays (HMDs) are finally emerging from the military environment, where they have typically been applied to *helmet-mounted* applications, such as night vision, avionics, and other areas for the purpose of augmenting a pilot's or soldier's vision. New applications include virtual reality, medical, and industrial-related uses.

HMD systems are dependent on several key technologies:

Display device. The display device is the source for the imagery that the user is looking at. This is often the most technically challenging part of an HMD, and it is one that is advancing quite rapidly. Current technologies include cathode ray tubes (CRTs), liquid crystal displays (LCDs), electroluminescent displays (ELs), field emission displays (FEDs), and others still in the laboratory. These display devices can be self-emissive (CRTs and FEDs), transmissive (LCDs), or reflective in their mode of operation. The transmissive and reflective displays require a separate means of illumination, whereas the transmissive devices use an integral form of backlight, such as a flat fluorescent lamp.

Drive electronics. Regardless of the display device used, there will inevitably be the requirement for drive electronics that convert the input video signal (NTSC, VGA, or other video format) to the proper format to generate displayed imagery on the display device. These drive electronics should be small in size, light in weight, and low in cost.

Mechanics. Independent of the optical design and the display device, there still must be a mechanical housing to hold them with other hardware and to support the entire system on the head. This includes some method for adjustments for focus and interpupillary distance. This must also be lightweight, low-cost, and rugged enough to protect the optics and display device.

Viewing optics. Given that we have a small, high-resolution display device such as a CRT or LCD, we require optics to view the display. If the eye could focus close enough to the display device, the viewing optics would not be necessary. However, this is not the case, as many people cannot focus their eyes closer than approximately 10 inches (25 cm). As with the other HMD technology areas, the viewing optics must be of suitable performance yet small, light in weight, and low in cost.

There are many parameters that must be considered when designing and configuring the viewing optics for an HMD. These, and some of the forms the optics may take, will be the subject of this chapter.

4.2   Fundamental Parameters

We must first consider the parameters of the user's vision. The range of human vision extends to over 200° in the horizontal dimension. The resolution at extreme angles is quite poor, as the eye detects only large objects in these regions. The eye is primarily sensitive to image motion in the outer periphery of the field, similar to the vision system of many insects. If you extend your hands out to your far left and right and flap your hands up and down while looking straight ahead, you can see your hands moving. You are usually unaware of objects in the outer periphery of your vision that are not moving.

The central portion of vision is known as the fovea, the region of high acuity. Here the eye can resolve approximately one minute of arc. This is equivalent to 0.0003 radian in angle.[1] This corresponds to the width of one of the horizontal bars (equal to the spaces between the bars) in the capital "E" on a standard eye chart for 20/20 vision, a frequency of about 30 line pairs per degree. A person can rotate his or her eyes over nearly ±45° so that the fovea can look directly at any object in this angular region. These are excellent guidelines to remember, and we will use them later in this chapter.

It is important to emphasize that not all HMD applications require resolution at the limit of human vision. An HMD that can only display objects of two to three minutes of arc or larger may still be quite suitable for many applications.

4.2.1   Resolution in a theater

As most of us are familiar with movie theaters, this is a very good place to start our discussion. Consider a movie theater where 35-mm film is projected in its horizontal (or landscape) format. A reasonably good lens will be able to provide a modulation of approximately 30 percent at 50 lp/mm. This means that a spot approximately 1/100 mm in extent will be resolvable on the film. Since the horizontal dimension of the image on the film is 36 mm wide, there are 3,600 horizontal resolution elements. Over an entire film format, there are 8,640,000 resolution elements. (For a good lens, image quality above 50 lp/mm will be present.) This film is then projected onto the screen at the front of the theater.

We now have a choice. We can sit in the front row, in which case the horizontal field of view will be enormous. Alternatively, we could sit in the middle of the theater, where we would have a somewhat rational

---

[1]If we multiply the distance to an object by 0.0003, we can determine the smallest feature that is resolvable by the eye. For example, at 3 m the eye can resolve a 1-mm object.

field of view of the screen, or we might sit in the last row, in which case the visual angle subtended by the screen would be quite small. Let us assume that we project the image onto a screen and view it over a full horizontal visual FOV of 40°. This means that we have 3,600 resolution elements horizontally spread over 40°, which equates to 90 resolution elements per degree, or $\frac{2}{3}$ minute of arc per resolution element. This is well matched to the eye's visual acuity of 1 minute of arc. While the above is not a rigorous treatment of the situation, it does give us a good idea as to what is going on. If we are sitting in the front row and the field of view is twice as large, the resolution will be degraded by a factor of two (1.3 minutes of arc per resolution element). Conversely, if we are sitting in the last row, the resolution will be improved accordingly.

Consider the situation where the user is viewing an HMD display with the aid of some form of viewing optics or magnifier from a close distance. Assume that the display (an LCD or similar device) has 512×512 pixels. For the same horizontal field of view of 40°, we find 12.8 resolution elements per degree, which equate to 4.7 minutes of arc per resolution element (4.2 mm at 3 m). While this is not too bad, it will take an image source with at least $1,000 \times 1,000$ pixels to produce imagery that begins to approach the limit of the human visual system.

By way of comparison, the HDTV standard has approximately 1,000 lines of vertical resolution. The goal for HDTV has always been imagery that is nearly indistinguishable from a 35-mm slide.

4.2.2 Pixel-based imagery

It should be clear from the previous section that for a given number of pixels, or resolution elements, the wider the field of view, the larger will be the angular subtense of the smallest resolution element. This is best illustrated by considering the parametric data shown in Fig. 4.1.

The abscissa shows the field of view (either horizontal or vertical) in degrees, and the ordinate shows minutes of arc. Each of the curves is labeled with a value that indicates the number of pixels. For example, if we have a 30° horizontal field of view and 420 horizontal pixels, each pixel would subtend 4.3 minutes of arc. If the number of horizontal pixels were increased to 640 (VGA resolution), each horizontal pixel would not subtend approximately 2.8 minutes of arc. Extending this further, with 1,000 horizontal pixels, each pixel would subtend slightly less than 2 minutes of arc and for 2,000 pixels the resolution would be slightly under 1 minute of arc.

As the FOV is increased, we are "stretching" the subtended size of the pixels. In many HMD applications, such as immersive virtual reality, we require extremely wide fields of view, and the subtended size of the pixels grows accordingly. For example, with a 70° FOV and

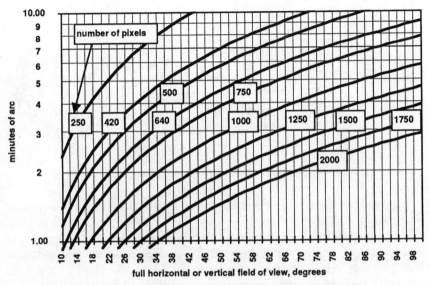

Figure 4.1.   Resolution as a function of number of lines or pixels and field of view.

a VGA level of resolution (640 × 480 pixels) each pixel now subtends about 6.5 minutes of arc, which is approximately equivalent to a spot $\frac{1}{4}$ inch in diameter at 3 m. These are big pixels!

It is best to specify a system based on its functional requirement or the needs and expectations of the user. If your HMD application is for an arcade with large fields of view, and if the users are nonengineering teenagers who are more interested in the experience than the resolution, then you can compromise your performance (at a lower cost). In this example with a 60° horizontal field of view, you would require nearly 2,000 horizontal pixels to bring the angular subtense per pixel to 2 minutes of arc. With 640 horizontal pixels the resolution is 5.6 minutes of arc per pixel. It is safe to say that your choice of display technology will be a driver in setting your specifications. To reiterate: The most important factors in specifying a system are the functional requirements and expectations of the user.

4.3   Basic Parameters of Head-Mounted Displays

A simple magnifying lens is a non-pupil-forming optical system. There is no well-defined exit pupil from the viewing optics at the position where the eye is located. If the eye were to move up or down from the position shown in Fig. 4.2, the light rays emitted from the object being viewed would still be able to enter the eye.

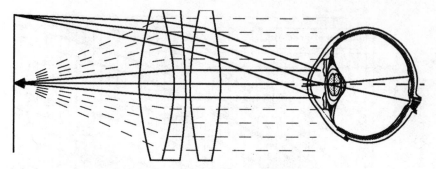

Figure 4.2. Optics of a magnifier, a non-pupil-forming system.

Alternatively, consider Fig. 4.3, where I show the optics of a simple astronomical telescope[2] used to view the moon or other objects. The front objective lens, shown as a two-element achromatic doublet, is imaged into the pupil of the eye. This location is simultaneously the exit pupil of the telescope as well as the entrance pupil of the eye. There is no light or light ray outside of this pupil location. If the eye were to move up or down, it would eventually move out of the area where there is light and the image would disappear. This is called a *pupil-forming system.* A good way to understand the difference between pupil-forming and non-pupil-forming systems is to place a white card at the location of the eye for both cases. With the non-pupil-forming system or magnifier, light will form on the card to the diameter of the magnifier. For the pupil-forming telescope, light will be present over a well-defined diameter. If the card is located at the exit pupil of the telescope, the diameter will be the exit pupil diameter.

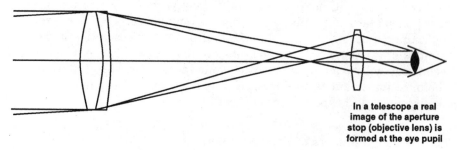

In a telescope a real image of the aperture stop (objective lens) is formed at the eye pupil

Figure 4.3. Optics of a simple pupil-forming astronomical telescope.

[2]A pair of binoculars is also a common pupil-forming optical design.

Figure 4.4 shows the basic parameters that are needed for HMD viewing optics. We know the entrance pupil diameter of the eye is approximately 7 mm in faint light and less than approximately 2.5 mm in bright light. It is therefore desirable that the exit pupil diameter of a viewing optical system in an HMD be at least 7 mm in diameter. The optic shown is not a pupil-forming system, and so there will be light outside of the desired pupil diameter, but a larger exit pupil eases HMD alignment. If good optical performance is only present over a 7-mm exit pupil, then the HMD must be aligned precisely to the eye. If the exit pupil diameter of the HMD viewing optics is enlarged to 12 mm, then the eye can move around within this region and you will still have adequate performance. This helps account for slippage of the HMD on the head. The larger the usable exit pupil diameter, the easier it will be to see the imagery.

Another fundamental parameter is *eye relief,* the clearance from the last component of the viewing optics to the cornea of the eye. In order to accommodate eyeglasses, the minimum eye relief should be at least 25 mm, with a larger eye relief more desirable. As the eye relief is increased, the diameter of the viewing optics increases in size (discussed later in this chapter).

There is a very interesting subtlety regarding pupils and eye relief. As noted earlier, the eye rotates about a point approximately 13 mm aft of the cornea. For wide-angle viewing systems, this point of eye rotation should coincide with the exit pupil of the viewing optics. This is shown in Fig. 4.4 for a straight-ahead view.

The entrance pupil of the eye is located approximately 3 mm aft of the cornea. If the exit pupil of the viewing optics system coincides with the entrance pupil of the eye, then the user will be able to see imagery from the entire field of view while looking straight ahead. With eye rotation, the user may lose the imagery off axis, because the point of

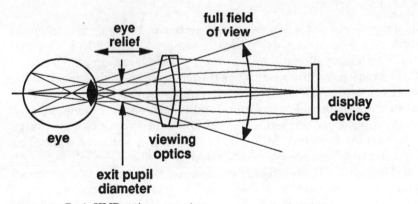

Figure 4.4. Basic HMD optics parameters.

eye rotation is approximately 10 mm aft of the eye's entrance pupil! If you ever have the opportunity to look through a very wide-angle (pupil-forming) telescope, you will be able to see imagery over the entire field of view when your eye's entrance pupil coincides with the exit pupil of the telescope. However, if you rotate your eye to look at the edge of the field, the imagery will likely disappear. If you move your eye closer to the viewing optics, you will be able to look to the edges of the field. At this position you may not be able to see the entire field when looking straight ahead—hence the dilemma we discussed previously, which can be overcome by enlarging the exit pupil. Fortunately, most HMD viewing optics systems are not pupil-forming, and the problem is less evident.

Field of view is another basic parameter of major importance. A large FOV contributes to the feeling of "immersion," though as previously discussed, resolution is reduced. For a display of 3:4 aspect ratio, the diagonal field of view is 25 percent larger than the horizontal field. While the horizontal field may be specified, imagery to the corners is usually mandatory. It is imperative to state what your field of view requirements are, and whether you are specifying the horizontal field or the diagonal field.

4.4 Performance Specifications for HMD Optics

As with any optical system, we can divide the performance specifications into at least two basic categories: (1) the first-order optical parameters and (2) those related to optical performance and image quality. Listed in Table 4.1 is a summary of all of the parameters that should be given some consideration in the design of HMD systems.

4.5 Magnification

The magnification of a lens is usually stated as 10 divided by the focal length of the magnifier expressed in inches. Thus a 2-inch (5.1 cm) focal length magnifying glass or viewing optic has a magnification of 5×. The reason for this seemingly strange equation is that most young people can focus their eyes on an object approximately 10 inches (25.4 cm) in front of their eyes. If they could focus on that same object 5 inches in front of their eyes, the object would appear twice as large in its angular subtense. At 2.5 inches the object would be 4 times as large, and so on—hence the relationship. If we place a simple lens with a focal length of 2 inches close to our eye and focus at infinity, we would be able to see the object 2 inches away. The object would now subtend a field of view approximately 5 times larger than that same object would appear to subtend at the normal 10-inch distance.

TABLE 4.1. **Typical HMD optical parameters**

Parameter	Typical specification
Display device full diagonal	30 mm
Full field of view	30°
Eye relief	≥ 25 mm for eyeglass computability
Exit pupil diameter	7 mm minimum, 10 mm (or better) goal
Resolution	3 minutes per element for video
Distortion	<5 percent
Dipvergence	5 arc minutes (non see-through system)
Convergence	1.5°
Magnification difference	1 percent between oculars
Brightness difference	10 percent
Best focus position, if fixed	Approximately 3 m
Color fringing	2 arc minutes maximum
Image quality	Integral of the MTF from 0–20 lp/degree (Z. Mouroulis, personal communication, 1994)

Dipvergence, or vertical alignment, is the angular difference in elevation between the light directed to the right eye and the light directed to the left eye.

Convergence is the apparent distance of the image from the observer corresponding to the horizontal angle subtended between the two eyes.

Color fringing is lateral color, the change in focal length as a function of wavelength. The user observes color fringes surrounding light/dark edges in the object being viewed.

The term *magnification* is used in many ways in optical systems, and Fig. 4.5 shows a summary of the many definitions of this term. Figure 4.5a shows a simple microscope that magnifies an object of height y to an image of height y'. The magnification is therefore y'/y.

If this image is formed on a CCD camera and the imagery is displayed on a monitor where the image height is now y'', the net magnification is y''/y. If this monitor is viewed by the human eye, and if the monitor were located across the room or across the street, it would appear quite small. If we were to view the monitor from 10 inches away, this would be the closest viewing distance that would be comfortable for a young person. We can therefore conclude that the net total magnification from the object to the eye is 10 inches divided by the distance to the monitor in inches times y''/y.

Figure 4.5b shows the magnification of a simple magnifying lens that was discussed earlier in Sec. 4.3.

Figure 4.5c shows a microscope—a pupil-forming system. Here the total magnification is the magnification of the objective lens (as in Fig. 4.5a) times the magnification of the eyepiece (as in Fig. 4.5b).

Another pupil-forming system is the astronomical telescope shown in Fig. 4.5d. The magnification is expressed as an angular relationship since the object is typically viewed at infinity. The magnification is θ/α, which can also be shown to be equivalent to D/d.

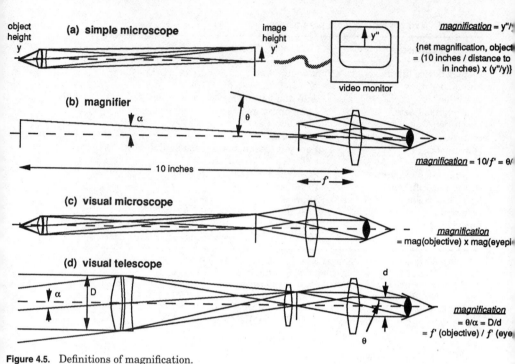

Figure 4.5. Definitions of magnification.

It is interesting that magnification is independent of eye relief. Figure 4.6 shows the eye looking through the same magnifier from different distances or eye reliefs. It is clear that the same size object of height *y* is viewed to cover a semi-field of view of angle θ in both cases. It is also clear that Fig. 4.6*b* requires the magnifier or viewing optics to be larger than would be required in Fig. 4.6*a*.

We can take this fundamental relationship and convert it to a parametric data set as shown in Fig. 4.7. The abscissa shows the full field of view in degrees, and the ordinate shows the viewing optics diameter in millimeters. This is plotted for various eye reliefs ranging from 5 mm to 40 mm. Thus, a full field of view of 30° requires approximately a 28 mm viewing optics diameter for a 40 mm eye relief. If the eye relief can be reduced to 10 mm, however, the viewing optics diameter decreases to approximately 12 mm. Note that these data are based on first-order optics and should be used only as a guideline.

4.6 Lens Aberrations

While the intent of this chapter is not to teach the reader the principles of lens design, it is important for cohesiveness to review the basics of aberrations and discuss how they apply to HMD optics.

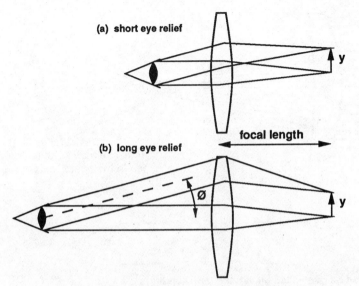

Figure 4.6. Two magnifiers with the same focal length and magnification and different eye reliefs require different lens diameters.

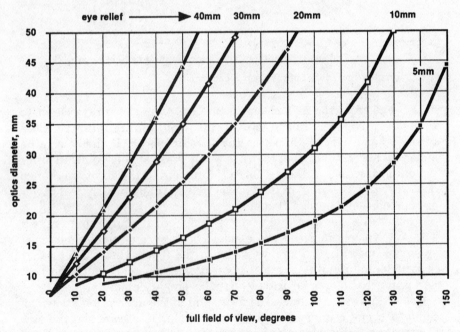

Figure 4.7. Approximate diameter of viewing optics as a function of field of view and eye relief.

Spherical aberration: Figure 4.8 shows the formation of spherical aberration. The lens shown in the upper part of the figure clearly does not create a perfect image, because the rays from the outer periphery

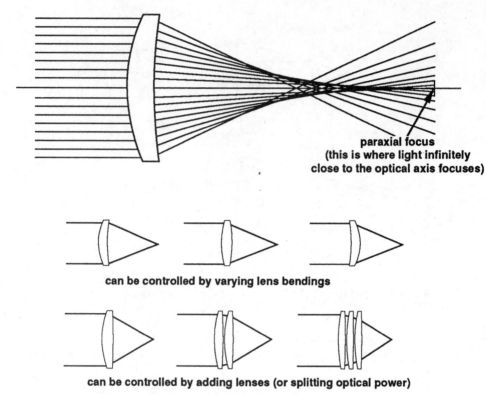

**paraxial focus
(this is where light infinitely
close to the optical axis focuses)**

can be controlled by varying lens bendings

can be controlled by adding lenses (or splitting optical power)

Figure 4.8. Formation and control of spherical aberration.

of the lens cross the axis closer to the lens than those rays closer to the optical axis. The image location for rays that are infinitely close to the optical axis is known as the *paraxial focus.* If there were no spherical aberration, all of the rays would cross at this point. In order to minimize spherical aberration, the designer has the ability to change or modify the "bending" of the lens as well as to split the optical power or add additional optical elements.

To illustrate the potential severity of spherical aberration, Fig. 4.9 shows a typical single optical element with an enormous amount of spherical aberration. The lower part of the figure shows a lens that is bent for minimum spherical aberration. This represents the best performance that can be achieved for this single element.

Coma: Figure 4.10 shows the formation of the aberration known as coma. A comatic image will look very similar to a comet in the sky. Coma is generally caused by high ray obliquities or angles of incidence on surfaces as is shown in the lower left portion of the figure. By altering the lens shape and/or the location of the aperture stop, the designer can increase the symmetry, thereby minimizing the coma.

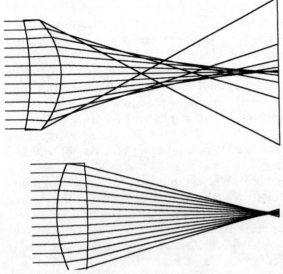

Figure 4.9. $f/2$ single-element lens with enormous amount of spherical aberration (top) and corrected for minimum spherical aberration (bottom).

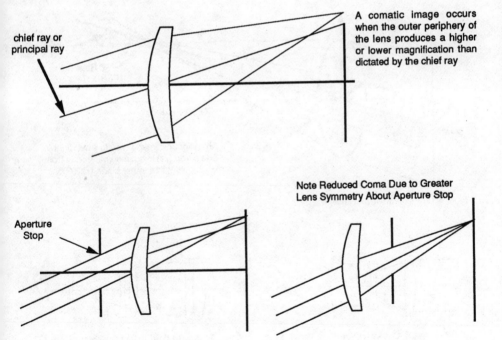

chief ray or principal ray

A comatic image occurs when the outer periphery of the lens produces a higher or lower magnification than dictated by the chief ray

Note Reduced Coma Due to Greater Lens Symmetry About Aperture Stop

Aperture Stop

Figure 4.10. Formation of the aberration of coma.

Astigmatism: Figure 4.11 shows the formation of astigmatism. An astigmatic image is one in which the rays in two orthogonal planes focus at different distances from the lens. As shown in the figure, the rays in the y-z plane focus closer to the lens than the rays in the x-z plane. As with many other aberrations, astigmatism can be corrected by selectively locating and bending lenses. Note that one of the deficiencies of the human visual system often encountered is astigmatism, which is caused by a nonspherical cornea. As the eye rotates to look in various directions, this aberration is constant. In an optical system, however, astigmatism (and other off-axis aberrations) increases with increasing FOV.

Field curvature: In the absence of astigmatism the image from any optical system will be formed on a curved surface known as the *Petzval surface.* The lens designer can very often flatten the field or correct the field curvature by locating negatively powered lens elements at positions within the system where the ray bundles are smaller as shown in the Cooke triplet lens in Fig. 4.12.

Distortion: This is an aberration in which a square or rectangular grid will image into a pincushion or alternatively a barrel-shaped

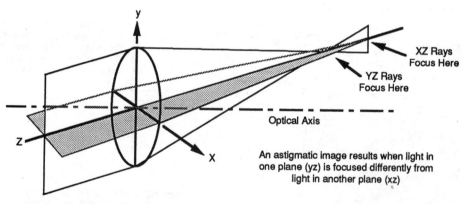

An astigmatic image results when light in one plane (yz) is focused differently from light in another plane (xz)

XZ Rays Focus Here

YZ Rays Focus Here

Optical Axis

Figure 4.11. Formation of astigmatism.

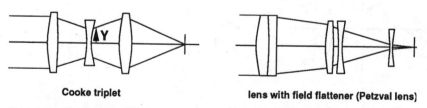

Cooke triplet

lens with field flattener (Petzval lens)

Figure 4.12. Correction of field curvature with a negative lens.

image. Figure 4.13 shows the formation of distortion. Figures 4.14 and 4.15 are plots of a grid pattern for various amounts of positive or pincushion distortion and negative or barrel distortion, respectively. Lines that we desire to be straight will bow or curve as is evident in

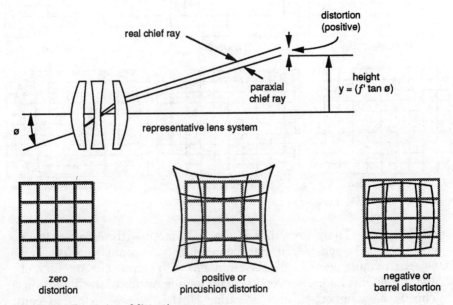

Figure 4.13. Formation of distortion.

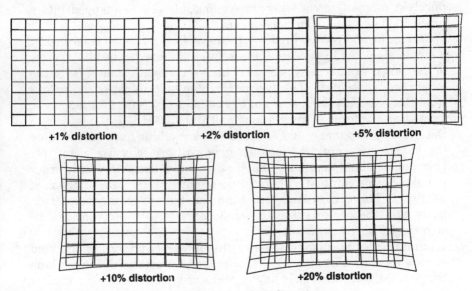

Figure 4.14. Distortion maps for positive or pincushion distortion.

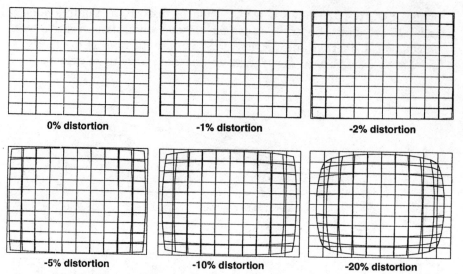

0% distortion   **-1% distortion**   **-2% distortion**

-5% distortion   **-10% distortion**   **-20% distortion**

Figure 4.15. Distortion maps for negative or barrel distortion.

these figures. There are other issues associated with distortion that also need to be considered. For example, the apparent velocity of an object traveling across a displayed image with significant amounts of distortion will change in proportion to the magnitude of the distortion. This is also important for binocular HMDs, where distortion can affect the alignment of an image observed by both eyes. Interestingly, distortion has nothing to do with image quality or resolution—it is purely a mapping error where a rectangular grid is mapped into a pincushion or a barrel form as shown.

Chromatic aberration: Figure 4.16 shows how an image is formed by a single element. In this case, the refractive index or bending power of the lens is greater for the shorter blue wavelength than it is for the longer red wagelength. This causes the blue to focus closer to the lens than the red. This chromatic shift is called *axial color.* If a display is located at the focus of such a lens and the viewer is looking in from the left, this will mean that the user will see a blue haze surrounding the image of a point. The lens designer will typically use negatively powered elements that have a higher dispersive characteristic in order to balance this chromatic aberration as shown in the lower portion of the figure. This causes the red and the blue light to come to a common focus, and the center green or yellow part of the spectrum will be slightly defocused as shown.

Lateral color is color fringing at the outer peripheries of the field of view as shown in Fig. 4.17. The image of a point will have a blue fringe or edge at larger field angles and a red fringe or edge at smaller

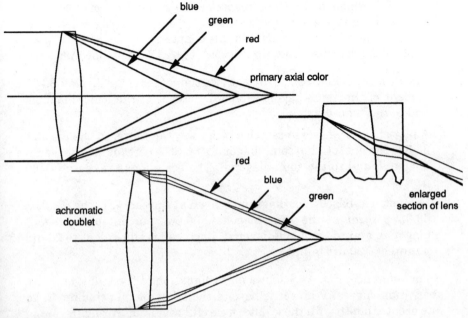

Figure 4.16.   Primary axial color and correction using achromatic doublet.

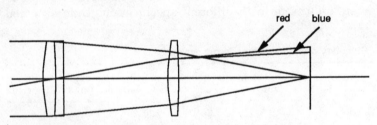

Figure 4.17.   Off-axis lateral color, which is seen as color fringing.

field angles. By using glasses of different dispersive characteristics the designer can minimize the lateral color.

4.7   Viewing Optics Designs

There are many optical design forms or configurations suitable for the viewing optics of HMDs. In order to summarize the vast number of possible solutions, I have taken some of the classical design forms in the literature (Smith, 1992, and Hopkins, 1962) as well as some custom designs and performed appropriate scaling of the data. The designs were scaled in focal length to view a 1-inch (25 mm) display full diagonal at their respective angular field of view. The resulting designs are summarized in Table 4.2. Figures 4A.1 though 4A.12 are found in the appendix to this chapter.

Each of the designs shows (in Figs. 4A.1–4A.12) the following:

A scale layout showing the ray paths to the center of the display, 33.33 percent of the semi-diagonal, 66.667 percent of the semi-diagonal, and to the semi-diagonal.

A plot of the transverse ray aberrations at the center of the display, 33.33 percent of the semi-diagonal, 66.667 percent of the semi-diagonal, and to the semi-diagonal. The scale for these data is 400 μm.

Geometrical-based spot diagrams shown at the center of the display, 33.33 percent of the semi-diagonal, 66.667 percent of the semi-diagonal, and to the semi-diagonal. Each box is equivalent to 20 arc minutes × 20 arc minutes.

For reference, Fig. 4.18 shows an explanation of the transverse ray aberration curves. While it is beyond the scope of this chapter to go into greater depth with these data, a careful review of this figure, along with each of the sample designs, will show how they compare with one another.

It is important to review how a system may be scaled in order to be usable for an application with different specifications than assumed

TABLE 4.2 Summary of optical design forms evaluated.

Figure number	Name	Focal length (mm)	Eye relief (mm)	Full field (degrees)	Exit pupil dia. (mm)	Distortion (per cent)
4A.1	Singlet	49.8	25.4	30	7	−3.7
4A.2	Doublet	48.9	25.4	30	7	−2.5
4A.3	Ramsden	48.3	22.6	30	9.52	−1.8
4A.4	Kellner	39.9	11.1	36	7.8	−2
4A.5	Orthoscopic	36.5	30.1	40	7.3	−4
4A.6	Plossl	37	31.1	40	7.3	−5.6
4A.7	Symmetrical	29.8	20.9	50	5.3	−8.4
4A.8	Erfle	24.1	17.8	60	4.7	−8.5
4A.9	Erfle-a	29	21.7	50	5.7	−6.1
4A.10	Berthele	24.6	16.9	60	4.8	−10.3
4A.11	Skidmore	18.4	9.1	80	4.6	−17.4
4A.12	Birdbath	45.7	—	40	6.4	−4.2

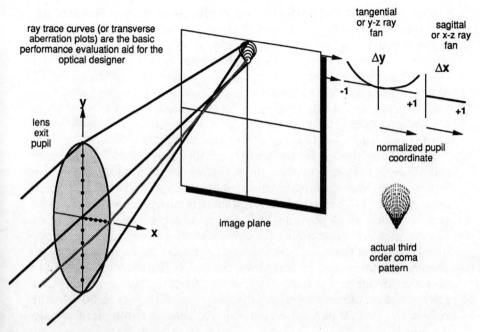

Figure 4.18. Explanation of ray trace curves.

here. As an example, we will look at the Plössl design in Fig. 4A.6. This design covers a 40° full diagonal field of view with the display device full diagonal of 1.0 inch.

Let us assume that you need a design for a 1.4-inch full diagonal, but you only need 30° full field of view. The first step is to determine the fractional field of view for a 30° full field of view, and this is simply 30°/40° or 0.75 of the field that the design covers in Fig. 4A.6. But this reduced field system will only image 0.75 of the original 1.0 inch full diagonal and 53.57 percent of the required 1.4-inch full diagonal, so we must scale up the system by 1.867 ×.

We can make use of the data in Fig. 4A.7 over 75 percent of its field of view, but the scale factor on the ray trace data needs to increase by 1.867×. The angular scale factor of 20 arc minutes × 20 arc minutes on the spot diagram remains the same as the blur size increases in direct proportion with the increase in focal length.

The aforementioned forms of designs are what might be considered *classical* in their configuration. They are direct view, do not contain any mirrors, flat or curved, and would not be considered applicable for see-through configurations. Flat-fold mirrors can be used to fold the light path between the optics and the display, but the image handedness of the display must be reversed from right-handed to left-handed due to the reflection.

One particular design form that could be used in a see-through application is the *Birdbath design* shown in Fig. 4A.12. In Fig. 4.19, I show two ways in which this could be done. The design uses a spherical mirror for all of its optical power. Since the pupil of the eye is the aperture stop, this design takes advantage of the powerful *Schmidt principle* where the off-axis aberrations of a spherical mirror with its stop at the center of curvature produces virtually no off-axis aberrations except field curvature. In most cases the accommodation of the eye is sufficient to take this into account.

Both of the design forms are nearly identical optically. The only difference is that the form on the right requires the final converging light to pass through the tilted beam splitter, which will introduce some astigmatism. In the right-hand system, the mirror is as close to 100 percent reflective as possible. In the system on the left, the mirror can be a beam splitter that allows for a see-through capability. The primary disadvantage of this system is that its theoretical throughput is only 25 percent maximum and is likely lower in reality. This low throughput is because the light must transmit or reflect from the beam-splitting fold mirror twice, and in the most advantageous case, with a 50 percent reflectivity and 50 percent transmissivity, the net throughput is only 25 percent.

Another interesting design form is shown in Fig. 4.20. Here light from a display device (a CRT is shown) is first collimated. It then enters an X-prism where the light is sent in two directions, left and right, ultimately reaching each eye. There is a throughput transmission problem in addition to the overall complexity of the design. However, this approach does offer the opportunity to view a single display with two eyes.

4.8   New Design Forms and Producibility Issues

Several relatively new technologies have become viable over the past few years. These include *aspheric surfaces, diffractive optics,* and *gradient index materials.* Lens designers know that aspheric surfaces can help to control lens aberrations, and this is advantageous with HMD optics because it can reduce the number of required lenses and improve the overall performance. The ultimate lens elements would likely be manufactured of injection-molded plastic or molded glass, with injection-molded plastic lens technology being quite mature.

Diffractive optics are beginning to play a role in HMD optics. The designer can use a diffraction grating with circular symmetry that has been etched or otherwise applied to the surface. Since a diffraction grating has the reverse chromatic dispersion properties of a bulk

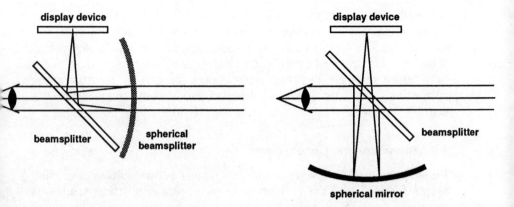

This system is sometimes known as the "birdbath"

Figure 4.19. Viewing optics with see-through capability using spherical mirror for focusing.

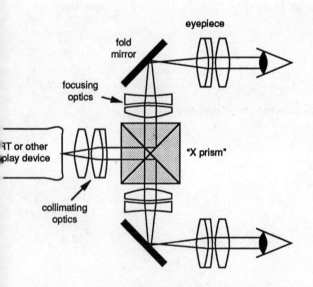

Note: Drawing for illustration and not to scale

Figure 4.20. Use of X-prism to direct light from one display device to both eyes.

glass or plastic lens element, a single lens with a diffractive surface can generally be made color-free or achromatic. The problem with diffractive optics is that they scatter light because of inaccuracies in the manufacturing process or light obliquities on the surface.

The other technology that may have a role in HMD optics is gradient index materials. This technology is still emerging and is somewhat costly.

Finally, we have the issue of producibility. As with any high-performance optical system, producibility—cost to manufacture and cost to assemble—is extremely important. To ensure an acceptable level of ultimate performance the design must be toleranced, with fabrication and assembly tolerances to each component. Because of the subjectivity of the human visual system, establishing the performance criteria for the tolerance analysis is not trivial.

4.9   Summary and Conclusions

I have presented design trade-offs and issues in this chapter regarding optics for HMDs. This subject can have more conflicting trade-offs than other optical design tasks because are we dealing with an image sensor—the human eye—that is exceptionally difficult to quantify as to its performance and image quality. The task is further complicated by the trade-offs of eye relief, exit pupil diameter, distortion, and lateral color. With a good understanding of the material in this chapter and the other chapters in this book, you should be able to proceed with confidence and success.

4.10   References

Hopkins, R. E. (1962). *Military Standardization Handbook, Optical Design* (MIL-HDBK-141).
Smith, W. J. (1992). *Modern Lens Design.* New York: McGraw-Hill.

Appendix: Optical Design Forms

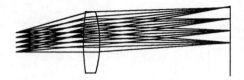

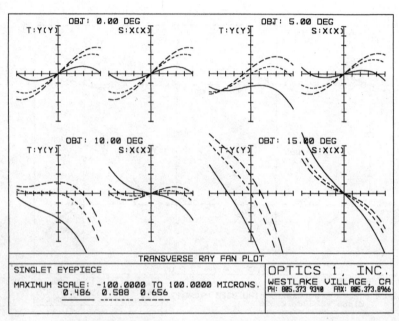

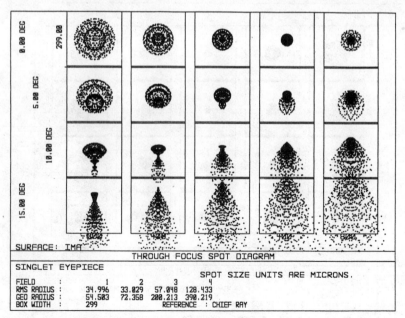

Figure 4A.1 Singlet eyepiece.

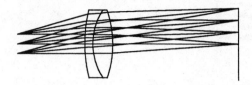

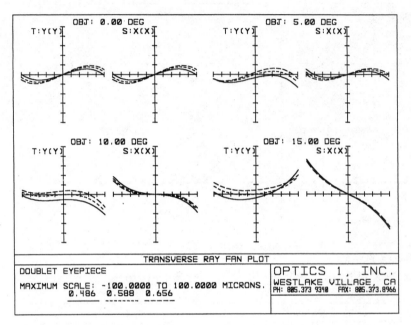

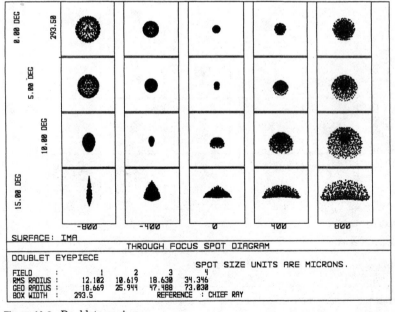

Figure 4A.2 Doublet eyepiece.

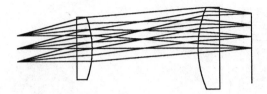

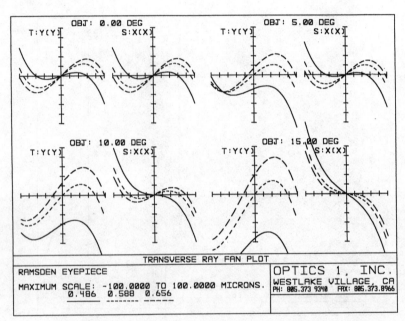

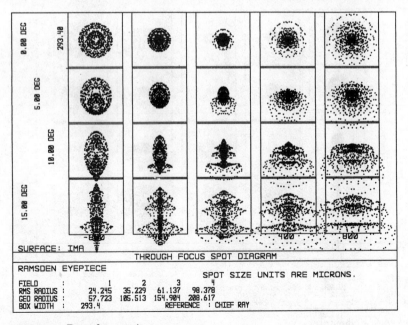

Figure 4A.3   Ramsden eyepiece.

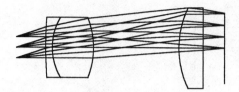

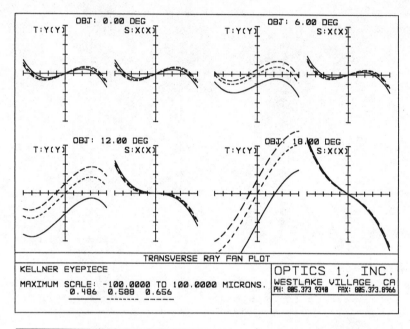

OBJ: 0.00 DEG
T:Y(Y)   S:X(X)

OBJ: 6.00 DEG
T:Y(Y)   S:X(X)

OBJ: 12.00 DEG
T:Y(Y)   S:X(X)

OBJ: 18.00 DEG
T:Y(Y)   S:X(X)

TRANSVERSE RAY FAN PLOT

KELLNER EYEPIECE

MAXIMUM SCALE: -100.0000 TO 100.0000 MICRONS.
0.486   0.588   0.656

OPTICS 1, INC.
WESTLAKE VILLAGE, CA
PH: 805.373.9340   FAX: 805.373.8966

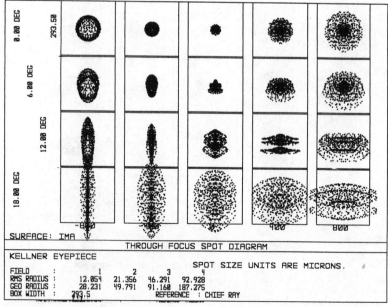

0.00 DEG   293.50

6.00 DEG

12.00 DEG

18.00 DEG

SURFACE: IMA

-800   -400   400   800

THROUGH FOCUS SPOT DIAGRAM

KELLNER EYEPIECE

SPOT SIZE UNITS ARE MICRONS.

FIELD :	1	2	3	4
RMS RADIUS :	12.054	21.356	46.291	92.928
GEO RADIUS :	28.231	49.791	91.160	187.275
BOX WIDTH :	293.5			

REFERENCE : CHIEF RAY

Figure 4A.4 Kellner eyepiece.

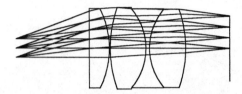

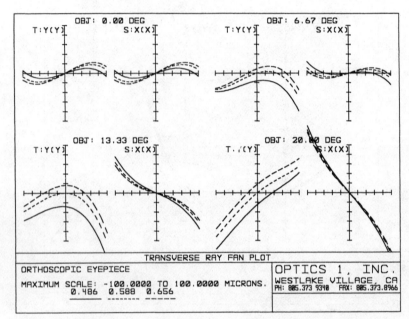

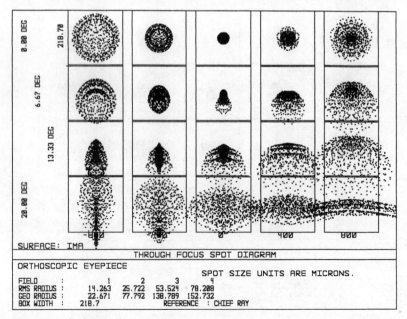

Figure 4A.5 Orthoscopic eyepiece.

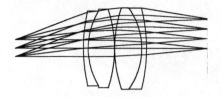

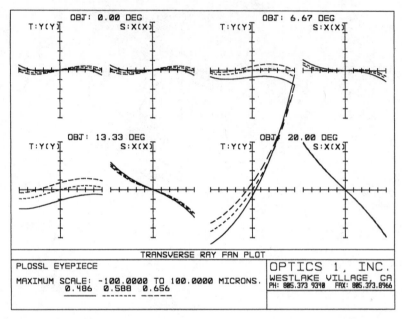

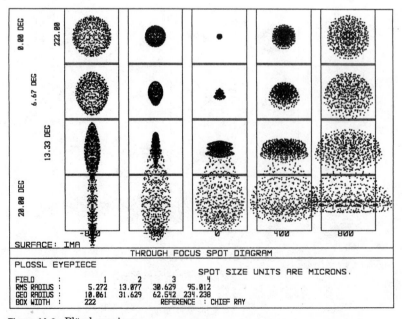

Figure 4A.6 Plössl eyepiece.

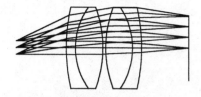

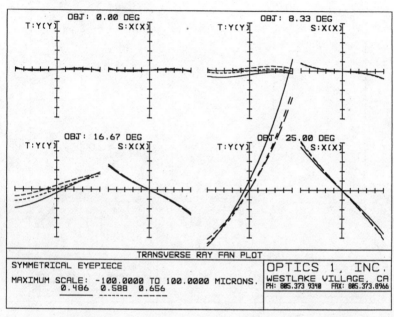

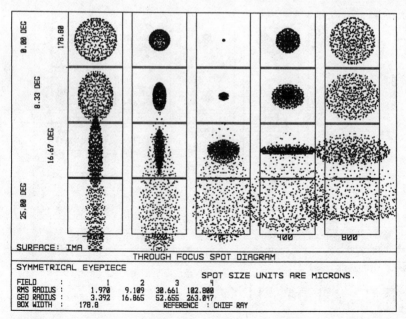

Figure 4A.7 Symmetrical eyepiece.

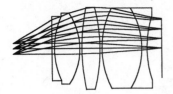

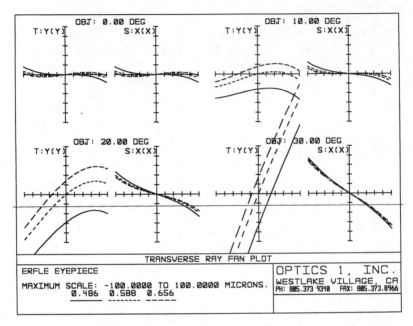

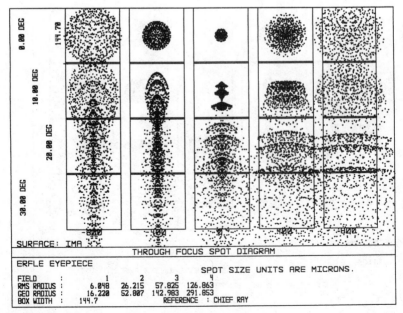

Figure 4A.8 Erfle eyepiece.

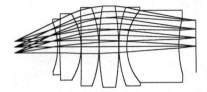

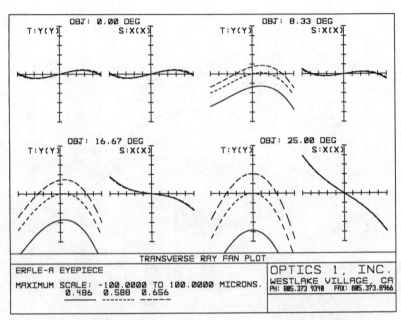

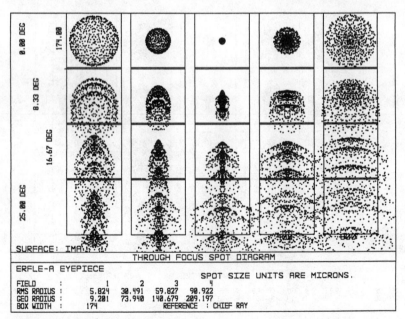

Figure 4A.9   Erfle-a eyepiece.

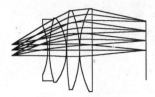

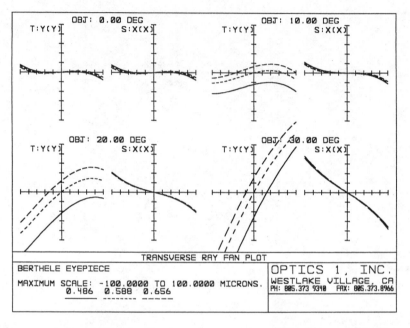

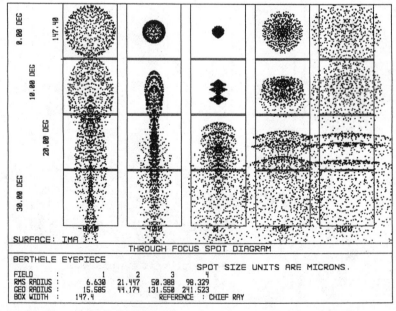

Figure 4A.10 Berthele eyepiece.

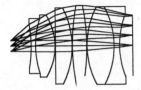

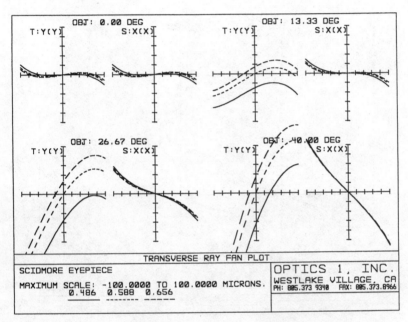

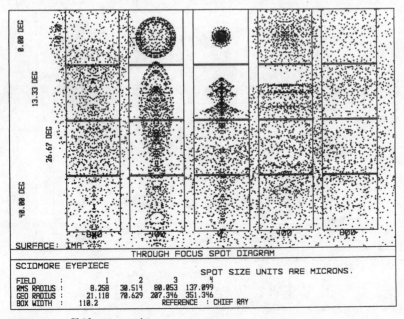

Figure 4A.11 Skidmore eyepiece.

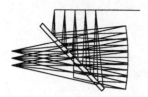

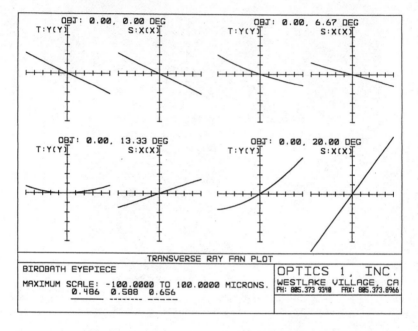

OBJ: 0.00, 0.00 DEG
T:Y(Y) S:X(X)

OBJ: 0.00, 6.67 DEG
T:Y(Y) S:X(X)

OBJ: 0.00, 13.33 DEG
T:Y(Y) S:X(X)

OBJ: 0.00, 20.00 DEG
T:Y(Y) S:X(X)

TRANSVERSE RAY FAN PLOT

BIRDBATH EYEPIECE

MAXIMUM SCALE: -100.0000 TO 100.0000 MICRONS.
0.486   0.588   0.656

OPTICS 1, INC.
WESTLAKE VILLAGE, CA
PH: 805.373.9340   FAX: 805.373.8966

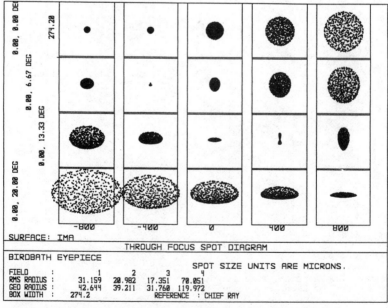

0.00, 0.00 DEG   274.20
0.00, 6.67 DEG
0.00, 13.33 DEG
0.00, 20.00 DEG

-800   -400   0   400   800

SURFACE: IMA

THROUGH FOCUS SPOT DIAGRAM

BIRDBATH EYEPIECE

SPOT SIZE UNITS ARE MICRONS.

FIELD	:	1	2	3	4
RMS RADIUS	:	31.159	20.982	17.351	70.051
GEO RADIUS	:	42.644	39.211	31.760	119.972
BOX WIDTH	:	274.2		REFERENCE : CHIEF RAY	

Figure 4A.12 Birdbath eyepiece.

Designing HMDs for Viewing Comfort

Kirk Moffitt

Head-mounted displays (HMDs) are complex viewing devices that can easily cause the user to experience eyestrain, headaches, disorientation, fatigue, and sickness. These symptoms of viewing discomfort are unacceptable if the potential benefits of HMDs are to be realized. This chapter defines HMD viewing comfort and user characteristics that are critical for designing a system that is comfortable to view. Design guidance is provided for achieving an acceptable level of HMD viewing comfort.

5.1 Preface

Head-mounted display (HMD) users want to obtain critical information, visualize a new design, or be immersed in a virtual environment. What users sometimes get is an uncomfortable viewing experience. The HMD-wearing user may see double imagery or blur, experience eyestrain and a headache, and feel disoriented and even sick (Stone, 1993). Alternatively, viewing comfort may be designed into the HMD to deliver the needed information, visualization, or immersive experience without the user suffering ill effects in the process.

Early HMDs using LCD image sources and directed at virtual-reality applications were notorious for offering low-quality images compounded by poor binocular alignment. The result frequently was disorientation, eyestrain, and a disappointed user. Early military HMDs using CRT image sources achieved higher-quality images (at a much higher monetary cost), but were criticized for problems involving binocular rivalry, ocular focus, and alignment. These systems also had weight and balance problems, and they put too much strain on the head and neck of the user.

As visual displays have matured, the image quality of commercial HMDs and military prototypes has improved. Obvious problems with alignment and focus have been corrected, but viewing comfort is still a challenge to the HMD designer. Specific problems include eyestrain resulting from misaligned binocular systems, fatigue resulting from a noncorrespondence between binocular vergence and visual accommodation in stereo HMDs, and feelings of unease and sickness resulting from a vestibular/visual conflict with immersive HMDs that simulate self-motion.

Viewing HMD imagery should be natural and comfortable. This requirement becomes more important as head-mounted displays penetrate more entertainment venues, engineering groups, and school curricula. HMD usage has been limited primarily to military aviators and test pilots, researchers, arcade users, and early adopters of cutting-edge technology. These initial users frequently overlook discomfort. With the advent of general applications and a wide range of users, HMDs need to become "bulletproof" in terms of viewing comfort. HMDs must conform to the viewing requirements of a diverse population.

The purpose of this chapter is to describe the factors underlying HMD viewing discomfort and to propose design requirements for alleviating these problems. Visual and ocular responses that directly affect viewing comfort take priority in this chapter, and perceptual conflicts that affect viewing comfort are also discussed. Related topics such as display appearance, mental stress, phobic responses, and head/neck strain are covered in other chapters of this book.

The intimate coupling of the HMD to the user makes a user-centered design critical. At the heart of this coupling is the response of the observer's visual and oculomotor system to the displayed imagery. A user-centered HMD design does not ask the user to adapt his eyes and visual system to peculiarities of the HMD. Rather, the HMD is designed in consideration of the user.

5.2   HMD Viewing Comfort

Natural scenes are easy to view. This ease of viewing can be contrasted with viewing displays such as computer monitors and devices such as stereo microscopes and HMDs. Placing HMD viewing comfort in the context of viewing natural scenes and a variety of displays can aid our understanding of the challenge of viewing comfort.

We are visually coupled to the natural world (e.g., Gibson, 1979). Scenes in the natural world are panoramic and extend to the horizon. The presence of the nose and the limitations of peripheral vision bound the image, but moving the head and body extends our view of the environment. It is easy to see the difference between close and distant objects, aided by the fact that most features are attached to the ground and nested. Distant objects are less clear than close objects, but that doesn't matter; close objects are more important. Objects are nested within other objects, and can be uncovered by moving the head or body or by pushing aside the close object. Of relevance for this discussion is that the observer of natural scenes experiences no visual stress and no eye or head pain.

Viewing HMDs can be very different from viewing the natural environment. The HMD image is obviously bounded, and differences between the left- and right-eye field of view (FOV) and imagery can be evident and distracting. Object distances are frequently ambiguous, and this makes exploring difficult. Most troublesome, there are many ways to misalign the HMD imagery, any of which can cause visual stress and eye and head pain.

Why is the viewing of natural vistas usually effortless and comfortable? Natural scenes are within the bounds of normal viewing—closer than infinity and greater than arm's length. Natural scenes are perfectly aligned. The images received by the right and left eyes have no vertical, rotation, distortion, or magnification difference. Horizontal parallax is due to the separation of the two eyes and gives us stereo vision. Furthermore, this horizontal variation corresponds to the focus of the image.

Our eyes automatically work together to facilitate the viewing of our environment. Shifting attention from a distant to a close object results in the *near response* or *near triad:* convergence of the eyes, accommodation of the lens, and constriction of the pupil. Convergence brings the object into central or focal vision, accommodation focuses the object to improve image quality, and pupil constriction increases the depth of field to compensate for dioptric errors caused by minor changes in linear distance. The convergence angle of the eyes is illustrated in Fig. 5.1.

All of these attributes of natural viewing can go awry in an HMD. For example, HMD imagery can be focused at the tip of the nose or beyond optical infinity. The left- and right-eye scenes can be misaligned in multiple ways. The binocular disparity that gives us stereo vision can be exaggerated or even incorrect. Finally, the vergence angle of the eyes to an imaged object need not correspond to the optical focus of that object. Any of these visual irregularities can create viewing discomfort.

How do HMDs compare to other displays in terms of viewing comfort? Viewing movies or television is usually effortless and comfortable, at least for moderate viewing periods. Staring at a screen for lengthy periods can lead to general feelings of visual fatigue that are preventable with frequent rest periods. One factor is that the observer has some control over viewing distance. The observer is also aware of the surrounding room, and can glance away from the display as needed. HMDs can be captive devices, with little opportunity for visual relief.

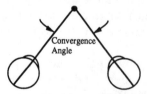

Figure 5.1. The two eyes fixating a target form the convergence angle. For an extensive treatment of the convergence angle and the computation of binocular disparity see Boff and Lincoln (1988, section 1.808) and Cormack and Fox (1985).

In contrast to movie screens and television, computer displays have routinely been associated with visual discomfort or *asthenopia*— including eyestrain, headaches, and blurred vision (e.g., Collins, Brown, Bowman, & Carkeet, 1990; Godnig & Hacunda, 1990; Sheedy, 1992). Godnig and Hacunda cite data showing that more than 50 percent of computer users experience visual discomfort. This percentage increases with more frequent computer use. Not surprisingly, this ocular discomfort is accompanied by other symptoms including back pain, shoulder pain, and fatigue.

High-quality binoculars are usually not associated with visual discomfort, although there are several critical adjustments. These adjustments include the distance between the exit pupils of the instrument and optical focus. The exit pupils of such instruments are small but easily placed within the eye's pupils. Improper focus can fatigue the observer, but the small depth of field of binoculars makes it easy to adjust focus for a clear and crisp image. In addition, binoculars are typically used for brief periods of time in combination with naked-eye viewing.

This is not quite the case with night-vision goggles (NVGs), when some NVG tasks require continuous viewing for long periods of time (e.g., Verona & Rash, 1989). NVGs for aviators are heavy and forward-balanced devices that are frequently helmet-mounted. Not surprisingly, there have been numerous reports of both neck and visual fatigue with NVGs. Incorrect setting of the lateral separation of the optical eyepieces has been implicated as one contributor to visual fatigue (Sheehy & Wilkinson, 1989).

Visual discomfort has also been associated with stereo microscopes (Helander, Grossmith, & Prabhu, 1991). This may be due to prolonged viewing periods, poor posture by the user, and the close working distance. Stereo microscope users have been found to focus the imagery several diopters too close, perhaps being influenced by the close working distance to the specimen.

A major difference between HMDs and these other displays and optical devices is the opportunity for misaligned and noncorrespondent imagery. HMDs have both complex eyepiece optics and electronic displays. Frequently, the optics are developed to support a wide field of view (WFOV), as well as for binocular viewing. The problems of optical distortion are combined with the need for an accurate interpupillary distance (IPD) setting. Electronic displays being imaged can vary along multiple dimensions, including magnification, rotation, horizontal and vertical offset, and distortion. With a binocular HMD, differences between the left and right displays in relation to any of these dimensions can create viewing discomfort. As with stereo microscopes, the user is well aware of a mechanism sitting in front of his eyes. This awareness can trigger a near response and over-accommodation.

The following sections of this chapter explore the role of the user in viewing and adapting to HMDs, their optical and display attributes and tolerances for eliciting viewing discomfort, the aftereffects of viewing HMDs, and HMD design recommendations to minimize discomfort. Although users may never confuse using an HMD with viewing and interacting with the natural world, HMD use should facilitate—not interfere with—the viewing experience.

5.3   User and HMD Characteristics

Users have visual and anatomical attributes that directly relate to HMD optical and mechanical characteristics. In some cases the HMD characteristics are fixed at one value while users exhibit a wide range of individual differences. In other cases users have fine sensitivity to visual parameters, whereas HMDs have difficulty meeting these tolerances. The emphasis of this section is on user attributes that affect HMD utility.

5.3.1   Visual acuity and eye relief

Visual acuity is sharpness of vision—the ability to resolve detail. Visual acuity can be degraded by refractive errors such as myopia, hyperopia, and astigmatism. With improvements in HMD image quality, good acuity becomes increasingly important. Eyeglass correction is one method of restoring visual acuity, making eyeglass compatibility a critical HMD attribute.

People typically take off their eyeglasses to use an optical instrument such as a microscope or telescope. Novice HMD users are frequently unsure about the need for eyeglass correction with HMDs. Few HMDs have focus adjustments, and most are focused between one meter and optical infinity with no provision for astigmatic correction.

Bifocal wearers are usually *presbyopes* who look through the top portion of their lenses to view distant objects and the bottom portion for near tasks such as reading a book. HMDs focused at optical infinity can be viewed by bifocal wearers looking straight ahead. If the HMD imagery is not head-stabilized, looking downward can result in viewing parts of the distant imagery through the portion of the optical correction meant for near viewing. With computer monitors such viewing has been linked to neck strain and viewing discomfort (e.g., Collins, Brown, Bowman, & Carkeet, 1990).

Eye relief is important to eyeglass wearers. Eyeglasses should fit comfortably underneath the HMD optics with no physical contact. The alternative to providing adequate eye relief is to provide refractive correction for each user, either as a focus adjustment or as an optical insert. Eye relief has been defined as the distance from the last optical element or mechanical component to either the exit pupil or the cornea of the eye. The exit pupil is located approximately 3 mm behind the front of the cornea. Eye relief for an observer wearing eyeglasses is shown in Fig. 5.2. An eye relief of 25 mm has been recommended, with a minimum of 15 mm needed for military-style eyewear (e.g., Farrell & Booth, 1975; MIL-STD-1472D,1989). These data use the exit pupil as the reference plane.

Viewing comfort for HMDs depends on adequate eye relief (Zwern, 1995). The HMD should not touch or press against the eye or brow, because HMD designs may depend on users wearing their normal optical correction. The nominal tolerance of 25 mm can be considered a minimum eye relief that will accommodate military eyewear but not all civilian styles. The key is to test eye and eyeglass compatibility using a mock-up early in the HMD design process. The designer should also develop general rules relating the optical focus of the display to the need for eyeglass correction.

5.3.2   Binocular balance and HMD alignment

Problems of binocular balance include *heterotropias* and *heterophorias*. Heterotropias are deviations of the eyes that prevent binocular fusion and stereo. Any problem with fusion interferes with the effective use of binocular HMDs. Heterophorias are horizontal, vertical, or rotational deviations of the eyes kept latent by binocular fusion. Latent visual problems of binocular balance can become evident during close viewing conditions or as a result of stress or fatigue. Oculomotor stress can be reduced by avoiding near focus and vergence settings. Heterophoria is kept latent by the binocular fusion in effect during normal viewing conditions.

Binocular HMDs do not provide a normal viewing condition. Some individuals with heterophoria may have difficulty viewing an HMD. This difficulty can involve problems with fusion and the perception

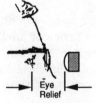

Figure 5.2.   HMD eye relief shown for an observer wearing eyeglasses.

of stereo, leading to diplopia, visual discomfort, and/or monocular suppression. This link is speculative, but could result from the limited FOV, texture, and depth of HMD imagery.

Stress and viewing conditions inherent in HMDs can alter the visual response, even when the HMD is perfectly aligned and functioning as intended. For example, "awareness of nearness" has been linked to an ocular convergence and inward accommodative response (Enright, 1987; Hokoda & Ciuffreda, 1983; Kotulak & Morse, 1995; Kotulak, Morse, & Wiley, 1994). Evidently, the presence of a structure such as a microscope tube near the eyes can inhibit the normal oculomotor response. It is likely that HMD optics could stimulate this response in some individuals.

Stress can also alter the ability to fuse binocular imagery and accurately focus. It has been my experience that many first-time viewers of HMDs and binocular devices exhibit stress and have difficulties ranging from disorientation to not being able to fuse the imagery. Initial exposure should be relaxed, with no performance pressure to immediately comprehend the imagery. Stress may also contribute to behavior such as staring at the screen with a decreased frequency of blinking. The result of infrequent blinking can be ocular dehydration, experienced as dry, red, and gritty eyes. Viewing comfort should be improved by reducing stress during HMD use.

5.3.3 Dark focus and vergence

Individuals can be characterized by relatively stable characteristics such as their *tonic accommodation* (dark or resting focus) and *vergence* (dark or resting vergence). A large range of individual differences can be found using these measures, with no agreed-upon link to normal or abnormal vision. Tonic accommodation and vergence define intermediate resting positions in the absence of spatial structure. These positions are intermediate between near and distant vision, and may be good settings for HMD imagery. This may be especially true for monocular systems, where instrument myopia (e.g., Hennessy, 1975; Owens & Leibowitz, 1983) is likely to be an important factor.

Observers exhibit a wide range of dark focuses and vergences. For example, Owens & Leibowitz (1980, 1983) report an average dark focus of 1.32 diopters (0.76 m) and an average dark vergence of 3.22° (1.16 m) from a sample of 60 college-age subjects. The range of dark focus extends from optical infinity to closer than 3 diopters (0.33 m). The range of dark vergence extends from zero degrees to as close as seven degrees (0.5 m). These averages and ranges are similar to those reported by other researchers.

There is some evidence that the dark focus or dark vergence can predict the best focus or vergence setting for a display. As an example of this research, Jaschinski-Kruza (1991) measured eyestrain at two computer monitor viewing distances: 50 and 100 cm. The more distant the dark vergence of the individual, the more likely was the report of eyestrain at the 50 cm viewing distance. When observers were asked to set their viewing distance based on comfort, they selected distances that appeared to be a compromise between the dark vergence and adequate character size.

5.3.4 IPD and exit pupil

The *interpupillary distance* (IPD) is another stable characteristic of the observer. Correspondence between the observer's IPD and the distance between the optical centers of a binocular HMD is critical to visual comfort. HMDs need to capture the IPD of the observer, either with eyepiece adjustments or with large optical exit pupils. If the observer's IPD is significantly different from the optical centers of a binocular HMD, then image deviation occurs—resulting in increased fusional demands on the observer. An IPD range of 50–74 mm captures most adult females and males (see Table 5.1).

If the eye and instrument pupils do not coincide, lateral offset of the imagery to compensate for optical displacement is not a satisfactory solution. Only the matching of the HMD optical centers and observer IPD achieves the desired outcome. HMD users may adapt to the vergence demand resulting from an incorrect IPD setting, but this is not desirable. The exit pupil of the optical system interacts with the IPD of the user and HMD adjustability to affect system usability and viewing comfort. A large HMD exit pupil fits a wider range of observer IPDs without adjustment. Unfortunately, a large HMD exit pupil usually requires a larger optical system and creates additional head-supported weight.

Binocular alignment is the attribute most critical to the viewing comfort of binocular HMDs. Providing an adequate range of optical separation for the IPDs of the target population is the first step toward ensuring binocular alignment. Several appropriate databases for adults are available, and data for children can be found in Diffrient,

TABLE 5.1 IPD Range

Range, mm	Source	Notes
56–71	Gordon et al. (1989)	5th–95th percentile f/m military
54–74	Gordon et al. (1989)	1st–99th percentile f/m military
50–73	MIL-STD-1472D	Binoculars eyepiece separation
51–68	Kroemer (1987)	5th–95th percentile f/m civilian

Tilley,~& Harman (1981). The other part of the IPD problem is arriving at a correct setting for each HMD user. Implications of incorrect IPD settings are described later in the section on binocular HMD tolerances.

5.3.5 Eye dominance

The importance of eye dominance is frequently stated, although there is little evidence to support any relationship to performance or comfort with either monocular or binocular HMDs. Eye dominance is typically defined as the eye used for sighting. A simple test is to point your finger at an object and close one eye. What happens? If the finger jumps to one side, your open eye is not dominant. If your finger remains pointed at the object, that is probably your dominant eye. Estimates of right-eye dominance range from 50 to 90 percent of the population (Crider, 1944). In an evaluation of a monocular HMD, Peli (1990) cites evidence that the dominant eye exhibits a small dominance in binocular rivalry, but that other HMD parameters are much more important.

5.3.6 Using HMDs to improve vision

From a different point of view, binocular HMD imagery could be developed to aid individuals with vision problems. The flexible alignment of HMDs that usually creates viewing problems could conceivably be used as a training aid for binocular imbalances. One HMD developed for visually impaired and legally blind users pairs a head-mounted camera with contrast-enhancement hardware to compensate for specific visual deficiencies (Siwoff, 1994).

5.3.7 Extent of eye movements, head tracking, and VOR

With naked-eye viewing, we use a combination of eye and head movements to survey our environment. An HMD system with head tracking can shift the HMD image in a direction opposite to head motion, creating the illusion of an unbounded world. Systems without head tracking yoke the HMD image to the head, requiring the observer to use eye movements to look at different locations. Most eye movements are less than 12° in extent, much less for tasks such as reading.

For systems without head tracking, eye movements are used to see different parts of the HMD image. Requiring the user to make large eye movements can quickly lead to fatigue. Although eye movements as large as 30° can be made without discomfort (Boff & Lincoln, 1988, section 1.207), sustaining this eccentric gaze is undesirable. HMD systems without head tracking should have a FOV that does not exceed approximately 24° horizontally and vertically. For HMD formats that position critical information near the edges of the display, the total FOV should be limited to less than about 18° when head tracking is not used.

Another problem with HMDs without head-tracking capability involves the *vestibulo-ocular reflex* (VOR). The VOR acts to stabilize the visual world by generating compensatory eye movements in response to head motion. Turning the head clockwise and the eyes counterclockwise maintains the gaze on a stable object. For systems without head tracking, HMD imagery is yoked to the head, negating the natural VOR. With the head and HMD imagery turning together, compensatory eye movements can create image motion and degraded image quality (Peli, 1995). VOR problems can also occur with head-tracking HMDs that have update rates (see Section 5.6) slower than the display refresh. Peli describes the case in which a moving object with a slow update rate can appear as a double or triple image, depending on the ratio of the refresh to update rates.

5.4   Binocular HMD Tolerances and Effects

In the natural world, viewing comfort is assumed and noticing double or unstable images is an odd and surprising occurrence. This can be contrasted with viewing binocular HMD imagery, where viewing is frequently painful, double images are common, and display artifacts are disturbing. If the eyes are binocularly aligned, fixated objects in the natural world create highly correspondent images on each retina in terms of location, size, shape, brightness, and color. The result is a single and stable image. The two retinal images created when viewing a binocular HMD may be quite different, resulting in double and/or unstable imagery.

If the HMD imagery misalignment is slight, the observer's visual system is capable of fusion into a single percept. With increasing misalignment, we can make small ocular adjustments to maintain image alignment. Increasing misalignment creates a thickening of the image and then an explicit double image, with possible suppression of one image to maintain single vision. The physiology of the oculomotor system determines the extent to which eye movements can compensate for image misalignment. Although we can make vertical, rotational, and divergent (beyond parallel) eye movements up to a few degrees, such ocular activity can lead to visual discomfort. This is especially likely in sustained viewing.

People show a great deal of variability in their response to misaligned or noncorrespondent imagery. Sensitive individuals may notice an immediate *tugging* of the eyes, followed by eye strain and headache.

Others will not notice any irregularity, due either to the vergence flexibility or to monocular suppression. Carpenter (1977) describes several individuals capable of voluntary control of vergence movements over a wide angular range. The effort involved in adapting to misaligned HMD imagery may be sustained for short periods of time with no problems, but may result in fatigue or eyestrain with longer viewing duration.

Alignment tolerance data come mostly from studies on binoculars and head-up displays (HUDs), with no known data being available specifically from HMDs. Alignment criteria range from discomfort to diplopia. These data apply to only one parameter at a time, as the combined effect of multiple misalignments is unknown. The best discussion to date related to HMD alignment comes from an Air Force technical report (Self, 1986). Self provides an excellent review of alignment tolerances, tracing the recommended numbers to the criteria that were used.

Thresholds or guidelines for eyestrain or visual fatigue are difficult to obtain. Measuring visual sensitivity as a function of target contrast or size can involve hundreds or thousands of trials that result in precise psychophysical data. The only danger from such experimentation is to the motivation of the observer. Exposing an observer to misaligned imagery is a different story. Once the observer experiences eyestrain and headache, which can last for several hours, further testing is jeopardized. This experimental difficulty, combined with an absence of theoretical interest in the topic among the scientific community, results in a lack of data directly related to eyestrain and visual fatigue.

Many of the tolerances discussed in this section are for binoculars or HUD symbology, devices that are not continuously used for lengthy periods. In contrast, many HMD applications require continuous use over minutes or hours. These applications include aviation sensor imagery, endoscopic surgery, and video games. This discrepancy in continuity and duration of use probably makes the cited tolerances too lenient.

Figure 5.3 shows a pair of eyeballs, each looking at a binocular image of a square target. Binocular fusion transforms each eye's perfectly aligned monocular images into the perception of a single binocular image. This binocular image lies in a direction that is intermediate between the two monocular images. Variations of this illustration will be used to show interesting HMD misalignments.

5.4.1 Vertical alignment

Although vertical misalignment is essentially nonexistent in the natural world, we have a limited ability to neurally fuse slightly misaligned

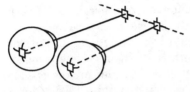

Figure 5.3. Visual fixation on a perfectly aligned
square target. Similar illustrations in this section
show various types of display misalignment. In
each case, foveal fixation is on the square target.
The cross indicates a nominal aligned position.

images and to make small vertical disjunctive eye movements to
maintain binocular fusion of slightly larger misalignments. In Fig. 5.4,
the right eye rotates downward to maintain fixation on the vertically
misaligned square target. This case is different from vertical disparity
in that the entire right-eye image is vertically misaligned.

Small amounts of vertical misalignment can be fused and perceived
as single images. Larger amounts of vertical misalignment can lead
to fatigue, eyestrain, diplopia, or monocular suppression. There has
been speculation that ocular fatigue and eyestrain can build up with
long-term viewing of even small amounts of vertical misalignment. Vertically misaligned imagery that can be initially fused may eventually
separate into diplopic images due to fatigue.

A wide range of absolute vertical misalignments that can be tolerated
were found in the literature. A combination of eye movements and
neural fusion is probably used to maintain single and/or comfortable
vision within this range of vertical misalignment. These data are based
on viewing with microscopes and binoculars, but devices that require
longer viewing times (such as HMDs) may require more stringent
alignment criteria.

The most exacting tolerance for absolute vertical misalignment is
found in the Self technical report (1986). Self cites a U.S. Navy training
text (Optical Man 3 & 2) that recommends the vertical misalignment
of the optical axes of binoculars barrels, not to exceed 2 minutes (min)
of visual angle. This text states that exceeding this tolerance can result
in eyestrain.

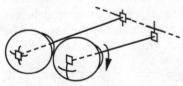

Figure 5.4. Absolute vertical misalignment. The
right eye rotates downward to fixate the square.

Tolerances of 5 to 10 min are recommended by Farrell and Booth (1975) and Jacobs (cited in Self, 1986). These data are based on microscopes and binoculars. No rationale is provided by Jacobs for an 8-min requirement. Farrell and Booth base their estimate on an unpublished study that had observers adjust a binocular microscope for adequate vertical alignment. Using the reported standard deviations, Farrell and Booth calculated the tolerance for vertical misalignment as 5 min for the most sensitive observer and 10 min for the average observer.

A tolerance for absolute vertical misalignment of 17 min is recommended by MIL-HDBK-141 (1962) for binocular instruments. An even larger tolerance of 34 min is recommended by Johnson (cited in Self, 1986). No supporting evidence is provided for either of these recommendations.

The tolerance given for absolute vertical disparity is mostly for viewing with binoculars, and should be considered an upper bound. For more extended HMD viewing, a conservative misalignment of 5 min is a good starting point. How can these alignment tolerances be verified?

Binocular viewing is not an acceptable method for HMD alignment due to large individual differences in binocular fusion and comfort thresholds. A more objective approach to binocular alignment is to separate the binocular imagery into its monocular components. One method is to display imagery that can be optically combined into a meaningful monocular image. For example, an "x" on one display and an "O" on the other can be combined into a monocular image of an x within an O. Two video cameras can also be used to combine imagery for monocular viewing, if camera-display resolution exceeds the binocular alignment requirements.

Relative vertical misalignment or binocular disparity is shown in Fig. 5.5. The square could be an object in the outside world and the circle a misaligned binocular HMD symbol. This type of imagery is used with binocular see-through aviation HMDs and augmented reality applications. The misalignment is relative between the two targets, and is referred to as *vertical binocular disparity*. Since vertical disjunctive eye movements cannot be used to fixate the circle and square target simultaneously, binocular fusion depends on neural activity. A small vertical disparity can be fused into a single percept, whereas larger amounts of disparity can create diplopia or monocular suppression. The range of binocular fusion with both vertical and horizontal disparity is called *Panum's fusional area*.

Tolerances for vertical binocular disparity, being dependent on neural fusion, are less than the corresponding absolute tolerances. Tsou, Grigsby, Jones, and Pierce (1995) state a recommended tolerance for

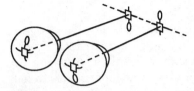

Figure 5.5. Relative vertical misalignment, or disparity. Seeing the circle as a single object requires neural binocular fusion.

transparent HMDs of 4 min. This recommendation is derived from current clinical tolerances and ranges of individual differences.

Self (1986) and Warren, Genco, and Connon (1984) cite head-up display (HUD) research by Gold and Hyman that was conducted at the Sperry-Rand Corporation in the early 1970s. These researchers had observers view dynamic HUD imagery against a static aerial view of terrain and buildings. After a 15-second observation period, observers rated the comfort level. A second study used static simulated HUD symbology viewed against a dynamic movie of the outside world. In the Gold research, a comfort tolerance of 3.4 min—less than the level that would cause diplopia—was recommended for vertical disparity. This recommendation should be considered tentative due to the small sample sizes that were used (three observers in one study, four in a second study) and the large individual differences that were observed.

Other data that can be used to derive tolerances for vertical binocular disparity come from a study by Duwaer and van den Brink (1981). These researchers used simple graphics to construct fixation markers and test stimuli. Observers indicated whether the test stimuli appeared *single, double,* or *neither single nor double*. The average binocular disparity for still seeing single ranged from 3 to 6 min, depending on the testing condition. As with the Gold studies, the range of individual differences was large. Thresholds differed between observers by up to a factor of 6 and varied with the amount of training by up to a factor of 2.5.

Duwaer and van den Brink also tested two observers on their ability to detect vertical disparity. One observer could detect vertical disparity in the fovea (centrally located) of only 2.2 min, though his singleness criterion was 7 min. The other observer had a lower threshold for detecting vertical disparity of only 1.6 min. Each observer reported using different criteria in detecting disparity. Although Duwaer and van den Brink reported large individual differences for vertical binocular disparity, the average tolerances (typically for diplopia) range from only about 3 to 6 min. Although we can detect smaller amounts of vertical disparity, limiting vertical disparity to this range should prevent viewing discomfort.

5.4.2 Horizontal alignment

Unlike the range of vertical movements, we have a wide range of horizontal vergence eye movements. This range is restricted by parallel axes, the near point, and image focus. We make convergent eye movements to look at close objects, and divergent movements to look at distant objects. The vergence eye movement is one component of the near triad, the other two being pupil dilation/constriction and visual accommodation. To view an object held in our hand, we converge our eyes, constrict our pupils, and accommodate (optically) inward. These actions serve to create an aligned image of the object, increase the depth of field, and bring the object into sharp focus. To shift attention to a distant object, we diverge our eyes, dilate our pupils, and accommodate outward. Figure 5.6 shows the left eye making a small diverging movement to maintain binocular fixation on the square.

Real-world objects are never more distant than optical infinity: parallel visual axes and zero diopters of focus. In addition, image focus and vergence demands are equivalent in the natural world. An object located one meter from the eyes requires one diopter of visual accommodation to accurately focus, and convergence equivalent to one meter to accurately fixate. These demands are in correspondence in the natural world but dissociated in a binocular HMD, where focus and vergence can take on independent values.

We have limited ability to diverge our eyes past parallel axes. This response is just not needed in the natural world. With HMDs, especially those nominally set to optical infinity, small amounts of divergent misalignment can be common. We also have a limit on convergence, frequently to the 10- to 20-cm region. This limit is commonly measured in conjunction with visual accommodation and pupil constriction to establish a *near point* of vision. Image vergence should not be closer than the near point of vergence and accommodation—10 to 20 cm. To avoid conflict with bifocal wearers and people with convergence insufficiency, focus and vergence should be outward from this near position. The most distant focus and vergence setting that should be considered is optical infinity.

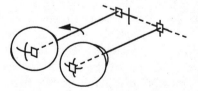

Figure 5.6. Absolute horizontal misalignment. The left eye is shown diverging to maintain binocular fixation on the square.

As with tolerances for vertical misalignment, much of the available data for absolute horizontal misalignment come from requirements for binoculars. In many cases, the rationale for the data is unknown. Jacobs (cited in Self, 1986) recommends a maximum divergence of 7.5 min and a maximum convergence of 22.5 min for the optical axes of the binocular tubes. Self also cites recommendations from the Army-Navy National Research Council of 1946 for a maximum divergence of 14 min and a maximum convergence of 28 min for binoculars. Johnson (cited in Self, 1986) allows much larger misalignments: 69 min divergence and 138 min convergence. Peli (1995) cites an ANSI standard of 17 min (0.5 prism diopter) of imbalance for ophthalmic prescriptions.

In natural scenes, object vergence and accommodation are correspondent. An object at one meter has a focus of one diopter. A distant object has a vergence approaching zero degrees and a focus approaching zero diopters. The vergence and accommodative systems are cross-linked, so that vergence demands drive accommodation and focus demands drive vergence. In optical viewing devices it is possible to dissociate the vergence and accommodation demands, in which case the cross-links are sending inappropriate commands.

Another approach to specifying absolute horizontal alignment requirements is to take focus or viewing distance into consideration. Farrell and Booth (1975) provide an envelope that relates the best viewing distance to convergence angle. Viewing distance can be equated to image focus. This envelope extends from a viewing distance of 0.5 to 2.0 diopters, or 2.0 to 0.5 meters. Using an average IPD of 63 mm, Farrell and Booth recommend that convergence be within 2.7° of the selected viewing distance (stopping at zero degrees).

Jones (1992), in a review of the interlinked role of accommodation and convergence in the fusion of displayed imagery, states that accommodation can vary about ±2 diopters independently of vergence. Similarly, vergence can vary about ±2 meter-angles (68 min) independently of accommodation. These are the limits of the oculomotor system before the imagery appears blurry or double. Jones states that a clinical rule of thumb is that no more than half this range (±1 diopter of accommodation, ±1 meter-angle (34 min) of vergence) provides for comfortable vision. These data should be limited to settings nearer than optical infinity or parallel optical axes.

The recommendation by Jones of ±1 diopter is probably too large for extended viewing. I have found ±0.25 diopter to be more desirable and ±0.5 diopter to be a maximum misalignment. For an HMD with a vergence setting of one meter, focus could range from 1.25 to 0.75 diopter without appreciable discomfort or eyestrain. Other researchers have proposed aligning HMDs in concert with the user's heterophoria, and they have developed a software calibration device for this alignment (Mon-Williams, Rushton, & Wann, 1995; Mon-Williams, Wann, & Rushton, 1993).

In nature, small amounts of horizontal misalignment or binocular disparity between objects provide us with the perception of stereoscopic depth (see Tyler (1983) for a review of binocular disparity). This viewing geometry results from the relative distances of objects with respect to our IPD. Horizontal binocular disparity is illustrated in Fig. 5.7.

Several researchers (e.g., Genco, 1983; Warren, Genco, & Connon, 1984; Gold, cited in Self, 1986) have provided horizontal alignment recommendations for head-up displays (HUDs) based on discomfort or diplopia thresholds. These are special cases in which HUD symbology is superimposed on outside-world scenery, and they were motivated by pilot reports of diplopia (e.g., Genco, 1983). This diplopia is caused by refraction of outside-world scenery by the curved windscreen of high-performance aircraft. A solution is to match the HUD optical vergence to this windscreen refraction, but the reader may ask why this is necessary.

We normally diverge and converge our eyes to look at objects at different distances. Furthermore, while we constantly maintain our ambient perception, we generally attend to only one object at a time. Why would the pilot not simply shift his attention and vergence between the HUD symbology and the outside world? One possibility is that the superimposed green symbology is directly in line with the outside-world object and accentuates any diplopia. Natural objects tend to be desaturated, opaque, and nested—not luminous, semi-transparent, and floating.

There are other cases in which a certain horizontal vergence setting is desired. For aerospace HUD and HMD applications, parallel or zero-degree vergence is usually required. This postures the eyes out of the cockpit in terms of both vergence and accommodation, in preparation for acquiring distant targets. While this setting can minimize the time needed to refocus the eyes, there is a danger in interrupting the natural pattern of glancing at objects at different distances. One

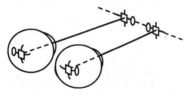

Figure 5.7.   Horizontal binocular disparity. Binocular fixation remains on the square, with the circle being horizontally convergent. The perceptual result from stereo vision is of the circle appearing closer than the square.

consequence of minimizing the need to change gaze and refocus is *HUD fixation,* where attention is trapped by the HUD symbology. For dome simulators, HMD symbology is frequently converged to the fixed distance of the simulator imagery.

Viewing discomfort can be noticed even when misalignments are insufficient to cause diplopia. Gibson (1980) found this discomfort to be associated with comments such as "uncomfortable to look at"... "confused"..."out of focus"..."unclear." The "confused" and "unclear" comments may be specific to binocular disparities that diverge graphics or symbology beyond optical infinity. That is, the outside world is presented at optical infinity and overlaid graphics are presented at an unnatural distance beyond optical infinity.

In the HUD study described in the previous section on vertical misalignment, Gold also measured viewing discomfort for horizontal misalignment. He reported that image divergence should not exceed 3.4 min, and image convergence should not exceed 8.6 min. A similar divergence requirement of 2.9 min was reported by Gibson (1980), and this level is significantly less than needed to elicit diplopia. Gibson also reported that his observers preferred that the HUD symbology be converged to 1.3 min anterior to the background scene.

An Air Force study in the early 1980s also provides recommendations for horizontal misalignment tolerances for HUDs (Genco, 1983; Warren, Genco, & Connon, 1984). In this study, observers fixated a distant real-world object through a third-story window and reported their perception of a briefly presented binocular symbol: single, double, or single and misaligned with the outside-world object (the latter indicating monocular suppression). The median disparity thresholds for diplopia, averaged over several conditions, were 4.1 min divergence and 8.8 min convergence. These recommendations are very close to those of Gold. Both 0.1- and 3.0-second exposures were tested by Warren et al., and there was more tolerance for disparity at the shorter exposure.

The data that have been cited indicate a tolerance for symbology or graphics of approximately 3 min divergence and 8 min convergence, relative to optical infinity reference imagery. This asymmetry may actually be due to a fixation disparity that is slightly closer than the image plane, perhaps corresponding to Gibson's recommended 1.3-min bias.

5.4.3 Accommodation/vergence dissociation with stereo displays

A related line of research involves vergence/accommodation imbalances found with stereo displays (see Chapter 9). Stereo displays have a fixed focus, yet they utilize binocular disparity to achieve depth. Shifting attention to a close object involves binocular convergence from the display, creating a decoupling of vergence and accommodation (Edinburgh Virtual Environment Laboratory, 1993; Mon-Williams, Rushton, & Wann, 1995). Yamazaki, Kamijo, and Fukuzumi (1990) reported symptoms such as eyestrain, dimness, and headache after having observers watch a one-hour movie on a stereoscopic monitor with field-sequential left-eye/right-eye fields. Another study linked fatigue from using a binocular stereoscopic viewer to a vergence/accommodation imbalance (Miyashita & Uchida, 1990).

Actual limits of fusion with stereoscopic displays were proposed by Yeh and Silverstein (1990). Ten subjects gave reports of a briefly presented stimulus being seen as single or double. Both *crossed* (the stimulus would appear in front of the display screen) and *uncrossed* (the stimulus would appear behind the display screen) disparities were tested at a viewing distance of 66 cm. The limits of fusion were approximately 27 min for crossed disparity and 24 min for uncrossed disparity. This translates into a tolerance of approximately ±0.1 diopter, which is more conservative than the recommendations cited in the previous section.

5.4.4 Rotational differences

Small amounts of HMD image rotation can be tolerated. With 100 percent overlapped left- and right-eye imagery, rotation can be countered with both sensory fusion and with small cyclo-rotation eye movements. A binocular rotation difference is illustrated in Fig. 5.8. One limit is vertical misalignment. As the images are rotated or clocked relative to each other, vertical misalignment increases with the radial distance from the image center. With partial-overlap imagery, rotation loses meaning and vertical misalignment becomes the correct assessment.

Studies have shown that observers can maintain fusion with relatively large amounts of image rotation. For example, Farrell and Booth

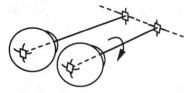

Figure 5.8. Binocular rotation, or clocking. The right image is rotated relative to the left image. The right eye may rotate or torque slightly to maintain binocular fusion.

(1975) cite data showing the ability to fuse binocular imagery with 5° to 15° of rotation difference. They also cite data showing that observers can correct rotational differences on the order of 1° to 2°.

In determining tolerances for image rotation, using vertical misalignment at an acceptable eccentricity is appropriate. The 5° to 15° rotations cited by Farrell and Booth are probably not acceptable for extended viewing. The 1° to 2° sensitivity of observers to binocular rotation differences may be the more relevant data, and could be used as an HMD alignment requirement.

5.4.5 Magnification difference

Although we can fuse magnification differences between the left and right images of up to 15 percent, sustained viewing of differences of only a few percent may cause visual fatigue. Magnification difference is illustrated in Fig. 5.9. Military standards for magnification differences based on binoculars specify an allowance of 5 percent (MIL-STD-1472D, 1989), 1 to 2 percent (MIL-HDBK-141, 1962), and 0.5 percent for sensitive observers (also MIL-STD-141). Tolerances for magnification are based on binoculars and are probably too large for HMDs. For lack of better data, the bottom range of binoculars tolerances (0.5 to 2 percent difference between the left and right image) should be acceptable for HMDs.

5.4.6 Luminance difference

Compared to binocular alignment and focus, a luminance difference between the left and right eyes is unlikely to be a major cause of HMD viewing discomfort. Tolerances based on binoculars are 5 percent (MIL-HDBK-1472D), 10 percent (MIL-HDBK-141), 10–25 percent (Self, 1986), and 25–50 percent (Farrell & Booth, 1975). Within this range, the visual system averages the luminances, and the apparent image brightness corresponds to the average luminance.

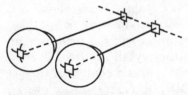

Figure 5.9. Binocular magnification difference. The right image is larger than the left image. Fusion into a single image is dependent on neural processes.

5.4.7   Changes in visual status

Viewing a display should not adversely affect the visual status of the observer. A stable visual status is one indicator of visual comfort and perhaps image alignment. Candidate vision tests include *phoria* (binocular balance), the near point, the far point, acuity, and binocular fusion.

The measures of visual status are complemented by observer reports of eye tugging, double vision, eyestrain, and head/neck ache. There are large individual differences in such reports, with some observers being relatively insensitive to large amounts of misalignment, whereas others find minimal amounts of misalignment to be intolerable. There are also large individual differences in reports of binocular rivalry, either with binocular partial-overlap displays or with monocular imagery displays.

With a mismatch between the optical centers of the HMD and the observer's IPD, the observer may exhibit adaptation to the increased vergence demand in only a few minutes (Piantanida, 1993). After removal of the HMD, visual discomfort could be experienced while the observer readapts to the perfect alignment of the real world. This readaptation should be complete in a few minutes. Piantanida points out an exception to this successful readaptation: the small proportion of people with intermittent *exophoria*. Fusional stress can cause an eye deviation in these people, triggering episodes of double vision. The most common cause of fusional stress is near-vision work. The fusional stress created by a mismatched HMD could also serve as a trigger for eye deviation.

Mon-Williams, Wann, and Rushton (1993) conducted an experiment with 20 observers, in which they found that using an HMD to explore a virtual environment for only ten minutes of time can result in temporary signs of binocular stress. These signs included distant visual acuity, distant heterophoria, and both distant and near associated phoria (binocular balance with fusion). The HMD was the VPL Eyephone LX, an early virtual-reality device with low resolution and fixed (65 mm) IPD. In a later study (Rushton, Mon-Williams, & Wann, 1994), these same researchers replicated their 1993 study with a higher-quality HMD—the Virtuality Visette™ 2000—and 50 observers.

Rushton et al. reported no visual acuity changes after 5-, 10- or 20-minute sessions. There were no significant signs of binocular imbalance. Six observers complained of either eyestrain, tired eyes, or disorientation. The most interesting finding was a change in the setting of the interocular distance of the HMD. Observers were allowed to set the HMD optical separation to achieve viewing comfort. Rushton's observers consistently set the optical separation to a smaller value than their IPD. This was attributed to a near response to the optics in front of the eyes.

5.5   Monocular HMD Tolerances and Effects

Monocular HMDs include an operational system for presenting infrared sensor imagery and symbology to helicopter aviators, several prototype symbology systems for military high-performance aircraft, and LED- and LCD-based devices for commercial and industrial use. Monocular HMDs usually have a smaller FOV than binocular systems and are not used for immersive experiences. The chief advantage of a monocular system is the absence of binocular alignment problems and the attendant eyestrain. The largest problems are perceptual irregularities related to left- and right-eye image conflicts.

Peli (1990) conducted a thorough analysis of the Private Eye™ monocular HMD. This particular display occludes the real world for one eye. Peli's analysis covers many of the viewing discomfort issues associated with monocular displays. The issues most relevant to viewing comfort are binocular rivalry and problems with monocular occlusion.

Peli tested three users on their ability to word-process a document with the computer screen presented to one eye using the Private Eye™ HMD and paper copy viewed by the other eye. One user superimposed the two images, another found it preferable to offset the images, and the third user found the setup to be uncomfortable and complained of eyestrain. Peli also measured the lateral phoria of the three observers after 45 minutes of HMD use on the word processing task. One of the three showed a small increase in exophoria with both distance and near vision.

With monocular HMDs that display video imagery, such as the Honeywell IHADSS used in the AH-64 Apache helicopter, a basic problem is binocular rivalry (e.g., Rash, Verona, & Crowley, 1990). The response to one eye viewing green video and the other eye viewing a dark cockpit and the outside world can be suppression of the eye viewing the dimmer cockpit and outside world. Viewing of these dissimilar images has proven especially fatiguing during lengthy missions. Voluntary switching of attention between eyes has met with difficulty among some aviators. Closing one eye can create additional fatigue and is not an effective solution.

5.6   Motion Effects

We explore our natural world using eye, head, and body movements. Changes in the visual scene correspond to these movements. Put yourself in the back seat of a moving automobile or the passenger seat on an airplane. The vehicle motion that stimulates your vestibular system is not always accompanied by the appropriate visual motion. In these cases much of the visual scene remains static, and the result can be motion sickness. Now consider the case in which there is little or no vestibular stimulation but highly salient cues to visual motion. That is what occurs in aircraft and driving simulators.

Simulator sickness in aircraft and driving simulators results in symptoms ranging from drowsiness to eyestrain to nausea. This sickness is caused by highly salient visual cues to motion being in conflict with an absence of vestibular cues to motion. Estimates of the proportion of simulator users who exhibit symptoms of simulator sickness range from 30 to 90 percent (Strauss, 1995). Simulator sickness can be contrasted with motion sickness, in which strong vestibular cues to motion have misleading visual cues to motion. Although motion sickness can culminate in vomiting, simulator sickness does not usually reach this stage. Cybersickness is an analog of simulator sickness that specifically applies to HMDs and immersive visual environments (Johnston, 1995; McCauley & Sharkey, 1992; Stanney, 1995).

Other problems with displaying scene motion on HMDs are lengthy lag times and long update rates. *Lag* is the time between when the head moves and when the visual scene is changed to reflect this movement. The *update rate* is the frequency with which each new visual frame is generated and displayed. Both long lag times and slow update rates have been implicated as contributing to simulator sickness and cybersickness (Biocca, 1992; Kalawsky, 1993; Pausch, Crea, & Conway, 1992). Although minimizing lag and increasing the update rate can improve well-being in a simulator or HMD, the fundamental visual-vestibular conflict remains intact.

Some researchers have reported data showing cybersickness to be common in immersive or WFOV virtual-reality applications (e.g., Regan & Price, 1994). After a 20-minute immersive VR experience with an HMD, 61 percent of their subjects reported some amount of malaise, including eyestrain, headache, dizziness, and nausea. Regan and Price also reported that 5 percent of their subjects withdrew from the study due to severe nausea or dizziness.

If these cybersickness symptoms sound bad, consider the worst-case scenario posed by Thomas Furness, Director of the Human Interface Technology Laboratory at the University of Washington: corruption of the vestibulo-ocular reflex can cause brain damage and LSD-like flashbacks (Gross, 1995; Johnson, 1995). Furness claims that the unnatural visual and vestibular stimulation found with HMDs can actually create new pathways in the brain with extended HMD use. During subsequent non-HMD activities the brain may unpredictably revert to these pathways, causing flashback reactions. Flashbacks and disorientation that outlast the time actually spent in the aircraft simulator have also been reported (Kennedy, Lane, Lilienthal, Berbaum, & Hettinger, 1992; Strauss, 1995).

Kennedy and Lilienthal (1995) discuss disturbances in balance and posture that can result from exposure to flight simulators and virtual-reality systems. These researchers have developed a video analysis of posture, and they found significant pre/post changes for participants in two WFOV dome flight simulators. The greatest disequilibrium was experienced by the co-pilot, seated 20° offset from the simulator design-eye point (the optimal position for a single observer). Kennedy and Lilienthal speculate that this disequilibrium has direct implications for the safety of simulator and WFOV users after they exit the system. No data were provided on the duration of these effects.

The content of the simulator or HMD scene and the user's behavior can determine the propensity to simulatorsickness or cybersickness. Freezing the action in mid-flight, backing up a dynamic scene, long exposures, requiring large head movements, and use when ill from the flu or a cold can all enhance the sickness problem (e.g., Frank, Kellogg, Kennedy, & McCauley, 1983; McCauley, 1984; Stoddard, 1989). Avoiding these actions can improve HMD comfort.

Improvements in HMD image quality will not solve the visual/vestibular problem. Existing motion bases are also not a solution. McCauley and Sharkey (1992) describe research at NASA Ames Research Center's Vertical Motion Simulator showing that pilots suffered just as much motion sickness on a motion base as on a fixed base.

One solution to the problem of cybersickness is to decrease the FOV. This suggestion runs counter to the desire to present the user with an immersive experience or peripheral information. HMD applications such as viewing movies or using a computer may be satisfied with a moderate FOV. While movies occasionally attempt to generate a sense of self-motion, many computer games and flight simulations give the user a strong sense of self-motion. These applications demand a WFOV and are the ones linked to cybersickness.

Another solution is to decrease the HMD's head-supported weight. DiZio and Lackner (1992) cite evidence that the added weight rearranges the relationships between motor commands for head movements and feedback from the head and neck. HMDs increase the effective weight of the head, increase the effective inertia of the head, and can shift the center of gravity of the head. DiZio and Lackner have observed that simply moving about while wearing such a device can elicit symptoms of motion sickness.

Other solutions to visual/vestibular conflicts involve see-through and look-around vision. Virtual I/O has developed a headset with a generous amount of look-around and look-under vision, and reports zero incidents of sickness (Buckert-Donelson, 1995). The stability provided by a view of the outside world as well as one's own body could provide the needed visual reference to prevent cybersickness. One value of an augmented-reality system, where synthetic imagery is combined with the real world, is to mitigate vertigo by providing a visual anchor to reality (Barfield, Rosenberg, & Lotens, 1995; Metzger, 1993).

5.7 Summary

Viewing comfort is a prerequisite for HMD usability. Devices that are uncomfortable, cause eyestrain and headaches, or create disorientation are unacceptable for general use. Our visual linkage to the natural world sets the standard for viewing comfort: unencumbered and panoramic viewing of perfectly aligned imagery. HMDs reside at the other end of the viewing spectrum. These are electronic displays that we view through complex optics, all mounted on our mobile heads. The challenge to the designer is to build viewing comfort into HMDs.

The utility of HMDs is partly determined by the characteristics of the observer. People with binocular vision imbalances or problems in fusing binocular imagery are not good candidates for general HMD usage. Certain characteristics of HMDs may create oculomotor stress, causing latent binocular deviances to become manifest. On the positive side, specifically designed HMDs may be an aid to individuals with poor visual acuity.

Not surprisingly, HMDs can give us eyestrain, disorientation, headaches, and sickness. Many of these problems stem from misalignment of the HMD for the observer. The good news is that we know the range of tolerances for most of these misalignments. A user-centered design puts the prevention of eyestrain and headaches, along with human performance and usability, at the center of the engineering process. Key factors in HMD design include adequate adjustability and an objective alignment method.

Monocular HMDs present a very unnatural view to the observer; each eye may see very different imagery. With video imagery, the result can be binocular rivalry and/or monocular suppression. Symbolic information has a better chance of success, especially with a see-through system. We can expect a wide range of individual differences in terms of the acceptability of monocular HMDs.

The problem of cybersickness resulting from a visual/vestibular conflict may prove to be the most difficult viewing comfort problem. The basic misalignment of these two sensory modalities cannot be corrected. Deletion of salient factors such as peripheral motion and the careful scripting of motion within the application can provide a partial solution. Providing see-through or look-around vision to the real world also has the potential to ameliorate cybersickness.

5.8 References

Barfield, W., Rosenberg, C., & Lotens, W. A. (1995). Augmented-reality displays. In W. Barfield & T. A. Furness (Eds.), *Virtual environments and advanced interface design*. New York: Oxford University Press.

Biocca, F. (1992). Will simulation sickness slow down the diffusion of virtual environment technology? *Presence, 1,* 334–343.

Boff, K. R., & Lincoln, J. E. (1988). *Engineering data compendium: Human perception and performance,* (Vol. I). Wright-Patterson Air Force Base, OH: Armstrong Aerospace Medical Research Laboratory.

Buckert-Donelson, A. (1995, May/June). Linden Rhoads, Virtual I/O. *VR World,* 35–36, 38–39.

Carpenter, R. H. S. (1977). *Movements of the eyes.* London: Pion.

Collins, M., Brown, B., Bowman, K., & Carkeet, A. (1990). Workstation variables and visual discomfort associated with VDTs. *Applied Ergonomics, 21,* 157–161.

Cormack, R., & Fox, R. (1985). The computation of disparity and depth in stereograms. *Perception & Psychophysics, 38,* 375–380.

Crider, B. (1944). A battery of tests for the dominant eye. *The Journal of General Psychology, 31,* 179–190.

Diffrient, N., Tilley, A. R., & Harman, D. (1981). *Humanscale 4/5/6.* Cambridge, MA: MIT Press.

DiZio, P., & Lackner, J. R. (1992). Spatial orientation, adaptation, and motion sickness in real and virtual environments. *Presence, 1,* 319–328.

Duwaer, A. L., & van den Brink, G. (1981). What is the diplopia threshold? *Perception & Psychophysics, 29,* 295–309.

Edinburgh Virtual Environment Laboratory (1993). What's wrong with your head mounted display? *CyberEdge Journal, Monograph 3.*

Enright, J. T. (1987). Perspective vergence: Oculomotor responses to line drawings. *Vision Research, 27,* 1513–1526.

Farrell, R. J., & Booth, J. M. (1975). *Design handbook for imagery interpretation equipment.* Seattle, WA: Boeing Aerospace Co.

Frank, L. H., Kellogg, R. S., Kennedy, R. S., & McCauley, M. E. (1983). Simulator sickness: A special case of the transformed perceptual world. In R. S. Jensen (Ed.), *Proceedings of the Second Symposium on Aviation Psychology* (pp. 587–596). Columbus, OH: The Ohio State University Department of Aviation.

Genco, L. V. (1983). Optical interactions of aircraft windscreens and HUDs producing diplopia. In W. L. Martin (Ed.), *Optical and human performance evaluation of HUD systems design* (Technical report AFAMRL-TR-83-095). Wright-Patterson Air Force Base, OH: Aerospace Medical Research Laboratory.

Gibson, C. P. (1980). Binocular disparity and head-up displays. *Human Factors, 22,* 435–444.

Gibson, J. J. (1979). *The ecological approach to visual perception.* Boston: Houghton Mifflin.

Godnig, E. G., & Hacunda, J. S. (1990). *Computers and visual stress.* Charlestown, RI: Seacoast Information Services.

Gordon, C. C., Churchill, T., Clauser, C. E., Bradtmiller, B., McConville, J. T., Tebbetts, I., & Walker, R. A. (1989). *1988 Anthropometric survey of U.S. Army personnel: Summary statistics interim report* (Technical report NATICK/TR-89/027). Natick, MA: U.S. Army Natick RD&E Center.

Gross, N. (1995, July 10). Seasick in cyberspace. *Business Week,* 110–111.

Helander, M. G., Grossmith, E. J., & Prabhu, P. (1991). Planning and implementation of microscope work. *Applied Ergonomics, 22,* 36–42.

Hennessy, R. T. (1975). Instrument myopia. *Journal of the Optical Society of America, 65,* 1114–1120.

Hokoda, S. C., & Ciuffreda, K. J. (1983). Theoretical and clinical importance of proximal vergence and accommodation. In C. M. Schor & K. J. Ciuffreda (Eds.), *Vergence eye movements: Basic and clinical aspects.* Boston: Butterworths.

Jaschinski-Kruza, W. (1991). Eyestrain in VDU users: Viewing distance and the resting position of ocular muscles. *Human Factors, 33,* 69–83.

Johnson, R. C. (1995, June 5). VR: Hazardous to your health? *Electronic Engineering Times,* issue 851, 1, 180.

Johnston, R. (1995, May/June). Losing everything to VR: Simulator sickness. *Virtual Reality Special Report,* 25–28.

Jones, R. (1992). Binocular fusion factors in visual displays. *SID 92 Digest,* 287–289.

Kalawsky, R. S. (1993). *The science of virtual reality and virtual environments.* Wokingham, England: Addison-Wesley.

Kennedy, R. S., Lane, N. E., Lilienthal, M. G., Berbaum, K. S., & Hettinger, L. J. (1992). Profile analysis of simulator sickness symptoms: Application to virtual environment systems. *Presence, 1,* 295–301.

Kennedy, R. S., & Lilienthal, M. G. (1995). Implications of balance disturbances following exposure to virtual reality systems. *Virtual Reality Annual International Symposium '95.* Los Alamitos, CA: IEEE Computer Society Press.

Kotulak, J. C., & Morse, S. E. (1995). The effect of perceived distance on accommodation under binocular steady-state conditions. *Vision Research, 35,* 791–795.

Kotulak, J. C., Morse, S. E., & Wiley, R. W. (1994). The effect of knowledge of object distance on accommodation during instrument viewing. *Perception, 23,* 671–679.

Kroemer, K. H. E. (1987). Engineering anthropometry. In G. Salvendy (Ed.), *Handbook of human factors.* New York: Wiley.

McCauley, M. E. (Ed.). (1984). *Research issues in simulator sickness: Proceedings of a workshop.* Washington, DC: National Academy Press.

McCauley, M. E., & Sharkey, T. J. (1992). Cybersickness: Perception of self-motion in virtual environments. *Presence, 1,* 311–318.

Metzger, P. J. (1993). Adding reality to the virtual. *Virtual Reality Annual International Symposium '93.* Los Alamitos, CA: IEEE Computer Society Press.

MIL-HDBK-141 (1962). *Optical design.* U.S. Department of Defense.

MIL-STD-1472D (1989). *Human engineering design criteria for military systems, equipment and facilities.* U.S. Department of Defense.

Miyashita, T., & Uchida, T. (1990). Cause of fatigue and its improvement in stereoscopic displays. *Proceedings of the Society for Information Display, 31,* 249–254.

Mon-Williams, M., Rushton, S., & Wann, J. P. (1995). Binocular vision in stereoscopic virtual-reality systems. *SID 95 Digest, 26,* 361–363.

Mon-Williams, M., Wann, J. P., & Rushton, S. (1993). Binocular vision in a virtual world: Visual deficits following the wearing of a head-mounted display. *Ophthalmic and Physiological Optics, 13,* 387–391.

Owens, D. A., & Leibowitz, H. W. (1980). Accommodation, convergence, and distance perception in low illumination. *Americal Journal of Optometry & Physiological Optics, 57,* 540–550.

Owens, D. A., & Leibowitz, H. W. (1983). Perceptual and motor consequences of tonic vergence. In C. M. Schor & K. J. Ciuffreda (Eds.), *Vergence eye movements: Basic and clinical aspects.* Boston: Butterworths.

Pausch, R., Crea, T., & Conway, M. (1992). A literature survey for virtual environments: Military flight simulator visual systems and simulator sickness. *Presence, 1,* 344–363.

Peli, E. (1990). Visual issues in the use of a head-mounted monocular display. *Optical Engineering, 29,* 883–892.

Peli, E. (1995, July). Real vision & virtual reality. *Optics & Photonics News,* 28–34.

Piantanida, T. (1993, November/December). Another look at HMD safety. *CyberEdge Journal, 9–10,* 12.

Rash, C. E., Verona, R. W., & Crowley, J. S. (1990). *Human factors and safety considerations of night vision systems flight using thermal imaging systems* (USAARL technical report 90-10). Ft. Rucker, AL: U.S. Army Aeromedical Research Laboratory.

Regan, E. C., & Price, K. R. (1994). The frequency of occurrence and severity of side-effects of immersion virtual reality. *Aviation, Space, and Environmental Medicine, 65,* 527–530.

Rushton, S., Mon-Williams, M., & Wann, J. P. (1994). Binocular vision in a bi-ocular world: New-generation head-mounted displays avoid causing visual deficit. *Displays, 15,* 255–260.

Self, H. C. (1986). *Critical tolerances for alignment and image differences for binocular helmet-mounted displays* (Technical Report AAMRL-TR-86-019). Wright-Patterson AFB, OH: Armstrong Aerospace Medical Research Laboratory.

Sheedy, J. E. (1992, April/May). VDTs and vision complaints: A survey. *Information Display,* 20–23.

Sheehy, J. B., & Wilkinson, M. (1989). Depth perception after prolonged usage of night vision goggles. *Aviation, Space, and Environmental Medicine, 60,* 573–579.

Siwoff, R. (1994, May/June). DEI: Digitally enhanced imager. *Virtual Reality World,* 63–65.

Stanney, K. (1995). Realizing the full potential of virtual reality: Human factors issues that could stand in the way. *Virtual Reality Annual International Symposium '95.* Los Alamitos, CA: IEEE Computer Society Press.

Stoddard, R. (1989, November/December). In search of the perfect ride. *Defense Computing,* 22–25.

Stone, B. (1993, October/November). Concerns raised about eye strain in VR systems. *Real Time Graphics,* 1–3, 6, 13.

Strauss, S. (1995, July). Cybersickness: The side effects of virtual reality. *Technology Review,* 14–16.

Tsou, B. H., Grigsby, S. S., Jones, R., & Pierce, G. E. (1995). Perspective projection and its tolerances in partial-overlap binocular head-mounted displays. *SID 95 Digest of Technical Papers, 26,* 364–367.

Tyler, C. W. (1983). Sensory processing of binocular disparity. In C. M. Schor & K. J. Ciuffreda (Eds.), *Vergence eye movements: Basic and clinical aspects.* Boston: Butterworths.

Verona, R. W., & Rash, C. E. (1989). Human factors and safety considerations of night vision systems flight. *Display Systems Optics II, Proceedings SPIE, 1117,* 2–12.

Wann, J. P., Rushton, S., & Mon-Williams, M. (1995). Natural problems for stereoscopic depth perception in virtual environments. *Vision Research, 35,* 2731–2736.

Warren, R., Genco, L. V., & Connon, T. R. (1984). *Horizontal diplopia thresholds for head-up displays* (Technical report AFAMRL-TR-84-018). Wright-Patterson Air Force Base, OH: Air Force Aerospace Medical Research Laboratory.

Yamazaki, T., Kamijo, K., & Fukuzumi, S. (1990). Quantitative evaluation of visual fatigue encountered in viewing stereoscopic 3D displays: Near-point distance and visual evoked potential study. *Proceedings of the Society for Information Display, 31,* 245–247.

Yeh, Y.-Y., & Silverstein, L. D. (1990). Limits of fusion and depth judgment in stereoscopic color displays. *Human Factors, 32,* 45–60.

Zwern, A. (1995, March/April). How to select the right HMD. *VR World,* 20–27.

HMD Head and Neck Biomechanics

Chris E. Perry

John R. Buhrman

Some neck injuries can be attributed to aircraft and ground vehicle accidents, sports and recreational activities, and routine and emergency military operations. The desire to use the head as a platform for new visual and auditory sensory technology may result in an increased potential for neck injury in both static and dynamic environments. This increased risk of neck injury is caused by the increase in total weight supported by the neck and the change in the head's center of gravity because of the addition of a head-mounted display (HMD). Designers of head-mounted information systems need criteria for allowable helmet-system mass properties (weight and center of gravity) that will not increase the risk of injury above acceptable limits. A research program at the U.S. Air Force (USAF) Armstrong Laboratory that reviewed accident statistics, current literature, and impact acceleration data showed that keeping the head-mounted mass below 4–4.5 lb (1.82–2.05 kg) and the center of gravity of the combined head and HMD as close to that of the bare head as possible will help to reduce the risk of injury. This research helped to establish interim USAF criteria for acceptable head-supported weight and center of gravity of HMDs used during ejection from aircraft. Continuing research efforts seek to refine and generalize these criteria by use of analytical methods that predict loads on the neck for various dynamic exposure conditions.

6.1 Introduction

Because of the rapid changes in technology, now more than ever people have to adapt to a continual supply of new hardware devices that enable us to interface with our environment better and faster. Of particular concern is the trend toward devices and systems that use the head as a platform. This trend is evident in both the military and civilian sectors. New Department of Defense (DoD) developments in head-mounted displays (HMDs) and helmet-mounted visually coupled systems such as night-vision goggles (NVGs) are being used to improve performance of ground troops and aircrew of fixed- and rotary-wing aircraft. Hospital operating rooms will use HMDs to enhance and speed surgery. HMDs are also currently being used in commercial virtual-reality systems for entertainment, engineering, and educational tasks. Ground vehicles (military, police, and industrial) are also candidates for HMD use by drivers and passengers requiring access to information from other compartments of the vehicle during movement. Unfortunately, extensive use of HMD systems could also lead to an increase in neck injuries.

Automobile crashes, falls, dynamic impacts, and other accidents are currently the most frequent cause of injury leading to death for people 44 years old or younger (Sances et al., 1981). A large body of knowledge exists in the area of head and neck dynamic response and injuries that has been derived from clinical evaluations; impact experiments on humans, animals, cadavers, and manikins; and from work with mathematical models. This chapter will attempt to narrow the focus of this information to HMDs and their effect on neck stresses in both static and dynamic environments. The focus will include neck anatomy, injury mechanisms, and mechanical responses. By knowing how the neck can be injured, what type of loading causes injury, and how the inertial properties (weight, center of gravity, and moment of inertia) of an HMD affect neck response and transmitted loads, HMD designers can develop new systems that are both advanced and safe.

6.2   Background: Basic Anatomy and Biomechanics

6.2.1   Basic anatomy

The anatomy of the head is ideally designed for its function. Not only does it protect the body's most vital organ, the brain, but it also enables us to interface with our environment by providing support for vision, hearing, smell, and speech. The neck is another important anatomical structure because it serves as the pathway between the body's command center, the brain, and the rest of the body. The head and neck can be vulnerable to injury when the environment applies severe stresses, such as those encountered in a fall or automobile accident. The additional load from HMDs can exacerbate this risk potential.

Numerous works have covered the gross anatomical structure of the head; they will not be covered in detail here. Of greater importance to the development of HMDs are the inertial properties of the head (weight, center of gravity, and moment of inertia). An excellent source of data comes from Beier, Schuller, Schuck, and Spann (1979), who performed measurements on twenty-one cadaver specimens. A summary of their results can be found in Table 6.1. Data are given in both English and SI units.

The center of gravity and moment of inertia data are defined relative to the head's anatomical axis system, which is defined by three vector axes: the positive y axis is a vector from the right to the left tragion, where the tragion is defined as the superior point on the flap of tissue anterior to the ear canal; the positive x axis is a vector normal to the y axis extending from the y axis to the infraorbitale (defined as the lowest point on the inferior bony ridge of the eye socket) and then translated to the midline on the face as defined by the sellion (indentation at top or bridge of nose); and the positive z axis is a vector normal to the intersection of the y axis and the x axis and projecting through the top of the head (superior). The intersection of the three axes defines the coordinate origin of the anatomical axis system. This is shown in the off-axis sketch of the head in Fig. 6.1.

The neck serves three primary purposes: to provide support to the head while allowing a wide range of motion such as rotation, lateral flexion (side to side), dorsal and ventral flexion and extension (chin to

TABLE 6.1   Average Inertial Properties of the Human Head

Head mass	9.47 lb	4.31 kg
Center of gravity (x, y, z)	$(0.33, 0.0, .123)$ in.	$(8.3, 0.0, 31.2)$ mm
Moment of inertia (I_x, I_y, I_z)	$(70.23, 76.03, 50.46)$ lb-in^2.	$(206, 223, 148)$ kg-cm^2

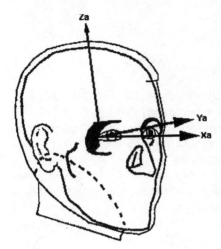

Figure 6.1. Head anatomical axis system.

chest and then back), or a combination of these movements; to provide a protective pathway for the major nerves from the brain to the rest of the body; and to provide a pathway for what we need to survive: air, food, and water. The cervical spine, as shown in Fig. 6.2 (shown with the skull and load vectors), is composed of seven cervical vertebrae separated by intervertebral discs and articulating joints.

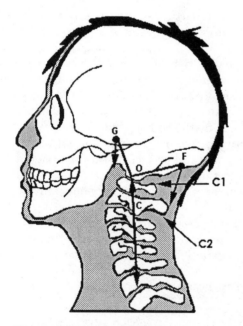

Figure 6.2. Anatomy of the neck.

The first cervical vertebra is the atlas (C1); the second cervical vertebra is the axis (C2). The atlas attaches to the base of the skull at the occipital condyles. The cervical vertebrae gradually increase in width from the axis down to the seventh cervical vertebra (C7). The average cross-sectional area of a cervical vertebra is approximately 0.51 in.^{2} (326 mm^2), (Sances et al., 1981). Assuming a normal upright posture, Fig. 6.2 also shows forces acting on the head and neck under static conditions, as previously noted by Anton (1987). It can be seen that the head is supported on the neck at the occipital condyle (approximately defined by point O and located approximately 1 in. down and 1 in. behind the previously defined y axis of the anatomical axis system) and that gravity acts at the center of mass of the head (approximately defined by point G and located approximately 1.33 in. forward and 2.23 in. up from the previously defined point O in Fig. 6.2). The gravitational vector induces a moment about the occipital condyle that is countered by the action of the posterior neck muscles, which act through point F. The resultant force from the head weight, any additional HMD, and the muscle action is reacted through the cervical vertebrae along vector C.

It is clear that wearing an HMD in any unusual or adverse environment—such as an automobile or other ground vehicle crash; an impact environment encountered during a jump, a fall, or ejection from an aircraft; or the sustained acceleration environment encountered during aircraft flight maneuvers—could affect both the effort required by the neck muscles to keep the head balanced and the loading of the cervical spine.

6.2.2 Basic biomechanics

The investigations of the mechanical properties of vertebrae date back to Messerer's work in the 1880's (Sances et al., 1981). The majority of the data in this area come from experimental work in the automotive research community. Most of the research involved cadaver specimens and related compressive loading in the neck in various degrees of flexion, extension, and lateral flexion to the degree of injury in the cervical spine.

Proceedings of the STAPP Car Crash Conference and SAE reports on human impact tolerance by Snyder (1970) and Brinn et al. (1986) are excellent sources of detailed information on the biomechanics of the neck and other body segments. Yamada (1973) investigated the biomechanical properties of various biological materials, and researchers in biomechanics have reviewed and referenced his work. The data in Table 6.2 have been adapted from his work as found in Sances et al. (1981).

TABLE 6.2   Average Failure Loads for the Human Neck

Load type	Cervical vertebrae	Cervical disk
Compressive breaking load	694 lb (3089 N)	705 lb (3138 N)
Tensile breaking load	270 lb (1199 N)	194 lb (863 N)
Torsional breaking moment	—	43.7 in-lb (5 N-m)

The data from the table seem to indicate that the neck is much stronger in a compressive state than in tensile loading; however, one must not overlook the fact that in compression, the primary structural members of the neck (vertebral bodies and intervertebral disks) tend to absorb the load, whereas in tension or a combination of tension and rotation, the ligaments and muscles of the neck help carry some of the load.

In the review by Sances et al. (1981), testing of human cervical cadaver spine specimens shows that the strength of the posterior and anterior ligaments range from approximately 40.5 lb (180 N) to 76 lb (340 N) before failure in tension. Additional research reported by Sances has also shown that when a spinal specimen (usually the first cervical vertebra through the first thoracic vertebra (C1–T1)) was exposed to a compressive force of 320 lb (1420 N), there was damage to the posterior ligaments of the cervical spine caused by C5 sliding forward over C6.

In terms of head rotation, average values of the range of motion of the head and neck with respect to the torso in flexion (positive or forward rotation of the head around the anatomical y axis) have been reported as 54°–72°; in extension (negative or rearward rotation of the head around the anatomical y axis) they have been reported as 39°–93°; in lateral bending (side-to-side rotation around the anatomical x axis) the average has been reported as approximately 59°; and in axial rotation (side-to-side rotation around the anatomical z axis) the average has been reported as approximately 128° (Sherk, 1989). Table 6.3 summarizes the data, and Fig. 6.3 (from Grierson and VanIngen-Dunn, 1995) shows the various anatomical motions of the neck.

To equate this three-dimensional motion of the head and neck with biomechanical loads or injury, studies have been conducted evaluating and comparing neck injury to the loads generated in the cervical

TABLE 6.3   Typical Head/neck Range of Motion

Head and neck motion type	Range of motion
Head/neck flexion	54°–72°
Head/neck extension	39°–93°
Head/neck lateral bending	Avg. of 59° (right to left)
Head/neck axial rotation	Avg. of 128° (right to left)

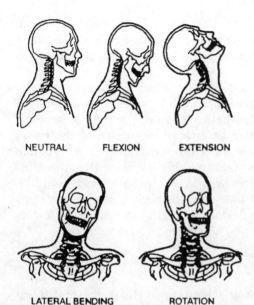

NEUTRAL   FLEXION   EXTENSION

LATERAL BENDING   ROTATION

Figure 6.3. Gross anatomical motions of the head and neck.

spine specimens during various degrees of flexion and extension. One commonly referenced study conducted by Selecki and Williams (1970) indicates that loads of approximately 450 lb (2003 N) generated during compression and extension (rearward head rotation) produced injuries (ligament and/or disk ruptures and vertebral fractures). Loads between 350 and 750 lb (1558 to 3338 N) generated during compression and flexion (forward head rotation) also produced damage to the cadaver neck. Loads of between 200 and 350 lb (890 and 1558 N) generated during lateral flexion produced some injury to the cervical spine. Sances et al. (1982) found that isolated human cadaver cervical spines fail at loads of 551 lb (2450 N) in tension and 420 lb (1868 N) in compression with slight flexion. Sances et al. (1981) reference tension studies in human cadaver spines that show ligament failure at loads of approximately 341 lb (1500 N); they also reference a study indicating that when a cervical spine specimen is subjected to shear loads, the ligaments begin to rupture at an average of approximately 245 lb (1088 N), and fracture of the odontoid process (upward projection of bone on the posterior portion of the axis; serves as a pivot point for the atlas and the skull) occurs at an average of approximately 278 lb (1226 N). However, data from a report by the Motor Vehicle Safety Systems Committee (1980) indicate a peak neck shear force of 437 lb (1944 N) without bone damage from dynamic tests with cadavers as a maximum tolerance value. Grierson and VanIngen-Dunn (1995) conducted a review of work showing that the cervical spine in pure compression can withstand up to 1000 lb (4450 N) before damage (fractures) occurs, but could only withstand up to 516 lb (2296 N) in a combined compression and flexion before some degree of fracture to the cervical spine results.

This wealth of information on the biomechanical properties of neck specimens strengthens the notion that neck injury is very dependent on the orientation of the head, whether the head and neck were directly loaded or indirectly loaded because of movement of the torso, and whether the neck was in a static (low-G condition) or dynamic environment. For example, the data in Table 6.2 show that the minimal compressive strength of the neck in a pure axial load would be approximately 700 lb (3115 N), but when the neck is also subjected to compression with extension or flexion, strength can drop to as low as 450 lb (2003 N) with extension and 350 lb (1558 N) with flexion.

6.3 Static Effects

When a person wears a helmet or some type of HMD, the first thing usually checked is comfort, then stability. However, comfort and stability are not necessarily compatible—a comfortable helmet might fall off your head when you move or bend over. If the weight and the center of mass of an HMD are taken into account, then an additional concern is how long can the HMD be worn until the neck becomes fatigued. This is particularly important to the physician who may wear an HMD for hours while performing delicate surgery, or an aviator who is required to wear an HMD for many hours in a rotary-wing or fixed-wing aircraft.

When the question becomes one of neck fatigue, then the strength of the neck muscles is important, because maintaining the head in a specific position (with or without an HMD) is a form of isometric exercise. According to Petrofsky and Phillips (1982), if isometric contractions of muscles are held at 10–15 percent of the muscle's maximum strength, then the muscle tension is nonfatiguing and can be held for an indefinite time period without fatiguing the muscles. When the sustained contractions begin to exceed 15 percent of the individual muscle's maximum strength or maximum voluntary contractions (MVC), the tension in the muscle will become fatiguing and the endurance time becomes a function of the tension or maximum strength. When the head is loaded by an HMD and the weight and center of gravity are altered, the muscles used to balance the head, as previously depicted by vector F in Fig. 6.2, will become affected.

Foust, Chaffin, Snyder, and Baum (1973) conducted a study of cervical muscle strength in approximately 180 subjects from 18 to 74 years of age (approximately equal numbers of males and females). Their overall results showed a dorsal (extensor muscle) strength of 29.6 lb

(131.8 N) and a ventral (flexor muscle) strength of 23.9 lb (106.2 N). Their study also examined the effect of physical parameters such as sex and age on the strength of the dorsal and ventral neck muscles. These data are average values representative of individuals in the 40th–60th percentile by height. Figures 6.4 and 6.5 show plots of neck muscle strength as functions of sex and age from this study.

The strength data from Foust et al. (1973) show that on average, dorsal muscles are stronger than ventral muscles, males have stronger necks than females, and beyond age 40, the strength of the neck decreases. These statements have practicable applications if an HMD is to be designed for a very specific target population.

The 1982 study by Petrofsky and Phillips used four subjects and evaluated the relationship between isometric strength and endurance for the neck muscles during dorsal (rearward), ventral (forward), and lateral (side-to-side) bending of the head. Figure 6.6 is a plot of a

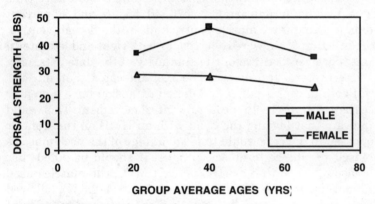

Figure 6.4. Mean strength of dorsal muscles.

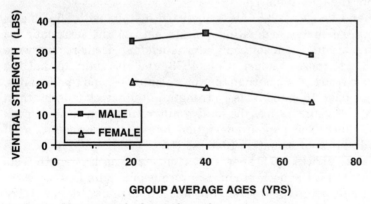

Figure 6.5. Mean strength of ventral muscles.

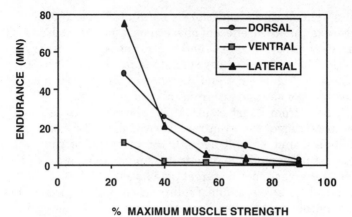

% MAXIMUM MUSCLE STRENGTH

Figure 6.6. Isometric strength and endurance relationship.

partial summary of the results showing the average endurance time until fatigue for muscle contractions at 25, 40, 55, 70, and 90 percent of the maximum voluntary muscle contraction in the dorsal, ventral, and lateral direction. Data were collected for both right and left lateral contractions; however, because of the similarity of the data sets, only the right lateral data are shown. The plot indicates that muscles that control both lateral bending and dorsal extension have a greater endurance than the muscles that control ventral movement. This would seem to indicate that loading the head with an HMD on the front or sides as opposed to the rear would take advantage of the neck muscles with increased resistance to muscle fatigue. It should be noted that this study also revealed that at a 90 percent MVC, the muscles used for ventral movement were the strongest at 25.3 lb (112.6 N), followed by the dorsal muscles at 19.8 lb (88.1 N) and the lateral muscles at 17.2 lb (76.4 N). The average MVC strength was 28.4 lb (126.3 N) for the ventral muscle, 22.7 lb (101 N) for the dorsal muscle, and 19.4 lb (86.2 N) for the lateral muscle.

Morris and Popper (1996) also conducted a study with 35 subjects (19 male, 16 female), evaluating the strength of the neck muscles controlling flexion, extension, and lateral bending. Their results were very similar to those of Foust et al. (1973) and also indicate that the dorsal muscles are stronger than the ventral muscles, and that females, on average, have 15 lb (68 N) less strength in their neck muscles than males. Table 6.4 summarizes the total group results from all three neck strength studies. The standard deviation for each measurement for all three studies is approximately 10 lb (44.5 N).

Mertz and Patrick (1971) show data from static neck strength tests that indicate that the neck produces a torque of 282 in.-lb (32.3 N-m) in resisting flexion (dorsal strength), and a maximum torque of 126 in.-lb (14.4 N-m) in resisting extension (ventral strength). These data

TABLE 6.4   Static Neck Strength Summary

Reference	Ventral strength	Dorsal strength	Lateral strength
Petrofsky and Phillips	28.4 lb†(126.3 N)	22.7 lb†(101 N)	19.4 lb (86.2 N)
Foust	23.9 lb (106.2 N)	29.6 lb (131.8 N)	—
Morris	17.5 lb (77.9 N)	30.8 lb (137.1 N)	18.5 lb (82.3 N)

†The mean data from Petrofsky and Phillips shows a slight difference from the other data sets—probably because only four subjects were used and the data may not portray an accurate picture of the difference between ventral- and dorsal-muscle strength.

enforce the position that dorsal muscles are stronger on average than ventral muscles; in addition, Petrofsky and Phillips' data indicate that dorsal muscles are also less resistant to fatigue when compared to ventral muscles. This statement also has very practical implications, since weight applied to any portion of the head may alter its center of gravity and may lead to additional stress on the neck muscles to correct for this new load and balance. This is of particular importance to HMD systems because they tend to shift the center of gravity of the head and HMD forward relative to the anatomical axis.

In a follow-up study to their work on neck strength and endurance, Phillips and Petrofsky (1983) conducted a series of experiments evaluating various head-mounted weight and center-of-gravity combinations and their relationship to neck muscle fatigue. Results indicated that for a low-weight HMD (approximately 3.0 lb (13.4 N)), the optimal location of the resultant center of gravity of the HMD for maximal resistance to fatigue would be approximately 2 in. (5 cm) in front of (along the anatomical x axis) the anatomical axis origin. For HMD weight around 5.0 lb (22.3 N), the center of gravity of the HMD may vary from 1 in. (2.5 cm) behind the anatomical axis to 2 in. (5 cm) in front of the anatomical axis origin. For heavier HMD systems (5–9 lb, 22.3–40.1 N), their data supported a center of gravity location of the HMD 1 in. aft of the anatomical axis origin for optimal neck muscle endurance. It should be noted that approximately 1 in. aft (dorsal) of the head anatomical axis origin and down (inferior) 1 in. in the z axis is the approximate location of the head-neck pivot point, or occipital condyles. These results are very important because they provide critical information relative to the relationship between various centers of gravity and HMD weight combinations that result in the optimization of neck muscle endurance. This information can then be used in current and future HMD design.

6.4   Dynamic Effects

Most of the data available describing the motion and acceleration of the head and neck in dynamic environments have been collected for the automotive industry and the DoD. Dynamic impacts to the head and neck can usually be divided into two categories: impact accelerations that occur when the head or neck comes into contact with an external structure, and abrupt accelerations of the whole body that cause the head and neck to accelerate because of torso acceleration. This section deals only with dynamic head and neck accelerations induced by torso acceleration.

6.4.1 Ground vehicle environment

In the ground vehicle environment the primary concern is the response of the head and neck during an impact simulating a vehicle (typically automobile) accident. By knowing the neck response in different impact or deceleration environments, automotive engineers can design safety systems (restraints, airbags, vehicle structural responses, etc.) that will not allow the head and neck to move or accelerate in such a way that may cause injury. Typical research studies evaluate the response of cadavers and live subjects in seated $-Gx$ impacts (a $-Gx$ impact acceleration represents what is produced in a front-end ground vehicle collision resulting in the vehicle structure being moved backward and the occupant being displaced forward into the seat belt). The response of the head is determined by measuring its linear acceleration and its angular acceleration or its rate of angular rotation. These measurements are made relative to the head's anatomical axis system (x-axis acceleration, z-axis acceleration, etc.). Forces in the neck are calculated using these measurements, basic dynamics, and knowing an estimate of the inertial properties of the tested headform. Unless otherwise noted, data are from tests involving volunteer subjects.

Typical flexion values about the occipital condyles without injury were approximately 213 in.-lb (24.4 N-m) in flexion and 403 in.-lb (46.1 N-m) in extension from tests conducted by Ewing and Thomas (1974) to answer neck loading concerns during aircraft ditching. Data from Mertz and Patrick (1967, 1971) show a maximum flexion value of 770 in.-lb (88.1 N-m) for one subject, which produced pain in the neck and along the thoracic spine. Additional tests showed a flexion value of approximately 522 in.-lb (59.7 N-m) for the initiation of neck pain. Cadaver tests conducted in 1971 by Mertz and Patrick showed maximum torque levels of between 1,547 in.-lb (170 N-m) and 1,660 in.-lb (190 N-m), and these were attained without bone damage as determined by x-ray analysis. This analysis, however, could not rule out soft-tissue (muscle and ligament) damage. Additional studies as cited by Sances et al. (1981) demonstrated lateral flexion values of 395 in.-lb (45.2 N-m) measured at the occipital condyles; these were achieved without injury. However, Sances et al. (1981) cite Gadd and

Culver, who conducted cadaver tests in which a torque value of 198 in.-lb (22.6 N-m) was measured in a combined forward and lateral flexion movement and produced minor injury to the neck. This seems to indicate that the neck is not as tolerant to combined motions (flexion, extension, rotation) as it is to pure neck motion or flexion in a single axis.

Limited data are available on the response of the head in a $-Gx$ impact environment when encumbered by an HMD. Perry (1990) conducted a series of $-Gx$ horizontal impacts (15-G peak, 45 msec rise time, 100 msec duration) with an instrumented manikin. The testing was initiated to determine the effects of HMD systems on biodynamic response during a simulation of a crash or emergency landing environment that could be experienced with a fixed-wing or rotary-wing aircraft. The helmet's weight varied from approximately 2.5 to 5.25 lb (1.14–2.39 kg). The manikin had a Hybrid III neck with a Denton six-axis load cell. The impacts generated tensile loads in forward flexion ranging from 195 to 308 lb (868–1371 N) and shear loads ranging from 380 to 470 lb (1691–2092 N), at a point on the manikin representing the occipital condyles. The tensile loads were below the previously referenced 341-lb (1500-N) ligament-damage threshold and the tensile cervical spine-failure threshold of 551 lb (2450 N). The shear load data indicated that three helmets with weights in excess of 4.5 lb (2.05 kg) produced loads equal to or greater than the 437-lb (1944-N) dynamic threshold limit. This may indicate that 4.5 lb is the upper HMD weight for the tested Gx impact acceleration.

6.4.2 Aerospace environment

The mission profiles of some military aircraft equipped with ejection seats are currently being expanded for more demanding operational scenarios. To improve pilot performance during these scenarios, a strong emphasis has been placed on the use of HMD or helmet-mounted visually coupled systems. These visually coupled systems can include integrated NVGs for increased visibility and helmet-mounted display systems for aircraft flight control and improved target designation for weapon delivery. By using the head and helmet as platforms for such systems, the total head-supported weight and combined head-helmet center of gravity may and often will be modified. If these systems are in use during aircraft emergency escape by rocket-powered ejection, or a crash, the potential exists for an increase in the rate of ejection-related neck injuries and fatalities caused by the head-supported weight. The addition of an HMD could not only affect crash impact and ejection-related neck injuries but could also affect pilot mission performance because of neck muscle fatigue, modification of the helmet's comfort and overall fit, and increased neck muscle strain during a high-G maneuver.

Research results could also be related to the impact experienced during a crash in either a fixed- or rotary-wing aircraft.

6.4.2.1 Sustained acceleration. The successful use of any type of HMD in the cockpit of a high-performance aircraft certainly depends on comfort, fit, and the environment in which the system is being used. The magnitude, direction, and duration of acceleration exposures on the crew members will affect the performance as the inertial properties of the HMD cause the neck muscles to readjust to the loading. Ultimately a balance must be achieved between improved aircraft performance, mission enhancement, and the physical stresses and potential safety decrement imposed on the crew member.

Research evaluating HMDs in a sustained G environment has been minimal. Darrah, Seavers, Wang, and Dew (1987) conducted a study relating limitations of aircrew in the cockpit while wearing HMDs. This study used a computer simulation to evaluate the homeostatic acceleration tolerance of the neck with several HMDs ranging in weight from 4.91 to 6.56 lb (2.23–2.98 kg) and with static moments relative to the occipital condyle ranging from approximately 77.01 to 179.69 in.-lb (8.81–20.56 N-m). The results indicated that +3.3 Gz (Gz indicates acceleration along the z axis of the anatomical coordinate system with the z axis oriented parallel to the spine) was the approximate acceleration the neck could withstand before the inertial properties of the lightest HMD tested would cause the head to pitch forward on the neck and possibly induce injury.

6.4.2.2 Ejection acceleration. The Dynamic Response Index, or DRI (Brinkley and Schaffer, 1971), is used to design ejection systems for the USAF to minimize injury and, in particular, spinal injury. The DRI is a spinal-injury indicator (thoracic and lumbar spine only, not cervical spine) that is currently used by the USAF to determine exposure limits for short-duration accelerations in the z axis produced by the ejection-seat catapult. It involves the use of a simple mechanical model to predict the probability of spinal injury, i.e., compression fractures of the vertebral body segments of the thoracic and lumbar spine. The model is a mechanical system composed of the common lumped-parameter elements: a mass, a spring, and a viscous damper. The mass represents the mass of the upper torso of the body, and the spring and viscous damper represent the dynamics of the spine. The DRI is calculated from the model using the following equations:

$$\ddot{S} = \ddot{\delta} + 2\zeta\omega_n\,\dot{\delta} + \omega_n^2\,\delta \tag{6.1}$$

$$\mathrm{DRI} = \frac{\omega_n^2\,\delta_{\max}}{g} \tag{6.2}$$

where $\ddot{\delta}$ = acceleration of the mass of the model (cm/sec²)
  $\dot{\delta}$ = velocity of the model mass (cm/sec)
  δ = compression of the model spring (cm)
  ζ = damping coefficient ratio (0.224) for the model
  ω_n = undamped natural frequency (52.9 rad/sec) of the model
  $\ddot{S}$ = input acceleration (cm/sec²) driving the model
  g = acceleration due to gravity (980 cm/sec²)

To estimate the probability of injury, i.e., probability of compression fracture of the spine for given DRI values, the spinal-injury rate for various operational ejection seats was plotted against the DRI value generated by the seat's catapult acceleration, as shown in Fig. 6.7. DRI values less than 18 predict spinal injury levels less than 5%. This relationship does not apply to the cervical spine; however, few cervical spine injuries have been observed in dynamic systems designed to meet a DRI value of 18.

Adding mass to the flight helmet was previously assumed to result in a corresponding increase in the risk of neck injury emergency escape. A criterion for the neck similar to the DRI has not been developed, in part because the weight variation of past and current operational flight helmets is comparatively narrow (approximately 3–4 lb) when compared to the range of seat accelerations (approximately 12 G–20 G) used to develop the DRI. However, because some potential visually coupled helmet system designs weigh up to 5 lb or more (Anton, 1987;

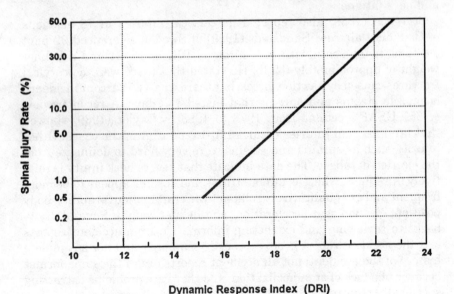

Dynamic Response Index (DRI)

Figure 6.7.   DRI as a function of probability of spinal injury.

Perry, 1993; Privitzer and Kaleps, 1990), criteria on HMD weight and center of gravity are required for development of future HMDs.

A search of the literature and accident data provides insight to the potential problems of using HMD in aircraft equipped with ejection seats. However, the reviews revealed insufficient quantitative data with which to build a relationship between helmet mass, center of gravity location, and neck injury probability.

Guill and Herd (1990) have reported a low incidence of neck fracture (1.7 percent) and a moderate rate of neck sprains and strains (12.2 percent) on ejection from Navy aircraft. They state that causal factors are many and varied, but make no attempt to relate head-mounted mass as a causal factor, except to conclude that "we are also of the opinion that a significant proportion of the serious ejection associated neck injuries are in fact likely to have been induced by the in-flight maneuvering/gyrating forces imposed upon the aircrew prior to ejection or during ejection. These, we believe, are especially significant and require consideration as helmets become the handy means for mounting sights and other needed equipment upon the aircrew" (p. 9-6). Anton (1990) was not able to find any severe neck injuries unless the head had struck part of the airframe. His consideration of helmet mass and center of gravity led him to recommend a helmet-mounted system less than 4 lb with a center of gravity near that of current helmets. This later recommendation was based largely on operational concerns cited in the 1983 laboratory study by Phillips and Petrofsky on helmet mass and neck fatigue.

Sturgeon (1988) also reported no cervical injuries in 78 ejections of Canadian aircrew. Sandstedt (1989) of Sweden referenced 92 pilot ejections using seats with an estimated DRI of 18–21 and a helmet weight of approximately 6.2 lb. He stated that there were no cervical fractures caused by ejection forces, but there was a 25 percent incidence of muscle strains and sprains that lasted 1–5 days. A recent review of 657 USAF ejections from 1978 to 1988 by Taylor (1989) showed that fatal cervical injuries occurred in 1.8 percent of the cases. Other injuries such as sprains and strains were very hard to delineate from the ejection database. The data indicate that severe neck injury is relatively infrequent whereas minor strains and sprains appear to be more frequent and are qualitatively related to aircraft maneuverability, body position, ejection seat acceleration, and helmet weight. Several drawbacks to reviewing and extracting information from accident reports are: the accident report is only a concise summary and not a complete history of the accident; not all accident reports are in the same format because of a lack of standardization; and there are problems extracting specifically required information from the accident report.

Other researchers such as Privitzer and Settecervi (1987), King, Nakhla, and Mital (1978), and Darrah (1987) have conducted modeling efforts to relate head-supported weight to neck forces induced by catapult accelerations. Privitzer and Settecervi report a safe maximum head-supported mass of 5.28 lb (2.4 kg), and King et al. report that vertebral disk and facet forces decrease as the head and helmet center of gravity shifts back with respect to the head's center of gravity by approximately 0.5 in. These modeling efforts show general trends toward greater forces and moments at the occipital condyles with increased mass, but they do not adequately quantify a relationship between model predictions and the potential for neck injury.

Several recent research studies evaluating human and manikin cervical neck response to a simulated ejection environment have been conducted and reported on by Perry (1993) and Buhrman, Perry, and Knox (1994). The tests were conducted on the USAF's Vertical Deceleration Tower (VDT) located in the Armstrong Laboratory at Wright-Patterson AFB, Ohio. The first study experimentally evaluated a weight and space mockup of a low-profile NVG, and the second study expanded the evaluation to include three prototype interim night integrated goggle head tracking system (I-NIGHTS) helmets with NVGs and three I-NIGHTS prototype helmets with NVGs and HMDs. These systems (including mask) ranged up to 6.64 lb (3.02 kg) total head-supported weight. The systems were tested at VDT impact acceleration levels of 10- and 15-G peak acceleration with a rise time of approximately 80 msec using an advanced dynamic anthropomorphic manikin (ADAM) developed for the USAF by Systems Research Laboratories, Inc. (Bartol et al., 1990). The VDT 10-G pulse produces DRI values similar to the ACES II ejection seat (DRI = 12–13), and the VDT 15-G pulse produces DRI values similar to the ejection seat in the B-52 aircraft (DRI = 17–19)(Perry, 1993).

Average acceleration response values measured with the manikin showed good correlation with the human data in the z axis and x axis, but the manikin rotational accelerations were approximately twice the human values because of the manikin's neck buckling during z-axis impact tests on the VDT. This highlights the fact that the primary design of the HYBRID III neck used with the manikin was to simulate head and neck response in an x-axis impact environment such as that experienced in automobile accidents.

Because of the z-axis correlations, z-axis compressive loading measured by the load cell in the manikin was assumed to be a good indicator of the loads in the human neck in the same acceleration environment (peak acceleration, helmet system, etc.). Figure 6.8 shows the peak z-axis head acceleration at various acceleration levels with a constant helmet-system weight of approximately 4.5 lb (2.05 kg) from tests with humans and manikins. The human data were used to calculate error bars, which show a ±1 standard deviation. Figure 6.9 shows the z-axis manikin peak loads at various helmet weights in

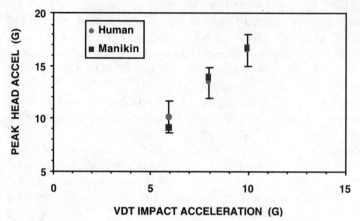

Figure 6.8. ADAM and human z-axis head acceleration at various impact acceleration levels.

10-G and 15-G impact environments. These data are an estimate of the peak compressive loads in the human cervical spine under the same dynamic environmental conditions.

The manikin compressive neck loads from Fig. 6.9 show that at VDT impact acceleration levels of 10 G, helmet weights of approximately 6.0 lb (2.73 kg) produce loads that are approximately 50 lb (223 N) greater than loads produced by current operational helmet systems (HGU-55/P, HGU-26/P) weighing approximately 3.5 lb (1.59 kg). The 6.0-lb helmet loads at 10 G are not greater than the 250-lb (1112-N) no-injury limit stated by Mertz and Patrick (1971) and used by the Navy (Weiss, Matson, and Mawn, 1989). This value is based

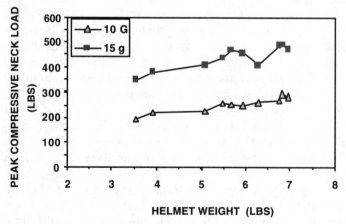

Figure 6.9. ADAM z-axis compressive neck loads from impacts with helmets of various weights.

on static neck strength and was considered to be the lower boundary for dynamic tolerance. Figure 6.9 also shows that the Swedish helmet of approximately 6 lb (2.73 kg) was producing peak compressive neck loads at the occipital condyles of at least 450 lb (2003 N) based on linear regression of the data. This assumption is based on the fact that the DRI of the Swedish seat is equal to or greater than the DRI from the B-52 ejection seat, which is approximated by the VDT 15-G impact pulse. Since no neck fractures were associated with the Swedish ejection seat, as indicated by Sandstedt, this 450-lb load may represent the upper dynamic tolerance limit of the cervical spine.

An additional study reported in Perry, Buhrman, and Knox (1993) and Perry and Buhrman (1996) evaluated a parametric shift in center of gravity and helmet weight using an HGU-55/P helmet modified to accommodate weights positioned in different locations. This allowed multiple combinations of head-supported weight, centers of gravity, and moments of inertia. Test conditions allowed changes in the helmet system's inertial properties along the positive and negative head x and z axes. The VDT used an impact level of 10 G to conduct all tests. Compressive, shear, and torque loads relative to the occipital condyle in the human subjects were calculated using the measured head accelerations (linear and angular) and an estimate of the combined inertial properties of the test subject's head and specific helmet system. It should be noted that the following values are from dynamic tests at 10 G on the VDT; there were no major injuries suffered from these loads, and less than 5 percent of the subjects reported sore necks lasting 1–3 days.

Data from the 1993 study by Perry, Buhrman, and Knox show that for helmet system weights varying from 3.0 to 7.5 lb (1.36–3.41 kg) with a combined center of gravity close to the head's anatomical axis origin, the calculated compressive load in the neck relative to the occipital condyle ranged from 184 to 275 lb (819–1224 N). The calculated shear load in the neck relative to the occipital condyle ranged from 41 to 74 lb (183–329 N). The calculated maximum torque around the anatomical y axis relative to the occipital condyle (M_y torque) ranged from 120 to 358 in.-lb (13.7–41 N-m). Figures 6.10–6.12 show these relationships. All data points represent an average response of approximately 12 subjects and are shown with error equal to 1 standard deviation.

When the helmet weight is shifted such that the helmet's y-axis moment of inertia I_y and its center of gravity are altered, the standard inertial weight (SIW) is a convenient method to analyze the changes in these parameters without the bias on center of gravity of analyzing the moment of inertia alone. This was reported by Perry and Buhrman (1996). The SIW is a dimensionless parameter that is a function of the weight and y-axis moment of inertia of both the HMD and the

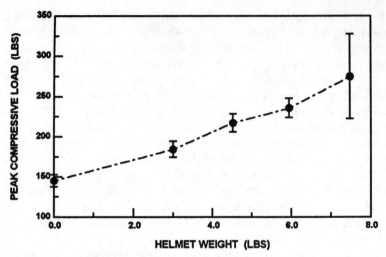

Figure 6.10. Estimated cervical compressive load at the occipital condyle as a function of helmet weight during 10-G impacts.

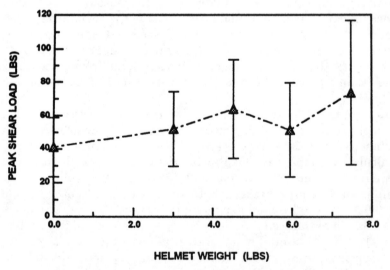

Figure 6.11. Estimated cervical shear load at the occipital condyle as a function of helmet weight during 10-G impacts.

reference headform (manikin or human). This parameter accounts for the primary inertial properties important in HMD design. Anthropometric regression equations (Clauser, McConville, and Young, 1969) estimate the weight of the human head and the y-axis moment of inertia—or the human head inertial properties can be estimated from

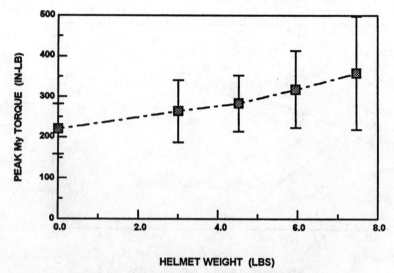

Figure 6.12. Estimated M_y torque at the occipital condyle as a function of helmet weight during 10-G impacts.

the data of Beier et al. (1979). The following equation calculates the SIW:

$$SIW = \frac{(HW)(HI)}{[(HW)(HI) + (H_D W)(H_D I)]} \qquad (6.3)$$

where HW = helmet system weight
    HI = helmet y-axis moment of inertia
    $H_D W$ = weight of head (or headform)
    $H_D I$ = y-axis moment of inertia of head (or headform)

The relationships between the SIW calculated as a function of the y-axis moment and the estimated loads in the neck relative to the occipital condyle are shown in Figs. 6.13–6.15. All data points represent an average response of twelve subjects and are shown with error equal to one standard deviation.

For this particular test series, the data show that as the weight of the helmet is increased and the moment about the y axis increases as indicated by the SIW value (because of the center of gravity of the head-helmet combination shifting forward along the anatomical x axis), the compression and shear loads increase and then start to decrease, as indicated by the second-order polynomial data trends ($r \cong 0.6$). The torque continues to rise in a linear fashion as the SIW value increases ($r \cong 0.94$), reaching a peak of approximately 490 in.-lb (56 N-m) at an SIW value of 0.6. It should be stated that a SIW value could be calculated as a function of the x-axis moment or z-axis moment, or

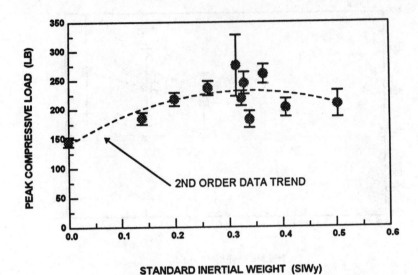

Figure 6.13. Estimated cervical compressive loads at the occipital condyle as a function of helmet inertial properties during 10-G impacts.

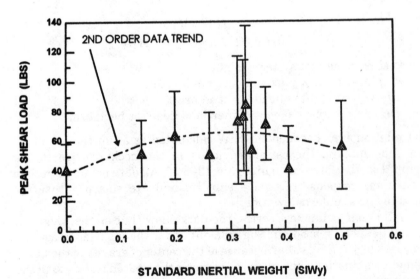

Figure 6.14. Estimated cervical shear loads at the occipital condyle as a function of helmet inertial properties at 10-G.

even a resultant of the three. However, the y-axis moment was chosen for this particular data set to reflect orientation of the visually coupled HMD system.

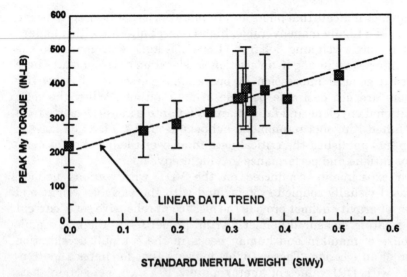

Figure 6.15. Estimated M_y torque at the occipital condyle as a function of helmet inertial properties during 10-G impacts.

6.5 Conclusions

There is a need for a methodology to relate environmental cause to probability, form, and severity of a neck injury. The neck will be injured, whether or not an HMD is involved, when the environment causes the neck to move (bend, flex, extend, compress, or some combination) beyond its normal limits and induce a load or combination of loads that exceed the structural integrity of the neck components (vertebrae, ligaments, and muscles).

A review of the literature and experimental data has brought forth the known physical properties of the neck, but only presented some general qualitative observations relative to helmet inertial properties and the response of the neck during static or dynamic impact environments. Data support the tendency for the neck to fatigue more as the neck-supported weight increases and as the center of gravity of the head or head-HMD combination shifts away from the occipital condyles. Statically, it can be reasonable to assume that HMDs should be symmetric and lighter, approximately 3.0 lb (1.43 kg) or less, with a maximum center of gravity in front of the head y axis of 2 in. (5.1 cm) to take advantage of the stronger and less fatiguing dorsal muscles in the neck. If sex and age are important population parameters for the design of the HMD, then minimal helmet inertial properties (mass, center of gravity, and moments of inertia) should be used to account for the potential decrease in neck muscle strength.

Data also indicate that pilots may be more at risk for neck sprains and strains in a highly maneuverable, high-G-onset aircraft with standard flight helmets weighing 3–3.5 lb (1.36–1.59 kg), as compared to the same injuries occurring in an ejection acceleration environment. There is also a general impression by operational personnel that heavier helmets are not desirable because of neck fatigue. When the total operational envelope and fatigue are taken into account, experts from the United Kingdom recommend helmets around 4.0 lb (1.82 kg). No group has published the critical data to answer the question of mass versus fatigue and performance in a flight envelope.

The experiments conducted on the VDT with various helmet-mounted visually coupled systems and with the variable-weight and center-of-gravity helmet provide the most comprehensive set of current data relating measured helmet inertial properties to the biodynamic response of manikin and human necks in the catapult acceleration phase of an ejection. These studies indicate that for impact environments with DRI values of approximately 18 (such as ejection seats with acceleration pulses ranging from 15- to 18-G peak in the z axis) or less, helmets should weigh less than 5.0 lb (2.27 kg) and have an SIW value of 0.20 or less relative to the head y axis. This will keep compressive neck loads below 420 lb (1780 N), which can be used as a threshold value for dynamic injury in cadavers but is a conservative value for injury in live human subjects. This SIW value is based on the reasonable assumption (shown in Fig. 6.9) that the impact environment with a DRI value of 18 produces manikin neck loads approximately 200 lb greater than the manikin neck loads in a 10-G impact acceleration, as commonly tested with the VDT at the Armstrong Laboratory. The SIW value of 0.20 produces a compressive neck load of approximately 220 lb at 10-G impacts (as shown in Fig. 6.7) and would therefore also produce neck loads of approximately 420 lb from impacts with DRI values of around 18. The SIW values, relative to the x axis and the z axis, should not exceed 0.44 and 0.43, respectively (these values are based on best available helmet data to date). It must be emphasized that all previously mentioned tests conducted with HMDs assume symmetry relative to the sagittal plane (right and left side) of the head.

The dynamic tests using the VDT also support HMD inertial-property criteria for catapult impact, which indicate that acceptable compression loads are seen for HMDs weighing less than 5.0 lb (2.27 kg) and have a combined ADAM head and HMD center of gravity within a box defined by -0.8–0.2 inches (-2.03–0.52 cm) along the head anatomical x axis, 0.5–1.5 in. (1.27–3.8 cm) along the head anatomical z axis, and -0.15–0.15 in. (-0.38–0.38 cm) along the head anatomical y axis. This is shown in Fig. 6.16. The zero-zero point on the plot represents

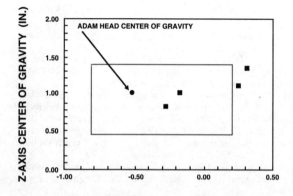

X-AXIS CENTER OF GRAVITY (IN.)

Figure 6.16. Weight and center-of-gravity criteria for HMD during catapult impact acceleration phase of ejection. The area within the box is considered safe.

the origin of the head's axis system discussed previously. The ADAM headform and the human head have approximate the same head axis system. The ADAM head center of gravity is shown for reference. The square data points represent previously tested helmets combined with the ADAM headform. If less than 5.0 lb (2.27 kg), the two helmets inside the box are acceptable, whereas the helmets outside the box are unacceptable.

The work of Petrofsky and Phillips (1982), Darrah (1987a), and Glaister (1988) indicated that flight helmets weighing 4 lb (1.82 kg) or less will be necessary to maintain performance for the required future mission durations; however, the critical experiments have not been done to finalize head-mounted mass limits and center-of-gravity locations based on fatigue and human performance in a sustained-G environment. In general, it can be recommended that helmet systems be lighter—3.5–4.0 lb (1.59–1.82 kg), in order to enhance overall pilot acceptance under in-flight conditions.

6.6 References

Aghina, J. C. F. M. (1985). Systematic radiographic examination of the spine for selection of F16 pilots: A preliminary report. *Advisory Group for Aerospace Research and Development Conference Proceedings, 396,* 41-1–41-4.

Andersen, H. (1988). Neck injury sustained during exposure to high-G forces in the F-16B. *Aviation, Space, and Environmental Medicine, 59,* 356–358.

Anton, D. J. (1990). Neck injury in advanced military environments. *Advisory Group for Aerospace Research and Development Conference Proceedings, 471,* K1-1–K1-7.

Anton, D. J. (1987). The biodynamic implications of helmet mounted devices. *Proceedings of Royal Aeronautical Society, London,*

Bartol, A. M., Hazen, V., Kolwalski, J., Murphy, B., & White, R. Jr. (1990). *Advanced dynamic anthropomorphic manikin (ADAM) final design report* (Technical Report AAMRL-TR-90-023). Wright-Patterson AFB, OH: Armstrong Laboratory.

Beier, G., Schuller, E., Schuck, M., & Spann, W. (1979, April). *Determination of physical data of the head: Center of Gravity and Moments of Inertia of Human Head* (Scientific Report No. 1). Institute of Forensic Medicine at University of Munich, Munich, West Germany.

Brinn, J., Eppinger, R., Hodgson, V. R., Melvin, J. W., Mertz, J. W., Pike, J., & Prasad, P. (1986). Human tolerance to impact conditions as related to motor vehicle design (SAE Information Report J885). New York: Society of Automotive Engineers.

Brinkley, J. W., & Schaffer, J. T. (1971). *Dynamic simulation techniques for the design of escape systems: current applications and future Air Force requirements* (Technical Report AMRL-TR-71-29). Wright-Patterson AFB, OH: Armstrong Laboratory.

Buhrman, J. R., Perry, C. E., & Knox, F. S. (1994). Human and manikin head/neck response to +Gz accelerations when encumbered by helmets of various weights. *Aviation, Space, and Environmental Medicine, 65*,1086–1090.

Clauser, C. E., McConville, J. T., & Young, J. W. (1969). *Weight, volume, and center of mass of segments of the human body* (Technical Report AMRL-TR-69-70). Wright-Patterson AFB, OH: Armstrong Laboratory.

Darrah, M. I. (1987). *Enhancement of head/neck computer model* (USAFSAM Contract Report for period 1 Feb. 1985 to 2 Aug. 1987). McDonnell Douglas Aircraft Co., St. Louis, MO.

Darrah, M. I., Seavers, C. R., Wang, A. J., & Dew, D. W. (1987). Acceleration loading tolerance of selected night vision goggle systems: a model analysis. *SAFE Journal, 17(2)*, 30–36.

Estep, C., & Schneck, D. J. (1992). *Modeling of the human head/neck system using rigid body dynamics* (Technical Report VPI-E-92-18). Virginia Polytechnic Institute and State University College of Engineering, VA.

Ewing, C. L., & Thomas, D. J. (1974). Torque versus angular displacement response of human head to −Gx impact acceleration. *Proceedings of 17th Stapp Car Crash Conference*. New York: Society of Automotive Engineers.

Foust, D. R., Chaffin, D. B., Snyder, R.G., & Baum, J. K. (1973). Cervical range of motion and dynamic response and strength of cervical muscle. Paper 730975, *Proceedings of the 17th Stapp Car Crash Conference*. New York: Society of Automobile Engineers.

Glaister, D. H. (1988). *The effects of off-axis loading on head mobility* (USAFSAM Memo Report). Brooks AFB, TX: USAF School of Aerospace Medicine.

Grierson, A. E., & VanIngen-Dunn, C. (1995). *Development of a comprehensive neck injury criterion for aircraft-related incidences* (Technical Report AL/CF-TR-1995-0038). Wright-Patterson AFB, OH: Armstrong Laboratory.

Guill, F. C., & Herd, G. R. (1990). Aircrew neck injuries: A new or an existing, misunderstood phenomenon? *Advisory Group for Aerospace Research and Development Conference Proceedings, 471,* 9-1–9-12.

King, A. I., Nakhla, S. S., & Mital, N. K. (1978). Simulation of head and neck response to −Gx and +Gz impacts. *Advisory Group for Aerospace Research and Development Conference Proceedings, A7,* 1–13.

Mertz, H. J., & Patrick, L. M. (1967). Investigation of the kinematics and kinetics of whiplash. *Proceedings of the 11th Stapp Car Crash Conference*. New York: Society of Automotive Engineers.

Mertz, H. J., & Patrick, L. M. (1971). Strength and response of the human neck. Paper 710855, *Proceedings of the 15th Stapp Car Crash Conference*. New York: Society of Automobile Engineers.

Morris, C. E., & Popper, S. E. (1996). Relationship between neck strength, anthropomorphic parameters, and gender with head motion under impact acceleration (Technical Report for Defense Women's Health Research Program Contract # 95MM5582). Wright Patterson AFB, OH: Armstrong Laboratory.

Motor Vehicle Safety Systems Testing Committee (1980). *Human tolerance to impact conditions as related to motor vehicle design* (SAE Information Report J885). Warrendale, PA: Society of Automotive Engineers.

Perry, C. E. (1990). *Horizontal accelerator tests of helmet-mounted visually coupled systems—preliminary summary report* (Internal Technical Memorandum). Wright-Patterson AFB, OH: Armstrong Laboratory, Escape and Impact Protection Branch.

Perry, C. E. (1993). *Evaluation of the effects of visually coupled systems on the biodynamic response to simulated ejection accelerations* (Technical Report AL/CF-TR-1993-0159). Wright-Patterson AFB, OH: Armstrong Laboratory.

Perry, C. E., Buhrman, J. R., & Knox, F. S. (1993). Biodynamic testing of helmet mounted systems. *Proceedings of the 37th Annual Human Factors and Ergonomics Society, 1,* 79–83.

Perry, C. E., Buhrman, J. R. (1996). Effect of helmet inertial properties on the biodynamics of the head and neck during +G$_z$ impact accelerations. *SAFE Journal, 26(2),* 34–41.

Petrofsky, J. S., Phillips, C. A. (1982). The strength–endurance relationship in skeletal muscle: Its application to helmet design. *Aviation, Space, and Environmental Medicine, 53(4),* 365–369.

Phillips, C. A., & Petrofsky, J. S. (1983). Neck muscle loading and fatigue: Systematic variation of headgear weights and center-of-gravity. *Aviation, Space, and Environmental Medicine, 54(10),* 901–905.

Privitzer, E., & Settecervi, J. J. (1987). Dynamic analysis of inertial loading effects of head mounted systems. *SAFE Journal, 17(2),* 16–32.

Privitzer, E., & Kaleps, I. (1990). Effect of head mounted devices on head-neck dynamic response to +Gz accelerations. *Advisory Group for Aerospace Research and Development Conference Proceedings, 471,* 13-1–13-14.

Sances, A., Weber, R. C., Larson, S. J., Cusick, J. S., Myklebust J. B., & Walsh, P. R. (1981). Bioengineering analysis of head and spine injuries. *CRC Critical Reviews in Bioengineering, 2,* 79–122.

Sances, A., Myklebust, J. B., Houterman, C., Weber, R., Lepkowski, J., Cusick, J., Larson, S., Ewing, C., Thomas, D., Weiss, M., Berger, M., Jessop, M. E., & Saltzberg, B. (1982). Head and spine injuries. *Advisory Group for Aerospace Research and Development Conference Proceedings, 322,* 13-1–13-34.

Sandstedt, P. (1989). Experiences of rocket seat ejections in the Swedish Air Force: 1967–1987. *Aviation, Space, and Environmental Medicine, 60,* 367–373.

Schall, D. G. (1989). Non-ejection cervical spinal injuries due to +Gx in high-performance aircraft. *Aviation, Space, and Environmental Medicine, 60,* 445–456.

Schultz, A. B., & Aston-Miller, J. A. (1991). Biomechanics of the human spine. In V. C. Mow and W. C. Hayes (Eds.), *Basic Orthopedic Biomechanics.* New York: Raven Press.

Selecki, B. R., & Williams, H. B. (1970). *Injuries to the cervical spine and cord in man.* Australian Medical Association Mervyn Archdall Medical Monograph #7. South Wales: Australian Medical Publishers.

Sherk, H. H. (1989). Physiology and biomechanics. In H. H. Sherk, E. J. Dunn, F. J. Eismont, J. W. Fielding, D. M. Long, K. Ono, L. Penning, R. Raynor (Eds.), *The Cervical Spine.* Philadelphia: J.B. Lippincott.

Shipley, B. W., Kaleps, I., & Baughn, D. J. (1983). Measurement of mass and cg location of helmet systems. *Proceedings of the 37th Annual Human Factors and Ergonomics Society, 1,* 74–78.

Snyder, R. G. (1970). State-of-the-art human impact tolerance. *International Automobile Safety Conference Compendium P-30.* New York: Society of Automotive Engineers.

Sturgeon, W. R. (1988). *Canadian Forces aircrew ejection, descent, and landing injuries, 1 Jan 75–31 Dec 87* (Technical Report DCIEM 88-RR-56). Downsview, Ont.: Defense and Civil Institute of Environmental Medicine.

Taylor, C. (1989). *Incidence of head and neck injuries in USAF accidents, 1978–1988* (Interoffice Technical Memo). Wright-Patterson AFB, OH: Armstrong Laboratory, Escape and Impact Protection Branch.

Vanderbeek, R. D. (1990). Prevalence of G-induced cervical injury in United States Air Force pilots. *Advisory Group for Aerospace Research and Development Conference Proceedings, 471,* 1-1–1-7.

Weiss, M.S., Matson, D. L., & Mawn, S. V. (1989). *Guidelines for safe human exposure to impact acceleration—Update A* (Technical Report NBDL-89R003). New Orleans, LA: Naval Biodynamics Laboratory.

Yamada, H. (1973). *Strength of Biological Materials.* Huntington, N.Y.: Robert E. Krieger.

Fitting to Maximize Performance of HMD Systems

Jennifer J. Whitestone

Kathleen M. Robinette

The fit of a head-mounted display (HMD) system is much more than its implied comfort. Fit affects stability, wearability, component placement, and ultimately, performance. This chapter examines anthropometric techniques such as using percentiles or relying on Frankfurt-plane measurements that are inappropriate for HMD or other helmet-system designs. These myths often lead to the wrong decisions concerning HMD designs. Instead, three-dimensional scanning and fit testing are described as methods that provide the designer with a foundation for building their HMD systems. In addition, future possibilities are defined as a means of working toward optimization of HMD fit.

7.1   Introduction

Fit is a tricky verb. We use it two ways when we talk about clothes or equipment: When a piece of equipment or clothing closely matches the dimensions of the body, we say, "That dress fits like a glove." When we talk about the process of measuring someone's size and adjusting equipment or clothing to match his or her dimensions, we say, "He's going to the tailor to be fitted for his tuxedo." Any nervous groom who has worn a rented tux, however, knows that being fitted doesn't mean the tuxedo will *fit*—going through the process of fitting doesn't ensure a good fit. The best way to ensure a good fit is to start with a good design.

A good design is particularly important for helmets. Now that helmet-mounted displays are common, a helmet doesn't just protect the head—it must also provide auditory and visual information to the wearer. The windows for this information can be quite small. A good fit therefore requires precise, three-dimensional placement of important components. A helmet that sits awkwardly on the wearer's head may not deliver all the information the wearer needs.

Comfort is obviously an important issue for helmet design: the most protective, information-packed helmet in the world is useless if it induces unbearable pain. But the performance requirements of modern helmets have been so expanded that comfort is now just one of many design objectives. A good fit now means more than comfort; it means the helmet protects the head, accomplishes all its information delivery requirements, and is comfortable to wear. Modern helmet design must balance performance and comfort to achieve a good fit.

Unfortunately, some common equipment design practices don't help to achieve a good fit; in fact, these practices make a good fit next to impossible. These myths can cause increased development costs and unnecessary inventory pileups. This chapter examines these myths and describes a design method that helps achieve the ultimate goal: helmets that fit.

Section 7.2 describes the four worst design practices, Section 7.3 discusses a design method that solves many problems, and Section 7.4 looks ahead to design practices and challenges of the future.

7.2   Anthropometric Myths: Methods That Don't Work

For helmet-mounted systems there are four major anthropometry practices that cause fit and performance problems and increased development and inventory costs: use of percentiles, use of the Frankfurt plane, reliance on a poorly defined concept of line of sight, and selection of sizes before design begins.

A key drawback to all these methods is that they do not use information from fit testing. Even detailed anthropometric data cannot meet designers' needs unless it is used in conjunction with fit-test results. Each faulty practice is discussed in a separate subsection below. Two of these, the use of percentiles and selecting sizes before design, are universal problems for nearly all engineering anthropometry applications.

7.2.1   Using percentiles: The impossible dream

The first attempts at characterizing populations using statistics relied on mean values, or "the average man" concept. One can imagine the customer requesting equipment or clothing "designed for the average guy." There are several problems with this approach, one of which was humorously illustrated by Damon, Stoudt, and McFarland (1966). They depict a man attempting to walk through a doorway designed for the average stature. A doorway of that size would be too short for 50 percent of the population! Daniels (1952) further demonstrates that there is no such thing as "the average man," because no one is average for all measurements.

The realization that the average, or fiftieth percentile, would exclude too many people led to the desire to characterize a range of people. The next statistic that became popular, therefore, was the percentile. Percentiles are univariate (one-variable) statistics that indicate the relative location of a variable's value with respect to the distribution of values for that variable. The fifth percentile value, for example, is the value for which 5 percent of the population is smaller and 95 percent is larger for one variable. The problem arises when such percentiles are used for more than one variable, because a percentile value for one variable is unrelated to a percentile for another. Robinette and McConville (1982) demonstrated that the sum of percentiles does not equal the percentile of the sum. They showed that with just eight percentile segments, the error at the fifth percentile was 150 mm for stature. In other words, no one can be a fifth percentile for all measurements—it is not physically or mathematically possible. So when designers use percentiles, the design incorporates proportions that do not exist in nature.

Zehner, Meindl, and Hudson (1992) further demonstrate that the range of people accommodated with percentiles is much less than desired. Figure 7.1 shows the actual percentage of people who are within the fifth to ninety-fifth percentile range for up to six head and face measurements. As can be seen, 90 percent of the population fall within the fifth to ninety-fifth percentile values for head length alone, but only 82 percent are within the range for both head length and head breadth, and only 78 percent for head length, head breadth, and pupil-to-vertex (top of head), until after six measurements only 57 percent of the population are within the fifth to ninety-fifth percentile range. Adding more measurements will further reduce this number until it approaches zero.

The more percentiles are used, the worse the problems become. There is almost no case, therefore, in which percentile values are appropriate for use in design or evaluation of HMDs. In other words,

So what is the continued attraction of percentiles? One explanation is that percentiles present the impression that a specific range of people will be accommodated if they are used. What actually is desired is to accommodate a large percentage of a given population. So in the definition of accommodation needs for an HMD, the customer should specify that desire. Instead of requesting that designers accommodate

HEAD LENGTH   90% OF POPULATION

HEAD BREADTH   82% OF POPULATION

PUPIL TO VERTEX   78% OF POPULATION

FACE BREADTH   69% OF POPULATION

FACE LENGTH   63% OF POPULATION

EAR TO VERTEX   57% OF POPULATION

Figure 7.1   The added reduction in accommodation with each successive application of fifth to ninety-fifth percentile values.

the first to ninety-ninth percentile, the customer should stipulate that designers must accommodate 98 percent of the population. Another explanation for the continued widespread use of percentiles is that percentiles are easy to list in tabular form in paper text. Multivariate data are impossible to portray for all applications in tabular form. Even books of bivariate plots require hundreds of thousands of pages to get every combination of every pair of variables in a single survey. The alternatives to percentiles require complete data sets with statistical analysis and visualization.

7.2.2 The Frankfurt plane: An oldie but not a goodie

The second flawed anthropometric practice is the use of the Frankfurt plane, or some similar anatomical landmark alignment systems that define "top of head" and "back of head." These systems came about in an attempt to establish a consistent way to measure skulls for comparison of species. The Frankfurt plane was established at a conference in Frankfurt, Germany, in 1884 (Ranke, 1884). Modern anthropologists have determined that these reference systems are actually inappropriate for their intended purposes. Furthermore, recent helmet studies have revealed that anatomical landmark orientations have little to do with the orientation of the head with respect to a helmet or an HMD. Measurements like "pupil to top of head" define the "top of head" as the highest vertical point with the head in the Frankfurt-plane orientation. The Frankfurt plane is defined in the Air Standardization Coordinating Committee air standard 61/83 as "a standard plane for orientation of the head. It is established by a line passing through the right tragion (the front of the ear) and the lowest point of the right eye socket." Figure 7.2 illustrates the problem using two subjects aligned in the Frankfurt plane. In this figure the same two subjects are shown aligned using the Frankfurt-plane alignment at the left and as the helmet is actually worn at the right.

As Fig. 7.2 demonstrates, the "top of head" defined by the Frankfurt plane does not coincide with the top of the head in the helmet for these two subjects. In fact, further investigation reveals that the location of the head in the helmet appears to be dependent on the contour of the cranium. The method used to establish the head-helmet relationship is described in Sec. 7.3.1.

These findings mean that designers who use traditional measurements for head and face equipment have unknown and unanticipated errors from their point of reference—axis system error—which can be quite large. The error caused by misalignment shown in Fig. 7.2 constitutes several inches in pupil location for just two subjects.

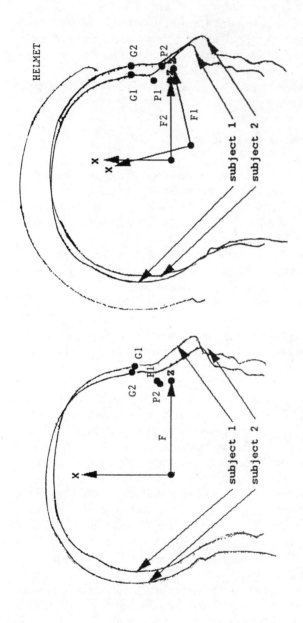

a) Traditional Measurement Alignment

b) Helmet Alignment

F = FRANKFURT PLANE BOTH SUBJECTS
G1 = GLABELLA FOR SUBJECT 1
G2 = GLABELLA FOR SUBJECT 2
P1 = PUPIL FOR SUBJECT 1
P2 = PUPIL FOR SUBJECT 2

F1 = FRANKFURT PLANE FOR SUBJECT 1
F2 = FRANKFURT PLANE FOR SUBJECT 2

Figure 7.2 Two subjects aligned according to the Frankfurt plane and within a helmet.

Before the arrival of advanced 3D surface-digitizing techniques, designers had little or no surface information pertaining to head alignment and no way of detecting errors. Efforts to design equipment to fit populations therefore led to at least two design mistakes. First, too many sizes were created because the added alignment error indicated that the population was more varied than it really was. Second, many sizes did not fit optimally because the added error caused odd proportions in size and shape, and the true size and shape variability were not represented.

It is not surprising, therefore, that in a test of the HGU-53/P U.S. Air Force (USAF) aircrew helmet, it was found that only two sizes (one for men and one for women) are required to fit the same percentage of the population as the original set of six sizes (Robinette, 1993). The size intended to be the largest, size 6, was improperly proportioned, so that even though it was larger in volume than the size 5, it did not fit larger heads. Furthermore, the two best sizes were not represented in the original set of six. At least 10 percent of the male aircrew population did not get a good fit in any size.

Another perhaps broader view of this same limitation is coordinate-system (or reference-frame) dependence. Even summarization and comparison of human shapes based on three-dimensional landmarks can be misleading if the analysis fails to remove dependency on the position in which the subject was measured. This is especially true since measurement systems typically report results in a coordinate reference system independent of the individual being assessed. This makes the measurement sensitive to one's location relative to the sensor reference frame. This dependency has been termed *observer inherence*. Cheverud, Lewis, Banchrach, and Lew (1983) and Lele and Richtsmeier (1991) devised methods to make comparisons between subjects using 3D landmarks. These methods resolve the observer inherence for landmark-type data sets for certain types of studies (e.g., growth studies and comparisons of species of primates), but are insufficient for applications in which contours and contour changes are critical variables.

7.2.3 Line of sight: More mystery than myth

The optimum performance of helmet-mounted optics and HMD systems requires a precise definition of the line of sight. The *line-of-sight angle* (LOSA), the relationship between the primary line of sight and the orientation of the head, is believed to be dependent on the position of the head, gravity, binocular vision, and fixation. There is, however, a range of definitions for line of sight. Even in the fairly small community that addresses HMD design issues, few researchers agree on the definition of line of sight.

Line of sight (LOS) is defined in the *Dictionary of Visual Science* (Schapero, Cline, and Hofstetter, 1960) as "the line connecting the point of fixation and center of the entrance pupil of the fixating eye." For helmet-mounted optical systems, LOS has been a very loosely defined concept. HMD designers must know the relative position of the LOS with respect to the helmet-mounted optics in order to position the optics coaxially with the LOS. Design of an HMD system assumes straight forward eye orientation; therefore, only movement in the sagittal plane (i.e., pitch) is considered. The pitch motion of the eye is controlled by the opposing force vectors of the superior and inferior rectus muscles. These antagonistic muscles roll the eye up and down to compensate for head position relative to a fixation point. The underlying assumption is that a relaxed eye position exists that causes the least fatigue to ocular muscles.

Line-of-sight angle (LOSA) has been evaluated and recorded in many studies relating to human factors design issues. It is referenced to either the horizontal plane or the Frankfurt plane. The angle is a positive value if the line is above the plane and a negative value if it is below the plane. A review of the literature reveals that a wide range of LOSA definitions exist. Figure 7.3 illustrates documented LOSA with respect to the horizontal plane. Experimental conditions for many LOSA studies are not consistent and are not documented with enough detail to recreate the experiment. In most studies, subjects stare at a visual target and the vector from their eye to the visual target is measured and recorded. Subjects are both standing and seated. Seat positions vary from upright (90°) to supine (180°). Content, size, and distance of the visual target vary considerably.

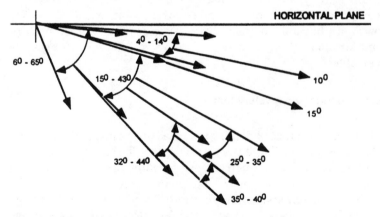

Figure 7.3   Line of sight angle with respect to horizontal plane.

Lehmann and Stier (1961) documented LOSA to be −38° with a standard deviation of ±6.3°. Morgan, Cook, Chapanis, and Lund (1963) recommend a LOSA of −15°, which includes a head rotation of approximately 5°. Van Cott and Kinkade (1972) define LOSA as −10°, whereas Stewart (1979) defines LOSA as a range from −35° to −40°. LOSA for Woodson (1981) is −10°, and for McCormick and Sanders (1982) it is −15°. Grandjean, Hunting, and Nishivama (1984) measured a LOSA of −4° to −14°, Ruhmann (1984) measured −25° to −35°, and Konz (1990) and Povlotsky and Dubrosky (1988) measured ranges from −6° to −65°. MIL-STD-1472D (*Human Engineering Design Criteria for Military Systems Equipment and Facilities,* 1989) states that a normal line of sight for an upright worker is −15° degrees from the horizon and −10° for a seated worker. Table 7.1 shows the same data in tabular form.

In a 1990 study sponsored by the Armstrong Laboratory (McMullin), the LOSA was determined for ten subjects using targets at a distance of 0.5, 1, and 2 m, as well as simulated infinity. Subjects were seated in a chair with five different seat positions. The angle of the line of sight in this study was measured relative to the seat back. The position of the head was not recorded. The results suggest that no significant differences were found by changing the image distance, and that therefore LOSA is not dependent upon the visual target distance. The subjects' LOSA was affected by seat position as the LOSA varied nonlinearly from −0.5° to −12° from an upright seat position to a supine position. Without information pertaining to the position of the head, however, the LOSA with respect to the seat-back angle is difficult to recreate or to map to specifications for helmet or head-mounted optics.

Hill and Kroemer (1986) define the LOSA with respect to the head as the angle between the Frankfurt plane and the line of sight. They

TABLE 7.1   Line-of-Sight Data in Tabular Form

Lehman and Stier (1961)	38° sd ±6.3°
Morgan, Cook, Chapanis, and Lund (1963)	−15°
Van Cott and Kinkade (1972)	−10°
Stewart (1979)	−35° to −40°
Woodson (1981)	−10°
McCormick and Saunders (1982)	15°
Grandjean, Hunting, and Nishivama (1984)	−4° to −14°
Ruhman (1984)	−25° to −35°
Konz (1990); Povlostky and Dubrosky (1988)	−6° to −65°

define a range of LOSA from 14° to −71°. In contrast to McMullin, they find that the image distance influences LOSA and they detect significant differences due to head position. However, Jampel and Shi (1992) demonstrate through the use of photographic and video analyses that the primary position of the eye is not dependent on the horizon, fixation, head position, or gravity. They also reference LOSA with respect to the Frankfurt plane, but find a range of approximately 0.5° to −8.5°. They propose that the primary position is not affected by passive mechanical forces but is the result of basic neurological oculomotor mechanisms.

Recently the Human Engineering Division at Wright-Patterson AFB sent a questionnaire to designers of HMDs to determine if a consensus of the definition of line of sight existed among developers of similar systems. Questions asked included, "When designing helmet-mounted optics, do you reference a preferred line of sight, and if so, what criteria are used? How is the line of sight referenced to the head?" The vendors reported line-of-sight definitions ranging from 15° to −15° with respect to the Frankfurt plane. Their information sources included MIL-STD-1472D, Dreyfuss (1960), and AFAL-TR-1992-00061. All vendors surveyed assumed that the seat-back angle significantly changes LOSA, but were unable to account for the angular difference. Likewise, the vendors believed that high-G maneuvers under which helmet systems shift position alter the pupil-optics vector. They also stressed that LOSA is a function of head attitude and is influenced by factors such as seat inclination, aircrew seating preferences (e.g., reclined versus alert), helmet mass properties, and mission environment. These vendors all requested additional information concerning line-of-sight issues. They communicated a need for procedures and test results to more accurately capture line-of-sight angle for varying seat positions and different types of HMD systems. Clearly, designers are eager for information relaying the "best" LOSA for different information presentation requirements depending on the mission type (field of view, position priority, clutter level, etc.).

7.2.4 Sizing before design: Building the cart without measuring the horse

Perhaps the most widely believed design myth is the idea that people can be grouped into sizes based on their anthropometry, regardless of the design and performance requirements of a system. It is a common misbelief, for example, that people naturally fall into three sizes: small, medium, and large. An engineer once complained about spending money on sizing because he believed "everyone knows what a size medium is." Sirvart Mellian (personal communication, 1995), a scientist at the U.S. Navy Clothing and Textile Research Facility, encountering this same response many times, has developed a quick way to refute this statement when speaking to a group. She asks everyone in the room who believes they are medium-sized to stand up. A wide variety of people stand, thereby demonstrating that all "mediums" are not alike.

The point is that anthropometry is not the only factor to consider when developing a sizing system. Design features and performance requirements of a system affect the amount of body-size variability the system can accommodate. For example, a helmet used for a virtual-reality (VR) game will not have to fit as tightly as an aircrew helmet that has to stay in place at 7 Gs. The VR helmet could therefore probably fit more people in a single size. Also, the VR helmet may need less adjustability than the aircrew helmet because precise placement of the display is less critical.

Quality of fit is the degree to which a head-mounted system can accommodate any individual in a population. The quality of fit for a population can be maximized in four ways: good proportioning and shaping of a single size; design features that broaden the accommodation range in a single size, such as adjustable straps or liners; adding sizes; and adding the ability to completely custom-make the system for individuals. Adding sizes, adding adjustable design features, and custom-making equipment can all affect cost. Methods for achieving a good quality fit for a population therefore are both performance and cost factors. Anthropometry and sizing are design trade-offs with other performance criteria, and it is usually cheaper to include them in the development process to optimize the design. This can perhaps be best understood by reviewing past sizing practices.

In the past, designers preset the number of sizes and the anthropometric measures to be accommodated in each size prior to development of the equipment. There have been two basic methods for doing this: starting with a base size and using grade rules to predict other sizes; and *anthropometric sizing*, where body measurements are used to somehow classify people to arrive at body sizes. The anthropometric sizing method appears to be more scientific and to account for more body-type diversity. In an early study by O'Brien and Shelton (1941), thousands of women were measured. They were then classified by body types, which were divided into discrete intervals called sizes. Measurements for each size were presented.

Randall, Damon, Benton, and Patt (1946) were among the first to use body measurements to derive sizing schemes for flight equipment after many fitting problems surfaced during World War II. Flight equipment was very specialized, with unique fit and performance requirements. There was no tailoring history on which to draw general rules for these new items. Anthropometric sizing became routine business for military organizations in several countries. Emanuel and Alexander (1957) used height and weight to separate people into six sizes for anti-G suits. The other relevant measures were estimated from means and standard deviations for the men who fell within an interval. The particular points used on the normal curve varied depending upon the measure. Similarly, Zeigen, Alexander, and Churchill (1960) used head circumference intervals to divide the anthropometry into sizes for helmets, and Emanuel, Alexander, and Churchill (1959) used face length and lip length to divide faces into size categories for an oxygen mask. Alexander, McConville, and Tebbetts (1979) and Tebbetts, McConville, and Alexander (1979) used height and weight to categorize men and women for flight suits.

Although these methods have become widespread, they are inherently flawed. People do not come in discrete sizes. If people naturally sorted into sizes, the same sizes would fit regardless of the design. T-shirts and dress shirts, for example, would come in the same sizes. Running tights and blue jeans would also come in the same sizes. More sophisticated anthropometric classification schemes are now being employed, such as principal-points analysis (Flury and Tarpey, 1993) and statistical clustering (Piecus, Smith, Standley, Volk, and Wildes, 1993; Corner and Robinson, 1996). Because the statistical analysis is more sophisticated, people are lulled into believing that these methods are better. In fact, they are not really different. No matter how sophisticated the methodology, if only anthropometry is used without data on performance and design criteria, then the analysis is still based on the misconception that classification of body sizes is sufficient to derive well-fitting item sizes.

Problems caused by preset sizing were highlighted in a program to develop a helmet with night-vision goggles (Blackwell and Robinette, 1993). With preset sizing systems a region of fit is implied, which the designer or engineer is said to "design to." In this case, three helmet manufacturers each used the same anthropometric and performance criteria to design a helmet system for size large. The helmet designs were very different, but all three helmet designs were supposed to fit a size large head. Figure 7.4 displays the results of a fit test conducted on the helmets. The helmets did not fit the same anthropometric region, nor did they fit the regions for which they were designed. Consequently the head sizes that achieved a good fit were not the same as the head sizes around which the helmets were designed, and they differed depending upon the design.

This development program also clearly demonstrates that improper sizing can have a major effect on performance testing of HMD systems.

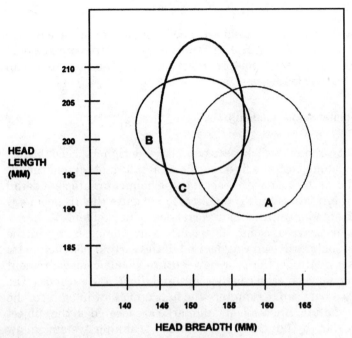

Figure 7.4   Fit regions for three size "large" helmets.

For example, the flight performance of these helmet systems was to be tested using a handful of test pilots, each of whom would fly with all three helmets. A pilot with 145 mm of head breadth and 200 mm of head length could be expected to rate helmet B as excellent, helmet C as marginal, and helmet A as a failure for optical performance of the attached night-vision goggles and for helmet stability. The subject's performance assessment is biased by the fit of the HMD system. Although helmet A could be the superior system, it would be impossible to determine this with comparative tests conducted according to the plan. Anthropometry and fit would disguise the true results. Worse yet, a poorer system could be erroneously judged as best.

When sizes are preset, the true range of fit within a single size is unknown, as are the variables that are most critical to fit in some cases. Without information about the range of fit in one size, how can designers determine the number of additional sizes needed? Fit information can be added from past experience with similar designs or by fit testing as part of the development process rather than just at the end of development. Fit testing at the end of development can only verify or refute some accommodation level rather than optimize the sizing.

7.3   Current Practices: Methods That Work

To avoid the four myths, designers need reliable fit data early in the development cycle. For HMDs, a prototype, a three-dimensional measurement device, and a comprehensive fit-test plan can provide all the information needed for design.

7.3.1   Three-dimensional scanning: Giving designers x-ray vision

To identify fit problems such as asymmetry, designers need fit-test results that include both the fit quantification information and the 3D spatial location of the head with respect to the helmet in sufficient detail to visualize the relationship. The ability to measure the 3D geometry permits accurate identification and correction of fit problems.

A method to arrive at some initial 3D forms for assisting in the early stages of helmet design was devised in the early 1990s (Robinette and Whitestone, 1992). Using a new anthropometric measurement technology, head geometry can be captured in a few seconds. The subject wears a bald cap to compress the hair and allow imaging of the shape of the cranium. Anatomical landmarks are located on the subject and distinguished by fiducials. Most surface scanning systems allow for identification of targets or fiducials within a data set. These targets must be larger than the resolution of the scan data to be visible in the data set. An example for a scan resolution of 1.5 mm is a round target, 4 mm in diameter, with an adhesive back to stick on the subject. Selection of the color of the dot is dependent on the scanning technology. These fiducials are identified in the scan and the coordinates determined to locate landmarks such as tragion (ears), inframalar (cheekbones), and other relevant features. Figure 7.5 shows both a photographic and an electronic image of a subject with fiducials. The Appendix to this chapter contains a listing of head and facial anatomical landmarks. Additional anatomical landmarks such as pupil location are located using image-processing software that allows the user to point and click on selected landmarks. Advanced post-processing methods are under development to automate the landmark identification process (Geisen, et al., 1995). Shown in Fig. 7.5 is a scanned image of the subject in surface mode (left) and wireframe mode (right). The landmarks are visible in the surfaced color scan file (shown here in monochrome) and are illustrated as an X in the wireframe mode.

The new anthropometric scanning technology provides much more information than could feasibly be used by designers, so the next challenge is to devise methods to distill the information into a useful form. One approach suggested is to start with bivariates of head measurements not dependent on the Frankfurt plane and select representatives from among the 3D scans (Robinette, 1992).

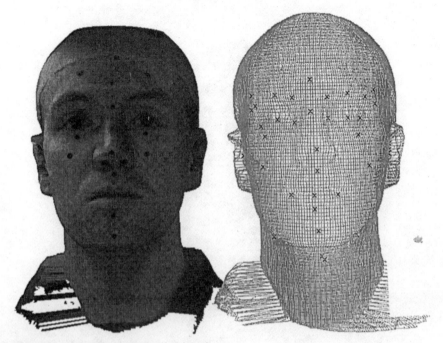

Figure 7.5 Scanned subjects with fiducials.

7.3.2 Feature envelopes: Marking the boundaries

Problems associated with measurement dependence on the Frankfurt plane and limitations with percentiles are alleviated with the use of feature envelopes. Feature envelopes describe the spatial location and orientation of areas of interest (i.e., features) with respect to a well-defined, easily duplicated coordinate system. For a given helmet system, this definition could include the range of pupil location along all three coordinate axes or the volume that contains the aggregate of all ears for a given population. These anthropometric design envelopes defined for an existing helmet are based on one critical factor: the relationship of the head to the helmet. Helmet systems do not fit the human head in exactly the same way across a sample of people. Figure 7.6 illustrates two subjects wearing the same helmet. The orientation of the head with respect to the helmet system is entirely dependent on the shape of the helmet, the liner system, and the added peripherals, such as optics or earcups. All of these components must be fitted optimally to the individual and, as a result, the helmet system sits on the head in a slightly different manner for everyone. In order to study these anthropometric design issues, accurate, high-resolution surface data is required. As equipment items are fitted more closely to the human body, it is essential to obtain the surface geometries of both the subject and

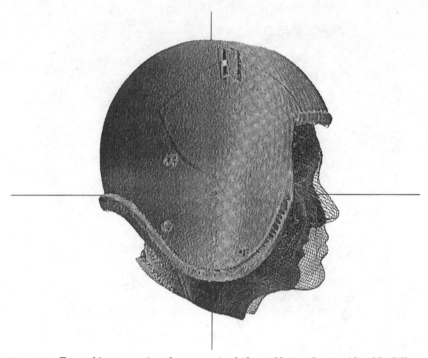

Figure 7.6 Two subjects wearing the same size helmet. Notice the considerable differences between the fits of the helmet on the two subjects.

the equipment. In this manner, the interface or definition of fit can be captured. This can be accomplished using a variety of surface scanning technologies and image-processing tools.

A prototype equipment item is needed for the acquisition and definition of feature envelopes. For helmet-mounted systems, the methodology to capture fit geometries requires a series of 3D scans of subjects with and without this equipment. The first of these scans is the unencumbered scan with fiducial points. After scanning the subject with fiducials, the helmet system is donned and the subject is rescanned. Surface areas common to both scans, in this case, on the forehead and eye regions, are used to register the two images (see Fig. 7.7). Registration can be performed by a least-squares fit of anatomical landmarks common to both scans or by a surface-matching approach (Bhatia, Fiehler, Smith, Commean, and Vannier, 1994). The resulting fused data set provides a view into the way in which the individual fits inside the helmet system. For identifying a population of subjects within a helmet system, a third step is required. A scan of the helmet alone is conducted. Registration of scan B and scan C is accomplished using the common surfaces found on the helmet system. By maintaining the transformation matrices from scan B to scan A and from scan C to scan B, and by eliminating scan B, the position of the subject

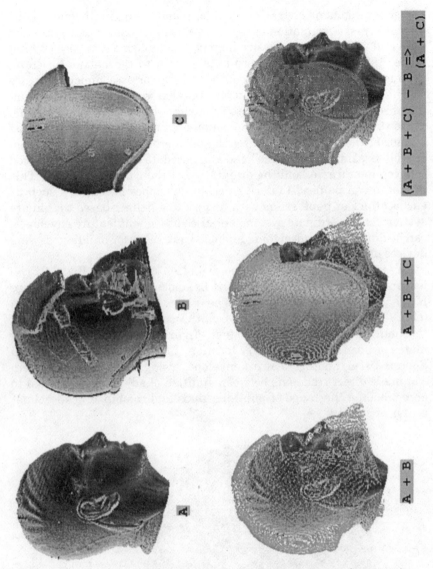

Figure 7.7 Sequence of registering three-dimensional scans to capture fit geometries.

underneath the helmet system can be visualized and recorded. This is a record of the fit geometry.

Figure 7.7 illustrates the sequence of registration: the helmeted figure on the lower left shows registration of scan A (unencumbered) and scan B (encumbered); the lower middle figure shows registration of scan A, scan B, and scan C; the lower right figure shows elimination of scan B, leaving the relationship of scan A and scan C.

For development of HMD technology using existing helmet shells as platforms, a population of representative personnel can be scanned with and without the same size helmet to establish a database (Whitestone, 1993). Assigning a coordinate system to the helmet based on symmetric, repeatably located points on it supports a common reference to which all scans can be registered. In other words, all of the encumbered scans are transformed into the helmet-based coordinate system. Since the relationship between encumbered and unencumbered scans is known, the unencumbered scans are referenced to this axis system as well. Essentially, all of the head and face data are thereby referenced to a common frame, and the orientation of their fit is preserved. This database can be used to derive design envelopes of features such as ear locations or pupil range with respect to a helmet-based coordinate system. This coordinate system, and the subsequent feature envelopes, can be duplicated by designers and used as a basis for building viewing devices or acoustic equipment.

An example of feature-envelope definition is shown in Fig. 7.8. The feature envelope for pupil locations has been defined for a size large HGU-55/P helmet. These data can be used with computer-aided design (CAD) software and used to drive anthropometric accommodation of the helmet-mounted optical system. Figure 7.8 also shows the estimates of head center-of-gravity location (Beier, Schuck, Schuller, and Spann, 1979), another feature envelope. Using these data, designers can make direct trade-offs between addition of adjustments needed to accommodate the range of pupil locations and combined head-helmet center of gravity.

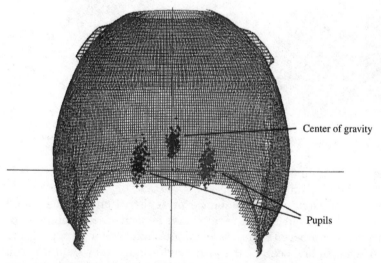

Figure 7.8 The pupil and head center-of-gravity feature envelopes for the size large HGU-55/P helmet.

7.3.3   Fit testing: the right data at the right time

Section 7.2.4 describes problems caused by defining sizes before establishing the design. One effective solution to these problems is to include fit testing in the development process rather than merely for verifying or refuting the success of the fit after development. This will help establish good item proportioning, establish the range of fit per size and the impact of design trade-offs, minimize the number of sizes, and optimize the design.

The advantage to this approach was demonstrated in a Navy program to develop a new women's uniform. In this study, an existing uniform was used as a prototype (Mellian, Ervin, and Robinette, 1990). Fit testing revealed that some sizes were unnecessary duplicates and that the diversity of sizes that were needed was not represented. A follow-up study (Robinette, Melian, and Ervin, 1990) describes the sizing scheme derived after the fit testing. This new sizing scheme reduced the number of major alterations from 75 percent to less than 1 percent without increasing the number of sizes. It also cut the time and cost of issuing the uniforms by 50 percent and reduced inventory waste. Millions of dollars were saved in the first purchase alone. The same kind of success is achievable for head-mounted devices. For an illustration of the fit-test process, see Fig. 7.9.

The first step in developmental fit-testing is to establish a prototype. The term *prototype* is used loosely here to mean any physical system that resembles the final design closely enough that it will help establish fit criteria. The prototype can be an existing system, an existing system modified to be similar, or a completely new system. The more the prototype reflects the new design concept, the more efficient and cost-effective the design will be, thereby reducing the risk of a poor fit quality or excessive number of sizes or adjustments. An example was provided in the previous section of using an existing helmet that described feature envelopes with respect to an existing helmet. These feature envelopes constitute fit information for prototype helmets (if these helmets are candidates for head-mounted displays).

The feature envelopes were derived from the spatial locations of the features with respect to the helmet when the wearer had a comfortable and stable fit. They geometrically quantify the fit for existing helmets.

An example of a modified existing helmet prototype would be to take one of the above prototypes and add integrated blocks that approximate the HMD size, shape, and weight. This modified helmet can then be tested for comfort, stability, and any other pertinent criteria, such as interference with other components, seats, or canopy. For those subjects who get an acceptable fit in the new configuration, feature envelopes can be created. This might be a desirable approach in the

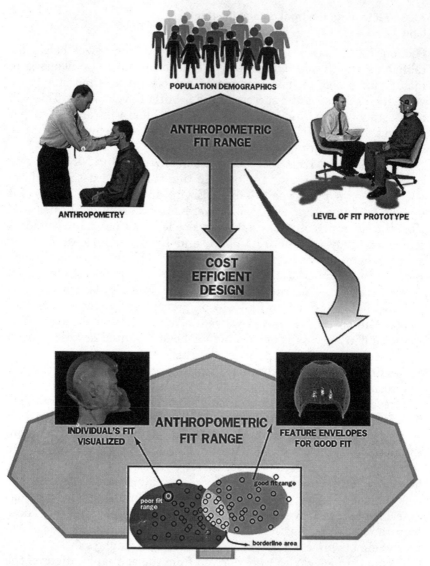

Figure 7.9 The fit test process.

instances in which there is a risk that the added weight and location of the HMD modules might change the location of the helmet's features.

The development of a new prototype helmet is becoming easier due to advanced anthropometric measurement devices that reduce the effort and time required to produce solid 3D models. In developing a pressure glove for a space suit, ILC Dover estimated that a 3D scan of a hand enabled them to reduce prototype production time by as much as 60 percent (Cadogen, 1990, personal communication). Using 3D forms in

3D CAD systems as solid models not only helps to create some of the first prototypes for testing—it also helps measure and control changes to the systems identified through fit testing. Using specially created prototypes may require extra time initially, but prototyping can speed up the overall development process.

There are two required components to the data collection during fit testing: measuring fit quality and anthropometry.

Measuring fit quality generally means more than just quantifying comfort. For HMDs, display stability is usually important, as is integration with other things, such as eyeglasses. These fit factors can affect each other. Determining the acceptable level of fit quality involves setting acceptable individual factor levels and the acceptable combination of factor levels. Setting the level that constitutes "too much" or "too little" usually necessitates assuming some risk of being wrong. These risks must be weighed against cost, time, and the particular conditions of the design. The following information gives some guidelines based on past experience with testing of helmet-mounted systems.

Fit-assessment questionnaires should be carefully worded so that the subject is not biased by the question. An example is provided below:

How does the system feel overall (circle one)?

A. Very comfortable

B. Moderately comfortable

C. Moderately uncomfortable

D. Severely uncomfortable

If C or D was circled, what was the cause (circle all that apply)?

A. General tightness everywhere

B. Chin strap scratch

C. Areas of tightness (please list)

D. Hot spots (color in region on pictures below)

In this example, hot spots are identified by having the subject color in the region of discomfort on the picture provided. However, hot spots can be better identified by having the subject point to trouble spots on the helmet while they are wearing it, marking the spots, and photographing or scanning the system with the hot-spot marks. With the 3D scanning option, the information can be transferred to a CAD system to help correct the system, or just visualized in 3D.

To assess comfort it is important to have the system worn for a length of time that represents how long it will eventually be continuously worn before the comfort assessment is made. If the expected wear time is 15 minutes or less, this is not much of a problem. If the expected wear time is several hours, a decision must be made based on cost, time, and availability of subjects for tests that would last a long time. In this case, research has shown (Blackwell and Robinette, 1993) that the system should be worn for at least 30 minutes for discomfort to become noticeable. After that point the risk of missing a poor-fitting system must be weighed against the difficulties of testing. Setting stringent comfort cut-off criteria can help minimize the risk. For example, any amount of discomfort might be set as the unacceptable point. This will increase the likelihood that the system will be deemed unacceptable when it may be tolerable, however. These are decisions that only the developer or experimenter can make, because they are limited by time and money.

Stability can be subjectively assessed and quantified, or a stability measure can sometimes be derived. Two methods were used in one helmet-fit study (Blackwell and Robinette, 1993). For the subjective quantification, visible movements of the helmet with respect to the head were rated by the investigators on a scale of 1 to 5 as the subject performed various head movements. For the more objective measurement, the movement was recorded in millimeters along the surface of the forehead when different forces were applied. With this method the location of the force applied was dependent on the helmet. Because the location of the applied force affects the movement level, this type of measurement can only be considered a relative measure of stability. Consequently the levels that constituted excellent, good, average, fair, and poor were subjective decisions. The decision about what was acceptable was based upon a combination of the two tests and a third question that asked the subject to rate his or her overall assessment of stability. If the subjects are experienced with similar systems, their subjective opinions can be the most valuable information of all.

Measuring anthropometry can include both traditional caliper and tape-measure measurements as well as 3D surface scanning. Although most traditional head and face measurements are dependent on a reference-axis system, there are some commonly measured ones that are not so dependent and that can be useful: head length, breadth, and circumference; bitragion breadth; and face length and breadth. Descriptions of these measures can be found in the Appendix. Measures that should not be used, no matter how attractive they may seem, include any measure to top of head, back of head, vertex, or wall; and any measure over the top of head or vertex. Other measurements to avoid are pupil to top of head, pupil to vertex, pupil to wall, pupil to back of head, and head height.

Three-dimensional surface scanning adds the measurement of contour and the ability to capture the relationship between the system and the person. The system–person relationship is referred to as the geometry of fit. This is obviously very valuable for quantifying the correlation between fit and anthropometry. If such technology is not available, much can be accomplished with the use of photographs (both encumbered and unencumbered, particularly from the side view). Transparencies or slides can be used to overlay the photos and measurements can be manually extracted, or digital cameras can be used to capture the information and the views can be overlaid in CAD systems or other graphics software packages to extract the measures.

Once the data are collected, the next step is to establish the range of anthropometry that is accommodated by a single size without changes. This range must then be compared to a sample of the end-user population to assess the proportion of the population accommodated by a single size without changes. This sample can be gathered as part of the fit test or might be acquired from some other source.

Next, the proportion accommodated in one size without change must be evaluated against different design mechanisms to achieve accommodation. One mechanism is to add a size by "growing" or "shrinking" the helmet in one or more directions. Another mechanism is to change the adjustment system so that it can move along a new axis or have greater movement along an existing axis. A third mechanism is to change the fitting system; for example, a helmet liner might be changed to a system of straps that provide greater stability and adjustability. These are all design trade-offs that require different amounts of time, effort, risk, and cost; the ability to quantify the amount of change needed in an area of the system in 3D space will greatly help to evaluate these trade-offs.

Finally, the design changes decided on should be tested to ensure that they actually do what they should. It is always possible that fixing one thing can cause a problem somewhere else. Design changes should also be verified with a fit test.

Although it is generally true that at least some of the population is accommodated by the prototype, there is the possibility that no one will be. The ability to quantify fit problems can also give direction to solving the problem. Furthermore, it is possible that all of the population is adequately accommodated, and then some. In other words, the system can be too accommodating. In this instance the system might be reduced in overall volume or weight or adjustment range without disaccommodating anyone—this can make it cheaper and better as well. Fit testing will not only assist in determining the best design decisions, but will also permit system optimization for population accommodation and cost. Fit testing with prototypes gives designers a clear view of fit problems early enough for them to take full advantage of the information.

7.4   Looking Ahead: Put Away Those Tape Measures

Computer simulations are used for earthquake studies, plane-crash investigations, and space exploration. If clothing and equipment designers can combine 3D data-collection technology with computer modeling and simulation software, anthropometry will leap into the twenty-first century. This chapter's final section examines some of the challenges facing the industry in the near future.

7.4.1   Defining line of sight

As the vendor survey shows, one of the most pressing needs for HMD designers is a definitive answer to the line-of-sight question. Acquisition and analysis of high-resolution surface topology of the human head with protective equipment is now possible. Although the exterior surface can be imaged, the primary line of sight is still invisible.

To solve this puzzle, an eye tracking system such as the RK416 Pupil Tracking System (ISCAN) can be used to determine the orientation of the primary line of sight with a global coordinate system. A surface scanning system can provide the orientation of the head with respect to the global system, thereby allowing the relationship of the head and primary line of sight to be established. The preferred line of sight can be evaluated for varying body postures from standing to sitting erect (as in an FB-111) to reclined as much as 34° (as in the F-16). Further, the relationship of the preferred line of sight to any selected head-mounted systems can be established using a surface scanning system. Line-of-sight angles measured and recorded using this procedure can be documented as a feature envelope.

7.4.2   Generic head alignment

The Frankfurt plane, as noted in Sec. 7.2.2, is not a useful design tool for aligning the head within a helmet system. As a matter of fact, it is completely irrelevant. What is evident, however, is the alignment of the back and top of the cranium to fit the helmet shell. By applying "lessons learned" from the interface databases of existing helmet systems, hypotheses can be formed to align the new helmet concept on the head. Known restrictions specific to the design concept will further refine orientation of the helmet and equipment with the head data. For instance, development of a helmet system to house active noise reduction (ANR) earcups may initially align the population of heads along the bitragion (ear) vector to restrict the size of the feature envelope for ear placement. Helmet-mounted systems will require that head scans originate from the pupil and rotate from this point to align the cranium. The important step in this concept is to assemble and apply all known relationships and restrictions and design adjustability into the system to account for the remaining variability.

7.4.3   Biofidelic computer-aided design head

With the design of protective equipment and workstations moving into the interactive computer-aided design (CAD) environment, there is a pressing need for an animated computer tool that accurately depicts the human body. Simply using range data to design an oxygen mask—the fit of which is dependent on soft tissue deformation—is limiting. To obtain the optimum man–machine interface, such a tool has to provide deformable surface representation of the body as it moves through its full range of motion. The ability of this tool to generate biofidelic animation requires that a linkage system be developed that is capable of reconstructing the complex articulations of the joints. The interaction of muscles and tendons as well as surface mechanical properties would have to be included for correct representation of the tissue deformation.

Many disciplines have addressed the 3D graphical representation of the head and face, including anthropometry, forensics, dermatology, physiology, engineering, computer graphics, medical computing, and surgery (Vannier, Marsh, and Warren, 1983). Software packages that allow for volume visualization of computed tomography (CT) and magnetic resonance imaging (MRI) scans are available, and the geometric accuracy of these images has been evaluated using test objects of known dimensions. Surface scan images from a database have been overlaid onto CT scans of skulls to ascertain the identity of unidentified persons and have been moderately successful, but these facial models do not represent true biomechanical mechanisms (i.e., muscle interaction) or true tissue deformation. There are numerous publications on the mechanical properties of skin and measured skin thicknesses and also correlations between skin properties and age and sex (Rhine and Moore, 1982). All available literature suggests that data exists, or that methods to measure and record parameters are available, for each of the components of a realistic representation of the head and face. The industry needs someone to step forward and combine these raw materials into a system that can truly automate anthropometric research.

7.5   Summary

This chapter discussed design techniques that have been misused and the specific problems created by these techniques during the development process. Attempting to define the sizing system, for instance, before prototyping the design is a common practice with many equipment items. However, this practice is inherently flawed, as the sizes cannot be adequately determined without fit-testing a prototype system. Similarly, reliance on percentiles and the Frankfurt plane as a frame of reference is apt to result in a poorly fitting HMD. Methods that do work, however, were described in a manner to allow the user to custom-fit his design process to the general technique, such as fit testing. Finally, challenges associated with developing a biofidelic model to provide the designer with yet more information regarding fit were discussed.

7.6 References

Air Standardization Coordinating Committee (ASCC) (1991). *A basis for common practices in the conduct of anthropometric surveys* (Air Standard No. 61/83). Washington, DC: Author.

Alexander, M., McConville, J. T., & Tebbetts, I. (1979). *Revised height/weight sizing programs for men's protective flight garments* (AMRL-TR-79-28). Wright-Patterson Air Force Base, OH: Aerospace Medical Research Laboratory.

Beier, G., Schuck, M., Schuller, E., & Spann, W. (1979). *Determination of physical data of the head.* Arlington, VA: Office of Naval Research.

Bhatia, G., Fiehler, G., Smith, K., Commean, P., & Vannier, M. (1994). A practical surface registration technique. *Photonics East 94, SPIE Sensor Fusion VII, 2355*, 135–146.

Blackwell, S. U., & Robinette, K. M. (1993). *Human integration evaluation of three helmet systems* (AL-TR-1993-0028). Wright-Patterson Air Force Base, OH: Armstrong Laboratory, Crew Systems Directorate, Human Engineering Division.

Cheverud, J., Lewis, J. L., Banchrach, W., & Lew, W. E. (1983). The measurement of form and variation in form: An application of three-dimensional quantitative morphology by finite-element methods. *American Journal of Physical Anthropology, 62*, 152–165.

Corner, B. D., and Robinson, D. G. (1996), A new method for quantifying and clustering human form based on spatial statistics and fuzzy logic. *American Journal of Physical Anthropology, Supplement 22*, pp. 89–90.

Damon, A., Stoudt, H. W., & McFarland, R. A. (1966). *The human body in equipment design.* Cambridge, MA: Harvard University Press.

Daniels, G. S. (1952). *The average man.* TN-WCRD 53-7 (AD 10 203). Wright Air Development Center, Wright-Patterson Air Force Base, OH.

Dreyfuss, H. (1960). *The measure of man: Human factors in design.* New York: Whitney Library of Design.

Emanuel, I., Alexander, M., & Churchill, E. (1959). *Anthropometric sizing and fit-test of the MC-1 Oral-Nasal Oxygen Mask* (WADC TR 58-505, AD 213 604). Wright-Patterson Air Force Base, OH: Wright Air Development Center.

Emanuel, I., & Alexander, M. (1957). *Height-weight sizing and fit-test of a cutaway G-suit, type CSU-3/P* (WADC-TR 57-432, AD 130 912). Wright-Patterson Air Force Base, OH: Wright Air Development Center.

Flury, B. D., & Tarpey, T. (1993). Representing a large collection of curves: A case for principal points. *The American Statistician, 47* (4), 304–306.

Geisen, G., Mason, G. P., Houston, V., Whitestone, J., McQuiston, B., & Beattie, A. (1995). Automatic detection, identification, and registration of anatomical landmarks. *Proceedings of the 39th Annual Meeting of the Human Factors and Ergonomics Society, 2*, 750–753.

Grandjean, E., Hunting, W., & Nishivama, K. (1984). Preferred VDT workstation settings, body posture and physical impairments. *Applied Ergonomics, 15* (2), 99–104.

Hill, S., & Kroemer, K. (1986). Preferred declination of the line of sight. *Human Factors, 29* (2), 127–134.

Human engineering design criteria for military systems equipment and facilities (MIL-STD-1472D). (1989). Washington, DC: U.S. Department of Defense.

Jampel, R. S., & Shi, D. X. (1992). The primary position of the eyes, the resetting saccade, and the transverse visual head plane. *Investigative Opthalmology and Visual Science, 33* (8), 2501–2510.

Konz, S. (1990). *Work design: Industrial ergonomics.* Worthington, OH: Publishing Horizons.

Lehmann, G., & Stier, F. (1961). Mensch und Geraet. In *Handbuch der gesamten Arbeitsmedizin,* (Vol. 1, pp. 718–788). Berlin: Urban und Schwarzenberg.

Lele, S., and Richtsmeier, J. T. (1991). Euclidian distance matrix analysis: A coordinate free approach for comparing biological shapes using landmark data. *American Journal of Physical Anthropology, 86*: 415–427.

McCormick, E., & Sanders, M. (1982). *Human factors in engineering and design* (5th ed.). New York: McGraw-Hill.

McMullin, D. L. (1990). Effects of image distance and seat inclination on the line-of-sight angle. In *Industrial Engineering and Operations Research* (pp. 26–45). Blacksburg, VA: Virginia Polytechnic Institute and State University.

Mellian, S. A., Ervin, C., & Robinette, K. M. (1990). *Sizing evaluation of Navy women's uniforms* (AL-TR-1991-0116). Wright-Patterson Air Force Base, OH: Armstrong Laboratory, Air Force Systems Command.

Morgan, C., Cook, J., Chapanis, A., & Lund, M. (Eds.). (1963). *Human engineering guide to equipment design.* New York: McGraw-Hill.

O'Brien, R., & Shelton, W. C. (1941). *Women's measurement for garment and pattern construction* (Miscellaneous Publication 454). Washington, DC: U.S. Government Printing Office.

Piccus, M. E., Smith, G. A., Standley, B. K., Volk, T. L., and L. B. Wildes, (1993), *Creation of prototype aircrew protection equipment based on face anthropometry,* Thesis, Air Force Institute of Technology, (AFIT/GSE/ENY/93D-2, AD-A273 865). Wright-Patterson AFB, OH.

Povlotsky, B., & Dubrosky, V. (1988). "Recommended" versus "preferred" in design and use of computer workstations. *Proceedings of the 32nd Annual Meeting of the Human Factors and Ergonomics Society, 2,* 501–505.

Randall, F. E., Damon, A, Benton, R., & Patt, D. (1946). *Human body size in military aircraft and personal equipment* (Technical Report 5501). Dayton, OH: Army Air Force, Air Material Command, Wright Field.

Ranke, J. (ed.) (1884). Verstandigung uber ein gemeinsames cranio-metrisches Verfahren (Frankfurter Verstandigung) [Standardization of a common head-measurement method]. *Archiver Anthropologie, 15,* 1–8.

Rhine, J. S., & Moore, C. E. (1982). *Facial reproduction tables of facial tissue thicknesses of American caucasoids in forensic anthropology.* Albuquerque, NM: Maxwell Museum Technical Series.

Robinette, K. M. (1992). Anthropometry for HMD design. *Proceedings of the SPIE, Aerospace Sensing International Symposium and Exhibition (1695), Helmet Mounted Displays III,* 138–145.

Robinette, K. M. (1993). Fit testing as a helmet development tool. *Proceedings of the 37th Annual Meeting of the Human Factors and Ergonomics Society, 1,* 69–73.

Robinette, K. M., and McConville, J. T., (1982), An alternative to percentile models, (SAE Technical Paper 810217), in *1981 SAE Transcriptions* (pp. 938–946), Society of Automotive Engineers, Warrendale, PA.

Robinette, K. M., Mellian, S., and Ervin, C. (1990). *Development of sizing systems for Navy women's uniforms,* Technical Report No. 183, Nacy Clothing and Textile Research Facility, Natick MA, and AL-TR-1991-0117, Armstrong Laboratory, Air Force System Command, Wright Patterson Air Force Base, OH.

Robinette, K. M., & Whitestone, J. J. (1992). *Methods for characterizing the human head for the design of helmets* (AL-TR-1992-0061). Wright-Patterson Air Force Base, OH: Armstrong Laboratory.

Ruhmann, H. P. (1984). Basic data for the design of consoles. In H. Schmidtke, (Ed.), *Ergonomic data for equipment design* (pp. 115–144). New York: Plenum.

Schapero, M., Cline, D., & Hofstetter, H. W. (Eds.). (1960). *Dictionary of Visual Science.* Philadelphia: Chilton.

Stewart, T. (1979). Eyestrain and visual display units: A review. *Displays, 1* (1), 25–32.

Tebbetts, I., McConville, J. T., & Alexander, M. (1979). *Height/weight sizing programs for women's protective garments* (AMRL-TR-79-35). Wright-Patterson Air Force Base, OH: Aerospace Medical Research Laboratory.

Van Cott, H. P., & Kinkade, R. G. (1972). *Human engineering guide to equipment design.* Washington, DC: U.S. Government Printing Office.

Whitestone, J. W., (1993). Design and evaluation of helmet systems using 3D data. *Proceedings of the 37th Annual Meeting of the Human Factors and Ergonomics Society, 1,* 64–68.

Vannier, M. W., Marsh, J. L., & Warren, J. O. (1983). Three dimensional computer graphics for craniofacial surgical planning and evaluation. *Computer Graphics, 17,* 263–274.

Woodson, W. E. (1981). *Human factors design handbook.* New York: McGraw-Hill.

Zehner, G. F., Meindl, R. S., & Hudson, J. A., (1992). *A multivariate anthropometric method for crew station design: Abridged* (AL-TR-1992-0164). Armstrong Laboratory, Wright Patterson Air Force Base, OH.

Zeigen, R. S., Alexander, M., & Churchill, E. (1960). *A head circumference sizing system for helmet design, including three-dimensional presentation of anthropometric data* (WADC-TR-60-631). Wright-Patterson Air Force Base, OH: Wright Air Development Center.

Appendix: Traditional Anthropometric Measures for the Head and Face with Minimal Axis System Dependence

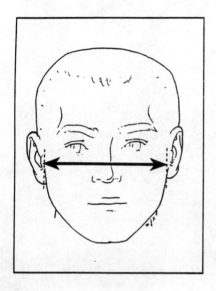

Figure 7A.1 Bizygomatic Breadth (Face Breadth): The straight line distance between the right and left zygion landmarks. This measurement is taken with a spreading caliper.
Landmark(s): Zygion (right and left)

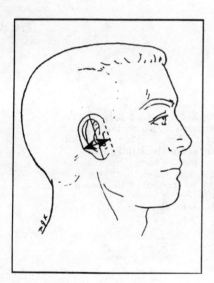

Figure 7A.2 Ear Breadth: The maximum breadth of the right ear perpendicular to its long axis. This measurement is taken with a sliding caliper.
Landmark(s): None

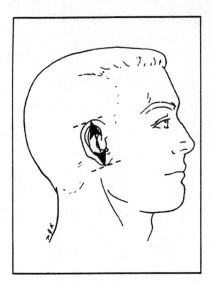

Figure 7A.3 Ear length: The length of the ear from its highest to lowest points on a line parallel to the long axis of the ear. This measurement is taken with a sliding caliper.
Landmark(s): None

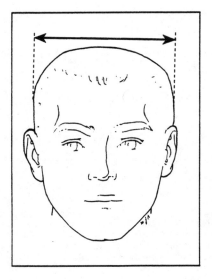

Figure 7A.4 Head Breadth: The maximum horizontal breadth of the head taken above the ears (usually slightly above and behind the ears). This measurement is taken from the rear of the subject with spreading calipers.
Landmark(s): None

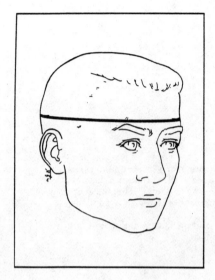

Figure 7A.5 Head circumference: The maximum circumference of the head above the browridges. This measurement is taken with a tape measure.
Landmark(s): None

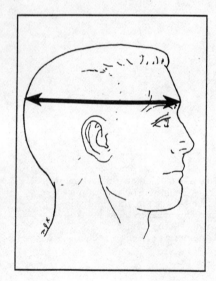

Figure 7A.6 Head Length: The distance from glabella to the most posterior point to the back of the head. This measurement is taken with a spreading caliper.
Landmark(s): Glabella

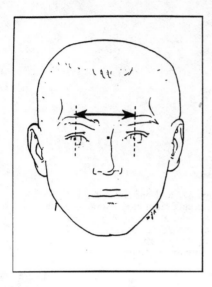

Figure 7A.7 Interpupillary Distance (IPD): Distance between the two pupils. This measurement is taken with a pupilometer.
Landmark(s): Pupil (right and left)

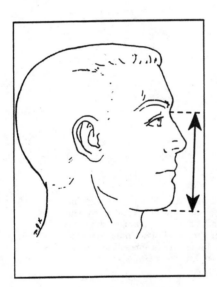

Figure 7A.8 Menton-Sellion Length (Face Length): The vertical distance between menton and sellion. This measurement is taken with a sliding caliper.
Landmark(s): Menton, sellion

Visual Requirements in HMDs: What Can We See and What Do We Need to See?

Elizabeth Thorpe Davis

To optimize the visual design of HMDs for usability and comfort, we must consider the capabilities and limitations of human visual perception, the relation between visual perception and the characteristics of the HMD's visual display, and the relation between visual perception and the performance of various tasks with either immersive or see-through HMDs. In this chapter I will examine the questions What can we see? *and* What do we need to see?

8.1   Simulated Visual Displays versus Real-World Perception

Often the simulated imagery of head-mounted displays (HMDs) is an attempt to mimic the visual information available in the real world. Yet the very nature of simulated displays implies that they can simulate information not readily available to our senses in real-world environments (e.g., the visual presentation of information from the infrared or ultraviolet spectra of electromagnetic radiation). They also can eliminate visual information normally available to our senses in the real world (e.g., reducing the spatial resolution of information or eliminating the 3D information normally provided by stereopsis). Sometimes information is eliminated because of current technological limitations in the hardware and software available for the HMDs. Sometimes information may be eliminated on purpose because the additional information is not cost-effective and/or adds no useful information for the tasks performed. That is, information is sometimes eliminated or added by purposeful design to enhance perception and performance with HMDs.

In optimizing the visual design of HMDs, we need to consider the questions *What can we see?* and *What do we need to see?*, as well as how these capabilities and requirements mesh with the characteristics of the visual display. In the first part of this chapter I will present these questions in terms of both the human visual system and the HMD's visual displays. Specific topics will include the perception of brightness, contrast, visual acuity and spatial resolution, temporal sensitivity and temporal resolution, the field of view (FOV), monoscopic versus biocular versus binocular views, and color versus monochrome. I will relate the human's visual capabilities to the characteristics of the HMD visual displays, describe limitations in achieving optimal display characteristics, and explain how these factors can affect performance with HMDs.

To obtain acceptable performance with HMDs, we need to mesh the human-computer interface design to exploit not only the capabilities and limitations of human visual perception, but also those of the HMD's visual display. As you will see, the capabilities and limitations of see-through and immersive HMDs can differ. For example, in augmented reality systems (e.g., see-through HMDs) we need to mesh the simulated visual display with the user's visual perception of the real world so that task performance is not hindered.

In the last part of this chapter I will present several tasks and situations in which HMDs may be beneficial, and I will also consider *What do we need to see?* in terms of several criteria for the usefulness of HMDs. That is, what characteristics of the HMDs and of human vision are necessary, or at least helpful, in performing these various tasks?

8.2 Characteristics of the Human Visual System and Their Relation to the Visual Displays of Immersive and See-Through HMDs

All of the factors considered here (e.g., brightness, contrast, spatial resolution, and field of view) are important for the design and use of HMDs. Many developers and users of HMDs consider the trade-off between spatial resolution and field of view (FOV) to be the most important (Task, 1991). Others consider the issue of time lag, which can cause spatial disorientation and even nausea, to be at least as important (Brooks, 1995). The order in which each of these factors is presented follows a logical sequence in terms of definitions, explanations, and visual information processing rather than the author's perceived importance of each factor.

8.2.1 Brightness and contrast

The human visual system can operate over an amazing range of luminance levels, from detecting the flicker of a candle several miles away on a dark, moonless night to perceiving objects in the dazzling midday sun of the desert (a luminance range of approximately 3×10^{-6} to 3×10^7 cd/m^2).[1] Most visual displays used in HMDs cannot match the high luminance levels or the large dynamic range of luminances that the human visual system can process. Although the human visual system can operate over a very large dynamic range, it also can discriminate very small differences in luminance. These luminance variations allow the human visual system to detect and recognize different objects in a scene. Rod photoreceptors mediate vision at scotopic nighttime luminance levels (3.14×10^{-6} to 0.314 cd/m^2), whereas cone photoreceptors mediate vision at photopic daylight luminance levels

---

[1]Damage can occur at luminance levels of about 10^8 cd/m^2 (e.g., looking at the sun). At somewhat lower luminance levels (e.g., from approximately 10^5 cd/m^2 and above), visual discomfort can occur. These luminance levels should be avoided.

(0.314 to 3.14×10^6 cd/m^2). Sensitivity to light also depends upon the retina's state of adaptation. After only a brief exposure to a higher luminance level, it can take almost an hour to dark-adapt for maximum scotopic sensitivity in a dimly illuminated environment. However, the human visual system can light-adapt to a brighter ambient luminance within a few minutes. The reader is referred to Hood and Finkelstein (1986) for a more detailed explanation of the human visual system's sensitivity to light.

Contrast is a measure of how much brighter or darker an object is compared to the mean luminance level or to the background against which it is presented. There are several methods for determining contrast. One formula for contrast is used when the luminance of an object is compared to the background luminance (e.g., for aperiodic patterns):[2]

$$C = \frac{L_{object} - L_{bg}}{L_{bg}} \qquad (8.1)$$

where L_{object} is the luminance of the object and L_{bg} is the luminance of the background. Notice that sometimes the luminance of the object may be less than that of the background, when the object is darker than the background. So contrast can have both positive and negative values for Eq. (8.1), indicating the polarity of the object with respect to the background, and is unbounded when L_{bg} is zero. Contrast perception is optimal above a mean luminance of 300 cd/m^2 (viz., in the photopic luminance range where rods are already saturated). At these luminance levels, objects with contrasts as low as 0.02 can be detected.

The necessary brightnesses and contrasts needed for head-mounted displays partially depend on whether the display is an immersive HMD (with only a simulated display) or a see-through HMD (with synthetic imagery or symbology superimposed upon a real-world visual scene). These two types of HMDs raise somewhat different problems. I will first discuss issues relevant for the immersive HMDs and then consider the special problems that a see-through HMD incurs.

---

[2]Another formula is the Michaelson contrast, used when the average luminance of the scene is constant and the patterns are periodic (e.g., a squarewave grating). This contrast is given by the following equation:

$$C = \frac{(L_{max} - L_{min})}{(L_{max} + L_{min})} = \frac{(L_{max} - L_{min})}{(2L_{mean})}$$

where L_{max} is the maximum luminance, L_{min} is the minimum luminance, and L_{mean} is the average luminance. Notice that this contrast value is always positive and would vary between 0 and 1. Also, for periodic patterns one can change the contrast of the pattern (e.g., by changing L_{max} and L_{min}) without changing the mean luminance. This is not true for aperiodic patterns such as squares or circles.

For immersive HMDs, the typical brightnesses and contrasts (or contrast ratios) for commercially available direct-view displays are shown in Table 8.1. As you can see, the monochrome CRT has the highest mean luminance and the electroluminescent display has the lowest mean luminance. Both displays operate within the photopic luminance range, although the electroluminescent display borders on the mesopic luminance range, where both rods and cones are active. None of the displays has a mean luminance that would cause discomfort or possible retinal damage. Although we could increase the overall mean luminance of the visual display by increasing the ambient luminance level, this would cause a veiling luminance that effectively reduces the contrasts of objects within the visual display.

The LED displays offer the largest contrast and could tolerate some veiling luminance. The color CRT displays, however, provide the lowest contrast. Adding a veiling luminance would not only reduce contrast but would also desaturate the colors in the display (see Sec. 8.2.6). For most applications a contrast between 9 and 24 is acceptable, although a 1980 report by AGARD cited a contrast of approximately 1000 as ideal (Padmos and Milders, 1992). Almost all of the contrasts shown in Table 8.1 are well above the minimum contrast required for detection of objects (excluding effects due to veiling luminance). Of course, even brighter displays with even higher contrasts could be produced with a laser display that paints images directly on the retinae, such as the one designed by Tom Furness. A recent review (Holmgren and Robinett, 1993) on the feasibility of scanned laser displays for HMDs or virtual environments in the near future is less optimistic.

TABLE 8.1 Typical Brightness and Contrast for Some Common Direct-View Displays[*]

Type of display	Brightness		Contrast ratio
	Average	Max possible	
Electroluminescent	102 cd/m^2	514 to 685 cd/m^2	9 (10:1)
LED	342.6 cd/m^2	3,426 cd/m^2 (single LED)	99 (100:1)
a.c. and d.c. plasma	103–171 cd/m^2	1,233 cd/m^2	19 (20:1)
Active matrix LCD	514 cd/m^2		16 or 31 (17:1 or 32:1)
CRT	822 cd/m^2 (color) 3,426 cd/m^2 (monochrome)		2 (3:1) 7 (8:1)

[*]Note: Contrast ratios are shown in parentheses.

For see-through HMDs, used in augmented reality systems, the graphics or symbology is superimposed upon a real-world scene. Thus, the real-world scene effectively provides a background against which the synthesized images and symbology are observed. At nighttime and under dim illumination, the effective background luminance provided by the real environment can be quite low, increasing the contrast and visibility of light symbols but decreasing those of dark symbols. On a bright, sunny day, however, the effective background luminance provided by the real environment can be quite high, drastically decreasing the contrast and visibility of even the brightest synthetic symbology. To offset this problem one could change the gain of the overall luminance from the real-world scene to match some constant value and, perhaps, simultaneously change the mean luminance and contrast gains in the synthetic imagery. The military currently uses a dark visor over the headset combiner to overcome the problem.

Another relatively straightforward approach would be to use light-sensitive filters over the headset combiner. These filters would uniformly become denser at higher luminance levels. This approach would reduce the wide diurnal fluctuation in the overall luminance of the real-world scene. (If there were spatially localized adaptation effects in the filters, however, the light-sensitive filters would reduce the contrast of some real-world objects. Moreover, depending on the time-lag of the filters, these light-sensitive filters might not be effective against sudden changes in light levels, such as the bright flash of a searchlight.) Additional measures may be necessary to fully compensate for the large dynamic range of effective background luminances provided by the real world. These additional measures could include changing the contrast and polarity of the synthetic symbology, depending on whether viewing is during daylight or nighttime hours.

8.2.2 Visual acuity and spatial resolution

Visual acuity is the ability to see fine details of the visual scene. For example, visual acuity allows one to see separate stars in the sky at night, to read signs along the roadside, or to identify a distant object. It is a type of threshold measure that is given in terms of the visual angle subtended by the smallest detectable object (i.e., the *proximal stimulus*).[3] Specifically, decimal visual acuity is the reciprocal of the threshold measure expressed in minutes of arc. Normal decimal acuity is 1.0, the ability to resolve a critical visual feature that subtends 1′

---

[3]The proximal stimulus refers to the stimulus image actually cast on the retina of the human observer's eye. We know that $\frac{1}{3}$ mm on the retina of the average adult male corresponds to approximately 1° of visual angle. Therefore, we can characterize the extent of the proximal stimulus in terms of visual angle (in units of degrees, minutes, and seconds of arc).

of arc. (This corresponds to a Snellen visual acuity of 20/20.) Characteristics of the visual display, however, often are treated as a distal stimulus, where physical measurements are made on the visual display itself. For example, pixel size on an HMD's CRT could be specified in fractions of a centimeter, independent of the human who observes the display. The relationship between distal and proximal stimuli is captured in the following equation:

$$S_\mathrm{p} = 57.3 \cdot 2 \arctan\left(\frac{S_\mathrm{d}}{2D}\right) \tag{8.2}$$

where S_p is the proximal stimulus on the retina of the human observer, S_d is the size of the distal stimulus presented on the visual display (e.g., in centimeters), D is the distance of the visual display from the nodal point of the observer's eye (specified in the same units as S_d), and 57.3 is a conversion factor to convert radians into degrees of visual angle. Thus, the size and spatial resolution of the proximal stimulus that the human perceives depends on both the size and spatial resolution of the distal, physical visual display as well as the viewing distance. The pixels of a display used in an immersive HMD always are viewed from a close distance. Therefore, the HMD will produce a larger proximal stimulus with much coarser spatial resolution than if that same visual display were viewed from a greater distance. For example, an active matrix LCD may have as many as 66 elements per cm (see Table 8.2). At a distance of 57.3 cm, each element of this display will subtend less than 1' of arc. But at a closer viewing distance of 5.73 cm, each element

TABLE 8.2   Typical Spatial Resolution and Display Size for Some Common Displays

Type of display	Spatial resolution (in elements per cm)	Display size	
		(in pixels)	(in mm)
Electroluminescent	8 to 20 (Research on 197 elements/cm)	240 × 320 485 × 645	152.4 25.4 × 25.4
LED	12 to 20	200 to 300	152.4*
a.c. plasma	12 to 24	1024 × 1024	430 × 430
d.c. plasma	12 to 24	Limited to 200 columns	
Active matrix LCD	34 to 66	1024 × 1024	160 × 160 or 300 × 300
CRT	20 lines per cm (color) 79 line pairs per cm (monochrome)	Resolution limited by phosphor granularity and spot size	

* Limited by power dissipation

will subtend about 9' of arc and, because the best human visual acuity is at least 1' of arc, the image may appear grainy, coarse, and blocky.

Optimal visual acuity.   Human visual acuity is best within the central 2° of the visual field at photopic levels of luminance, and for high-contrast objects that are either stationary or moving slowly.[4] Objects viewed within the central 2° of the visual field fall on the fovea of the human retina, which contains only cone photoreceptors.

Cones found in the central fovea are slender and tightly packed in an orderly hexagonal array so that the distance between the center of two adjacent cones is only 2.2 μm or approximately 23" of arc. Thus, humans can detect an object with a fundamental spatial frequency of 60 cyc/deg, the Nyquist limit for the cone mosaic of the central fovea.[5] Yet, the foveal cones also can detect the offset of two high-contrast lines that differ by only 2" of arc (Klein and Levi, 1985). Because this vernier offset is much smaller than the diameter of a single cone, the detection of offset or of vernier acuity is classified as a *hyperacuity.* In terms of the physical display, this means that jaggy lines in the visual display can be detected by the human visual system if the lines are of high contrast, the offsets are 2" of arc or greater, and they are presented within the central 2° of the visual field. (These jaggy lines are caused by spatial aliasing that occurs in the visual display due to the poor spatial resolution of the display.) Immersive HMDs that use wide fields of view (WFOV) are especially susceptible to jaggy and distorted edges in the images, because of the trade-off between FOV

---

[4]Dynamic visual acuity (DVA) usually is worse than static visual acuity and will decrease as a function of the object's velocity. This deterioration occurs whether the motion is caused by movement of the object or by the observer's movement. Thus, the poor spatial resolution for HMD imagery is less perceptible for moving objects and images than it is for stationary ones. The relation of dynamic acuity to target velocity can be described by the following semi-empirical equation:

$$y = a + bx^3$$

where a is the static visual acuity, x is the target velocity in deg/sec, and b is a parametric measure of dynamic visual acuity (Miller and Ludvigh, 1962). Notice that dynamic visual acuity is not significantly correlated with static visual acuity. For example, two observers could have the same value for static visual acuity, but if the value of the b parameter is very different, then their dynamic visual acuities would be very different from each other. Finally, there is an exception to this semi-empirical equation: If a target is only moving at a velocity of 2.5 deg/sec or slower, then dynamic visual acuity will be very similar to static visual acuity (Westheimer and McKee, 1975).

[5]Fortunately, the upper cutoff for the modulation transfer function of the eye's optics also is approximately 60 cyc/deg; the effective contrast of higher spatial frequencies is reduced to zero. This is fortunate because otherwise, spatial aliasing would occur for visual patterns with spatial frequencies higher than 60 cyc/deg. This could be tested using a laser interferometer to bypass most of the eye's optics (Williams, 1985). If lasers are used to paint images directly on the retina, one should consider these potential aliasing effects that can occur within the human visual system.

and spatial resolution. Some of these trade-off problems and potential solutions will be discussed later in more detail.

Effects of mean luminance. Changes in mean luminance result in changes in the light adaptation of the human visual system (Hood and Finkelstein, 1986) and thus affect the ability to detect and recognize objects. Visual acuity is worse at mesopic luminance levels, when both rods and cones function, than at photopic levels. Visual acuity is the worst at scotopic levels, however, when only rods function (e.g., by moonlight it would be difficult to read this chapter or to decipher fine details of objects). In fact, as the ancient astronomers realized, under scotopic luminance the best visual acuity is obtained approximately 4° from the point of fixation (because there are no rods in the fovea).

The effect of mean luminance on visual acuity and on sensitivity to the entire range of visible spatial frequencies is captured in the set of contrast sensitivity functions (CSFs)[6] shown in Fig. 8.1. These CSFs can be used to calculate the minimum contrast that can be used in a display that has a given mean luminance. Notice that at the highest mean luminance shown (300 cd/m^2) the high spatial-frequency cutoff is 60 cyc/deg (corresponding to the best obtainable visual acuity), the overall contrast sensitivity is at its maximum, and the visual system is more sensitive to intermediate spatial frequencies than to lower or higher spatial frequencies (i.e., the CSF is bandpass). As the mean luminance decreases both the high spatial-frequency cutoff and the peak sensitivity shift to lower spatial frequencies, and the CSF becomes more low-pass in character.

Because visual acuity capability is worse at lower average luminance levels, coarser spatial resolution can be tolerated at these lower luminances. Also, more contrast is needed to detect objects at these levels. This is an advantage when using a see-through HMD during nighttime or an immersive HMD with a visual display that does not have a very high ambient mean luminance. For a see-through HMD

---

[6]Because the human visual system is thought to perform a crude Fourier analysis of the visual scene and because by Fourier analysis any complex visual scene can be broken down into sinusoidal gratings of the appropriate spatial frequency (in cyc/deg), orientation, etc., the use of contrast sensitivity functions (CSFs) is a particularly appropriate way to analyze human spatial vision (Graham, 1989). CSFs not only provide an estimate of visual acuity, but also provide additional information about sensitivity to the visible range of spatial frequencies. The high spatial-frequency cutoff requires the maximum contrast for detection and corresponds to the human observer's visual acuity. Moreover, as shown by the bandpass CSFs in Fig. 8.1, human observers often are most sensitive to intermediate spatial frequencies (i.e., they require less contrast to detect them). Although CSFs are used to determine the *least* amount of contrast necessary to detect a grating, contrast matching functions also can be obtained at higher, suprathreshold contrasts. Whereas the CSF often has an inverted "U" shape, the shape of the contrast matching function becomes relatively flat as contrast is increased and mean luminance is held constant.

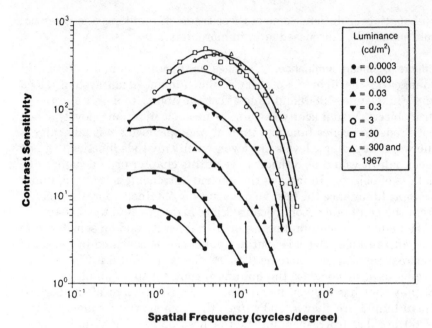

Spatial Frequency (cycles/degree)

Figure 8.1. Contrast sensitivity as a function of spatial frequency for several different levels of mean retinal illuminance (shown in Trolands) for a given observer. The spatial pattern was a sinewave grating (4.5° wide by 8.5° high) with a monochrome light of 525 nm. An artificial pupil with a diameter of 2 mm was used. To convert retinal illuminance to luminance (measured in cd/m²) multiply the retinal illuminance by the area of the artificial pupil. (Source: F. L. Van Nes & M. A. Bouman, (1967). Spatial modulation transfer in the human eye. *Journal of the Optical Society of America, 57.* Reprinted with permission of the Optical Society of America.)

on a bright, sunny day or an immersive HMD with a relatively high ambient, mean luminance, however, the required spatial resolution of the display must be much better to match the visual acuity capabilities of the human visual system.

Effects of visual field location. Visual acuity becomes worse as one moves from the central 2° of the visual field to the periphery. This loss in visual acuity occurs because there are both rod and cone photoreceptors in the periphery and because there is greater spatial pooling of information in the periphery than in the fovea. The decline in photopic visual acuity for eccentricities ±20° from fixation can be estimated with the following equation:[7]

---

[7]At very low scotopic luminance levels (e.g., 0.006 cd/m² and below), visual acuity is approximately constant from 4° out to 30° of eccentricity (i.e., a decimal acuity of 0.1 or less) (Mandelbaum and Sloan, 1947). For higher scotopic luminance levels, however, visual acuity is best at 4° of eccentricity and rapidly declines with increasing eccentricity. This latter function is very different from the one described in the text at photopic luminance levels.

$$MA(E) = \frac{MA(O)}{[1 + (E/E_2)]} \qquad (8.3)$$

where MA(O) is the minimum visual angle that can be resolved at fixation (viz., the center of the visual field), MA(E) is the minimum visual angle that can be resolved at an eccentricity of E degrees, and E_2 is the eccentricity at which the resolution has changed by a factor of two. The value of E_2 depends upon the specific task (e.g., grating resolution or vernier offset), the specific meridian (e.g., along the horizontal or vertical axis through the fovea), and upon individual differences. For example, E_2 has a value between 0.3° and 0.9° for vernier acuity or tasks involving judgments of spatial offset, as in the perception of jaggy lines in HMDs (Levi, Klein, and Aitsebaomo, 1985). Yet E_2 has a value between 1.5° and 4° for grating resolution (Swanson and Wilson, 1985). At eccentricities of greater than 20°, visual acuity deteriorates at a rate faster than predicted by Eq. (8.3).

One consequence of this retinal inhomogeneity is that worse spatial resolution can be tolerated in the periphery than can be tolerated within ±1° of fixation. If one could either monitor eye position within an HMD[8] or ensure that the user only uses *straight-ahead* viewing, then one could use finer spatial resolution tiling within the central 2° and coarser tiling beyond this region. This tiling technique may coerce the user to scan or search the visual scene by using head movements, which are slower than the more natural eye movements.

Effects of age. Because spatial vision capabilities change over the life span of an individual, one should consider the users' ages when designing immersive or see-through HMDs. From age 30, both visual acuity and sensitivity to intermediate and high spatial frequencies decreases as a function of age (Fig 8.2). These age differences are more apparent at lower luminance levels than at higher levels. Some of these losses are due to optical factors, such as clouding of the lens, and can be mostly bypassed with a laser display. Some losses, however, are due to neural factors and cannot be readily overcome. (A happy consequence is that the spatial resolution limitations of a given HMD may be less of a problem for an older user than for a younger user.) Moreover, the lens of the older adult gradually loses its ability to accommodate (i.e., to change shape) so that it cannot produce a sharp image of nearby objects on the retina, resulting in presbyopia. (See Fig. 8.3 for a schematic of losses in accommodation range as a function of age.) Problems caused by presbyopia may be overcome by using collimated HMDs,

---

[8]Mary Hayhoe and Dana Ballard at Rochester University in New York are developing an HMD in which eye movements can be monitored. A prototype HMD is currently in use.

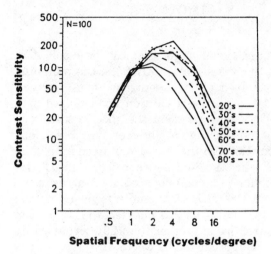

Figure 8.2. Contrast sensitivity as a function of age for 100 observers (from 19 to 87 years old). Stimuli were stationary sinewave gratings with a mean luminance of 103 cd/m². (Source: C. Owsley, R. Sekuler, and P. Siemsen, (1983). Contrast sensitivity throughout adulthood. *Vision Research, 23,* Copyright 1983 by Pergamon Press. Reprint permission pending.)

Accommodation Changes with Age

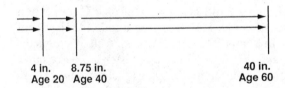

4 in.   8.75 in.   40 in.
Age 20   Age 40   Age 60

Presbyopia -- farsighted because of loss in accommodation with age

Figure 8.3. Changes in the near-point of accommodation as a function of age. At age 20, the typical human can accommodate to an object only 10.16 cm (4 inches) away. By age 40, the near-point of accommodation is 22.23 cm (8.75 inches), and by age 60, it has increased to 101.6 cm (40 inches). Presbyopia is caused by this loss in accommodation of the lens as a function of age, resulting in farsightedness.

so that simulated objects are focused at optical infinity. Moreover, the decreased pupil size, caused either by senile miosis or by the presence of higher luminance levels, also results in greater depth of field and fewer optical distortions, and can partially compensate for some of the losses in accommodation experienced by older people. The decreased pupil size, as well as clouding of the ocular media and lens, also causes a reduction in retinal illuminance (a measure of the amount of light reaching the retina). In sum, vision in older adults suffers from reduced retinal illuminance, reduced visual acuity, and poor accommodation to nearby objects—factors that should be considered in designing HMDs for older users.

Effects of interactivity. An interactive environment can compensate for poor performance due to poor HMD spatial resolution. An interactive environment could be created by an immersive HMD used with head-tracking to create a virtual environment. Smets and Overbreke (1995) found that when the camera's view of a visual scene was yoked to the user's head movements (the active condition) the user could solve a visual puzzle most quickly. But if the camera presented only a single viewpoint (the still condition) or scanned the scene independently of the user's head movements (the passive condition), puzzle-solving performance was significantly slower. These results suggest an ecological approach to the design and use of HMDs (e.g., Gibson, 1979): Performance with an HMD partially depends on interactions between the visual quality of the image (e.g., the spatial resolution) and the user's active interaction with and manipulation of the environment.

8.2.3 Critical flicker fusion (CFF), temporal resolution, and motion

Critical flicker fusion (CFF) is the lowest rate at which the user will perceive a fused image in a visual display that is flickering on and off. The temporal frequency of the CFF is measured in cycles per second (Hz). For temporal frequencies lower than the CFF the display can appear to flicker. The CFF depends on several factors: overall mean luminance, level of the observer's light or dark adaptation, spatial-frequency content of the image, location within the user's visual field, and certain technical considerations (e.g., the phosphor persistence of a CRT). The highest CFF is obtained for foveally viewed images of high mean luminance (photopic levels) presented to a light-adapted eye. In this case, CFF can even be *above* 60 Hz (Travis, 1990).

The user's sensory CFF has implications for the necessary temporal resolution of the display. Both frame rate and update rate affect the temporal resolution of the display. *Frame rate* is a hardware-controlled variable that determines how many images each eye sees per second, whereas *update rate* is the rate at which changes in the image are updated. The update rate depends upon the computational complexity of an image and can be no faster than the frame rate. If the temporal resolution of the visual display is too low, it can hinder task performance or cause illusory motion artifacts. In fact, a low temporal resolution may even cause nausea while using an immersive HMD (with a wide field of view) or a see-through HMD in which the real and virtual images do not temporally mesh together in an appropriate manner (Piantanida, Boman, and Gille, 1995).

Effects of mean luminance. The effects of mean luminance on CFF and on sensitivity to a range of temporal frequency modulations are shown in Fig. 8.4 (deLange, 1958). These functions were obtained by sinusoidally modulating the time-varying luminance of a 2° disk with blurred edges. The functions shown here could provide an estimate of the frame rate and update rate necessary to perceive a fused and smoothly changing visual image. The CFF is the high-frequency cutoff shown in each of these functions; it corresponds to a contrast sensitivity of 0.01 or a contrast amplitude modulation of 100 percent. Notice that as the mean luminance level decreases, the CFF and overall sensitivity for detecting temporal modulation also decrease. These results have implications for HMD design and use. If an immersive HMD has a display with a low overall luminance level or if a see-through HMD is used under dim ambient light conditions, then the temporal resolution of the display need not be as good as in those used with a high mean or ambient luminance. That is, slower frame rates and update rates

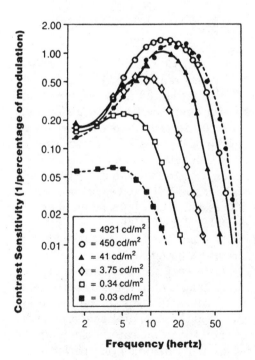

Figure 8.4. Temporal contrast sensitivity as a function of temporal frequency (measured in hertz) for several different levels of background retinal illuminance (shown in Trolands) for a given observer. The temporal modulation was sinusoidal and the target was a 2° disk presented on a large surround. (Source: H. de Lange, (1958). Research into the dynamic nature of the human fovea-cortex systems with intermittent and modulated light. I. Attenuation characteristics with white and colored light. *Journal of the Optical Society of America, 1958, 48.* Reprinted with permission of the Optical Society of America.)

will mesh with the reduced CFF and temporal sensitivity of the human visual system. Conversely, HMDs that have high mean luminances also require much better temporal resolution to match the CFF and temporal sensitivity of the human visual system.

Effects of object size or spatial frequency content. In general, CFF decreases as a function of the spatial frequency content of the visual display (Robson, 1966). Displays that have coarse spatial resolution or large, blocky images (e.g., most immersive WFOV HMDs) may require better temporal resolution than displays with a finer spatial resolution (e.g., HDTV viewed from a distance of a few meters).

Effects of visual field location. CFF also tends to decrease as a function of eccentricity within the visual field (Graham, 1965). This relationship has been obtained at photopic luminance levels where cone function dominates, as well as at dimmer luminance levels (and when the retina is in a dark-adapted state) where rod function dominates (Boff & Lincoln, 1988). An opaque HMD with a narrow FOV and a high overall luminance will predominantly stimulate foveal cones. The display may need a relatively fast frame rate to match the user's CFF under these conditions. Conversely, an immersive WFOV HMD with a dimmer overall luminance may predominantly stimulate rods and cones in the peripheral retina. Under these conditions, a display with a slower frame rate possibly may be sufficient to match the user's CFF.

Effects of phosphor persistence. Perceived flicker in a display can also be influenced by certain technical characteristics of the visual display in addition to the frame rate. For example, phosphor persistence of a CRT can affect the perceived flicker in a display that has a fixed frame rate (Boff and Lincoln, 1988). When several different phosphors were tested (viz., P1, P4, P7, P12, P20, P28, and P31), the P20 phosphor produced the *least* perceived flicker and the P31 produced the *most*. That is, the persistence of the illuminating elements used in an HMD can reduce the necessary frame rate of the display. Unfortunately, the persistence of these illuminating elements also can cause *motion blur* in the changing image if the decay lasts noticeably longer than the duration of a single frame or the update rate.

Effects of update rate on motion perception. The faster the angular speed of moving virtual objects presented in the HMD, the higher the update rate (and frame rate) must be to avoid image shaking or jerky movement (Padmos and Milders, 1992). The maximum displacement per frame that gives an impression of smooth, continuous motion is approximately 15' of arc (Braddick, 1974). The minimum update rate (in frames per second) is the virtual object's angular speed (in arcmin/sec) divided by 15. Thus, slowly moving objects within the visual image can be adequately handled by an update rate of 30 Hz, but very fast-moving objects cannot. The update rate of an HMD will constrain how fast virtual objects can move and also will constrain how quickly the user can move his or her head within an immersive HMD.

Effects of sensor lag in head-tracking. The user's head position is monitored in an immersive HMD and the visual scene is updated according to the head position. In this case, the update rate depends on sensor lag in determining the head position as well as on the computational complexity of the visual imagery. These two factors are additive. The result may be that the change in the visual imagery noticeably lags behind the actual head movement, unlike perception in the real environment. Consequently, this perceptible time lag could give rise to the perception of jerky motion and to illusory motion perception (objects seem to be moving when they are supposed to be stationary). These perceptible time lags also produce conflict between the vestibular and proprioceptive systems, which signal a head movement, and the visual system, which has not yet registered a corresponding change in the visual image.

8.2.4 Field of view (FOV)

No existing HMD achieves the wide field of view of the human visual system operating in a real environment. One of the most pressing challenges now facing designers and developers of HMDs is to *simultaneously* provide the user with a wide field of view *and* good spatial resolution. The trade-off between the display's FOV and spatial resolution is a crucial issue. A related issue for the design and development of HMDs is how wide the FOV must be to competently perform tasks with either an immersive or a see-through HMD. I will consider both of these issues.

The instantaneous field of view is defined as the sensor's field of view *without* any movement. For humans, the instantaneous monocular FOV is about 160° in the horizontal direction and about 120° in the vertical direction. The FOV is wider on the temporal side (about 100°) than it is on the nasal side (about 60°) because the nose blocks part of the FOV. The instantaneous binocular field of view for humans is almost 200° of visual angle in the horizontal direction. Although both horizontal and vertical FOVs matter, the horizontal FOV is often emphasized because it is considered more important.

Trade-offs between FOV and spatial resolution. The horizontal FOV of current HMDs ranges from 22.5° for General Reality's HMDs to 155°

for Kaiser Electro-Optics' full-immersion HMD-1 (Anonymous, 1995). Not only do these HMDs have more narrow horizontal and vertical FOVs than the human FOV, but their spatial resolutions are poor relative to human visual acuity. That is, the pixel's angular resolution is significantly worse than 1' of arc (where 1' corresponds to a Snellen visual acuity of 20/20). To achieve the WFOVs of immersive HMDs there is a corresponding perceptible decrease in the spatial resolution of the display and/or a noticeable increase in the weight and complexity of the HMD. Both of these factors affect comfort and performance. In order to achieve a wider FOV with a fixed number of pixels, the pixel size is magnified, but the resulting spatial resolution may decline to worse than 20/200 VA (the criterion for legal blindness). The decrease in spatial resolution caused by magnification is especially problematic for color displays, because a *color pixel* usually consists of an RGB triad or a triplet of pixels. Thus, monochrome and gray-scale displays, with their simple, single pixels, may provide better spatial resolution when magnified to create a WFOV HMD.

Some other techniques to increase the FOV without a significant decrease in spatial resolution include the following:

1. Tile multiple displays (e.g., multiple LCDs, as Kaiser Electro-Optics has done)

2. Mixed resolution to provide a higher spatial resolution within the central 2° to 5° of vision (assuming *line-of-sight* viewing) and a lower spatial resolution outside this region (e.g., CAE Electronics' high-priced fiber-optic HMD)

3. Binocular HMDs with only partial overlap

4. Design of a different and better display element

Each of these approaches still has technological hurdles to overcome and each can significantly increase the complexity, cost, and weight of the HMD. Yet we want to minimize the amount of weight carried by the user's head and to increase the comfort of using an HMD. Minimizing the HMD's weight and increasing its comfort are especially important ergonomic considerations for long-term use of HMDs, although less important in situations where a user will don the HMD for only a short period of time. Of course, one could maintain good spatial resolution and minimize the HMD's weight by limiting the FOV. There are other perception and performance issues involved with this alternative.

The effect of FOV on task performance with HMDs. Two related questions about the necessary size of FOVs are (1) What is the minimum FOV necessary for acceptable performance? and (2) What effects do smaller FOVs have on perception and performance? Too small a field of view increases the number of head movements the user must make to determine where things are located and interferes with situation awareness. Moreover, peripheral vision can help in ego-orientation, locomotion, and reaching performance (Dichgans, 1977, cited in Alfano and Michel, 1990). A WFOV display can produce better orientation within the environment and a stronger sense of self-motion (Padmos and Milders, 1992). Bodily discomfort, dizziness, unsteadiness, and disorientation are reported for subjects moving around with a restricted FOV, although WFOVs can increase simulator sickness (Padmos and Milders, 1992). Peripheral vision also helps the user to establish visual position constancy and to understand events that occur over a panoramic visual field (Wallach and Bacon, 1976).

The minimum acceptable FOV depends on the task to be performed and on the field size available in the real environmental situation. The latter may be especially true with a head-up display (HUD) or see-through HMD display in which the display's virtual imagery or symbology is superimposed on the real environment. In this case, we need to mesh the intradisplay FOV (within the HMD) and the paradisplay FOV (outside the HMD). For applications involving symbology and alphanumerics, good foveal resolution is needed and the minimum intradisplay FOV for the HUD or see-through HMD is 15° to 30° (Wells and Haas, 1992). Often the intradisplay FOV is from 20° to 60°, but the paradisplay FOV is even wider because the user can "see around" the device (Rolland, Holloway, and Fuchs, 1994). Although parts of the HMD device may obscure part of the view in such open-design, see-through HMDs, the paradisplay FOV is almost as wide as the human's natural, instantaneous binocular FOV. When symbology or alphanumerics are superimposed on the real-world image, the larger paradisplay FOV causes no problem for the smaller intradisplay FOV (and vice versa).

For opaque, immersive HMDs with imagery, it is often the case that the larger the FOV, the better the performance. Some examples of better performance with larger FOVs are dive bomb delivery, orientation within the environment, and movement through the virtual environment (AGARD, 1987, cited in Padmos and Milders, 1992; Foley, 1987; Osgood and Wells, 1991; Proffitt and Kaiser, 1986). Alfano and Michel (1990) studied the role of peripheral vision in competent performance of visuomotor activities such as walking, reaching, and forming a cognitive map of a room while wearing goggles that restricted the field of view to 9°, 14°, 22°, or 60°. Each restricted FOV resulted in some perceptual and performance deficits, with the 9° and 14° FOVs producing the greatest deficits. Piantanida, Boman, Larimer, Gille, and Reed (1992) also found an inverse relation between field size (i.e., 28°, 41°, 53°, and 100° horizontal FOV) and performance. Search time for locating a target (either among distracters or without distracters) was longer with the smaller FOVs. Moreover, Sandor and Leger (1991) reported that tracking task performance was moderately impaired with a restricted FOV of either 20° or 70° in the horizontal direction, probably because visuomotor tracking requires head stability.

These studies suggest that a horizontal FOV much wider than 60° or 70° would be better for using opaque HMDs in tasks that involve spatial orientation, navigating through the environment, and manipulating objects. Yet Padmos and Milders (1992) report that a 50° horizontal by 40° vertical FOV is adequate for performing takeoffs and landings in flight simulators or for lane-changing, merging, and turning in driving simulators. The optimal or minimal FOV for opaque HMDs with imagery remains an unresolved issue.

Eye, head, and body movements can increase the effective FOV of an HMD. Normally such movements operate synergistically to scan a scene. However, small FOVs force the user to limit the amount of eye movement[9] and to make more head and body movements to view the environment. These head and body movements require more time and are less efficient than saccadic eye movements. They also require more updates of the virtual imagery, with resulting potential time-lag problems (see Sec. 8.2.3). For more efficient scanning of a visual scene with a restricted FOV, one possibility is to use auditory spatial localization (e.g., Wenzel, 1991; Wenzel, Wightman, and Foster, 1988). In real-world scenarios, the auditory system can alert the visual system where to look next; audition can do the same within the virtual environment of an HMD.

8.2.5 Binocular HMDs versus monocular or biocular HMDs

In a binocular HMD a slightly different image is presented to the right eye from that presented to the left eye. To achieve wider FOVs designers of HMDs may use partially overlapping binocular displays, so that binocular vision is available within the central overlapping region but only monocular vision is available in the periphery. This type of display has both advantages and disadvantages compared to binocular HMDs with complete overlap. In general, binocular HMDs also have advantages and disadvantages compared to monocular or

---

[9]In unrestricted, real-world environments the oculomotor range of eye movements is ±55° in humans (Sandor and Leger, 1991). But such eye movements within an HMD display can cause problems. Most HMDs are designed for *line-of-sight* or *optical axis* eye viewing and are not designed to monitor eye position within the head. Moreover, eye movements of ±55° within an HMD would be very uncomfortable and should not be required of HMD users; instead a ±15° range of eye movements is more reasonable. Researchers at Rochester University, however, are currently using an HMD prototype that can monitor eye position.

biocular HMDs. In a monocular HMD the image is presented to one eye and the other eye views the real environment. In a biocular HMD both eyes view the *same* image. The advantages and disadvantages of these different HMDs are discussed in the following sections, as are some characteristics of human stereopsis relevant to the design and use of binocular HMDs. I will also consider some tasks or situations for which these different HMDs are well suited and some for which they are not.

Stereopsis and fusion. For most humans, the region of binocular overlap where stereopsis can occur is 120°. A monocular visual field of approximately 35° flanks the region of binocular overlap on each side. Stereopsis results from fusion of the two slightly different views that our laterally displaced eyes receive in viewing a real or simulated 3D environment (Arditi, 1986; Davis and Hodges, 1995; Schor, 1987; Tyler, 1983). It can provide an enhanced perception of depth and three-dimensionality for a visual scene, as compared to that provided by monoscopic views (Yeh and Silverstein, 1992).[10] Stereopsis can provide very fine depth information within a meter of the observer (Surdick et al., 1994), as well as relative distance information for an object located at a much greater distance. For example, stereopsis allows us to distinguish an object one mile away from another object at optical infinity (Boff and Lincoln, 1988; Tyler, 1983; Wickens, 1990).

Retinal disparity and interpupillary distance. Retinal disparity underlies the perception of stereoscopic depth among objects in 3D space. Retinal disparity occurs when the retinal image of an object does not fall on corresponding points in the two eyes.[11] (see Fig. 8.5). A negative retinal disparity causes an object to appear closer than the point of fixation, whereas a positive retinal disparity causes it to appear farther away. The greater the retinal disparity between two objects, the greater the perceived depth or distance between them. Retinal disparity depends on the real (or simulated) distances of the objects from the observer as well as on the interpupillary distance, as shown in the following equation:

$$\delta = \frac{id}{D_1(D_1 + d)} \tag{8.4}$$

---

[10]Note that many monocular cues to depth exist, such as occlusion of objects, texture gradients, and absolute motion parallax.

[11]Corresponding points of the two retinae are defined as having the same vertical and horizontal distance from the center of each fovea (e.g., Arditi, 1986; Davis and Hodges, 1995; Tyler, 1983). For example, if an observer fixates an object in the distance, the image will fall in the center of the fovea for each eye (i.e., on the corresponding points).

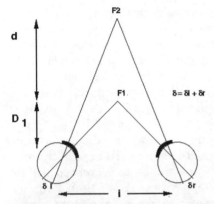

Figure 8.5. Two eyes with an interpupillary distance of i are fixated on point F_1, at a distance D_1 from the interocular axis. In this example, the fixated object at point F_1 would cast an image on corresponding points of the retinae (the center of the fovea for this example). For an object at point F_2, however, at a distance of $(D_1 + d)$ from the interocular axis, the image does not fall on corresponding points of the two retinae. Instead, the retinal image of F_2 in the left eye lies at a different position and distance from the center of the fovea than does the retinal image of F_2 in the right eye. This difference creates a retinal disparity of δ, which is the sum of the retinal disparities in the left and right eyes (i.e., $\delta = \delta_l + \delta_r$). (See the text for another equation to estimate retinal disparity, δ.) Disparities measured from the center of fovea outward are negative and, conversely, those measured from the center of the fovea inward are positive. In this example, δ has a negative retinal disparity; this corresponds to an *uncrossed* retinal disparity and indicates that the object at F_2 is farther away than the point of fixation at F_1. If δ had a positive disparity, this would correspond to a *crossed* retinal disparity and would indicate that the object is closer than the point of fixation at F_1.

where δ is retinal disparity (in radians), i is the interpupillary distance (IPD), D_1 is the distance of the fixated object from the observer, and d is the depth or distance between the two objects (so that $D_1 + d$ is the distance of the second object from the observer). In a computer graphics display, retinal disparity is created from the binocular parallax of objects in the image.

The average IPD is approximately 63 mm, with a range of 53 mm to 73 mm (Kalawsky, 1993). The larger the IPD, the larger the retinal disparity that results from viewing two objects at different depth planes (Eq. 8.4). In general, the optical design of a binocular or biocular HMD should have an adjustment for the user's IPD. The optics of the binocular (or biocular) HMD could be adjusted mechanically to align with the pupil of each eye. Alternatively, if the exit pupil is big enough, the binocular parallax of the computer graphics or the offset of the two sensors can be adjusted for the user's IPD. (Note that in an immersive HMD one could enhance stereoscopic depth perception by increasing the *virtual* IPD of the user. Thus, small depths or distances among objects should appear much more salient.)

Sensory fusion and diplopia. If the images in the two eyes are similar, each binocularly viewed object may appear fused into the percept of a single object. *Fine stereopsis* corresponds to the range of retinal disparities over which the percept of an object appears fused. This range is known as *Panum's fusion area* (e.g., Davis and Hodges, 1995). It is ellipsoidal in shape so that more horizontal disparity (e.g., 10' to 20' at fixation) than vertical disparity (e.g., 2.5' to 3.5' at fixation) can be tolerated. Panum's fusion area is affected by a number of display characteristics so that it increases as the display's luminance is decreased; objects are presented at more peripheral locations; and the spatial frequency content of the image is decreased (below 2.5 cyc/deg). If the binocular retinal disparity is larger than Panum's area, one perceives a double image (diplopia) instead of a fused, single image. Although some judgments of stereoscopic depth are still possible with diplopic images, diplopia can cause feelings of nausea, eye strain, and performance deficits while using HMDs, especially if the HMD is used for a long period of time (Kalawsky, 1993; Piantanida, Boman, and Gille, 1995).

Binocular rivalry, monocular suppression, and binocular luster.   If the left and right monocular images fall on corresponding points of the two retinae but are very different in spatial characteristics, then binocular rivalry may result. That is, instead of perceiving both monocular images simultaneously, the monocular images may be perceived in alteration (i.e., the right monocular image is suppressed while the left monocular image is perceived, and vice versa). Increased contrast of one of these monocular images will increase the prevalence of the corresponding eye over the fellow eye (Blake, 1977; Levelt, 1968). Transient, alternating suppression can become continuous suppression of one monocular image by the other if one eye dominates the other. Suppression often occurs when the spatial stimulus presented to one eye is sufficiently different in orientation, length, or thickness from that presented to the other eye (Levelt, 1968). Binocular luster often occurs, however, when an area has a uniform luminance and the luminance for one eye is different from that for the fellow eye (e.g., of opposite polarity). In that case, a lustrous or shimmering surface of indeterminate depth is perceived.

Binocular HMDs with partial versus complete overlap.   Using a binocular display with only partial binocular overlap can result in a wider FOV while maintaining an adequate spatial resolution and reduced weight and size, as compared to a binocular HMD with complete binocular overlap and the same FOV (Melzer and Moffitt, 1989; Melzer and Moffitt, 1991). HMDs with only partial binocular overlap have flanking monocular regions of the display, resulting in very different spatial and luminance configurations for the right and left eyes. The region of partial binocular overlap also is considerably smaller than the 120° of binocular overlap for natural human stereoscopic vision. Both of these characteristics cause problems for the HMDs with only partial binocular overlap as compared to HMDs with complete binocular overlap. That is, binocular rivalry and monocular suppression can occur within the flanking monocular regions of the display. Moving or temporally modulating image contours are particularly effective in producing binocular rivalry and monocular suppression (Tyler, 1983).

If the "wrong" eye is suppressed in a binocular HMD with partial overlap, a moonlike crescent shape is perceived (*luning*) in the flanking monocular region (e.g., Melzer and Moffitt, 1989; 1991). This luning can result in fragmentation of the visual field—the FOV appears to be fragmented into three distinct regions. To overcome the luning effect Melzer and Moffitt suggest either creating dark contour lines to separate monocular and binocular regions of the image or blurring the edges between these regions. Blurring may result in less fragmentation of the visual field. Moreover, Melzer and Moffit have reported that *convergent* binocular HMDs result in less luning than do divergent binocular HMDs. This finding has been replicated and validated by others (e.g., Klymenko and Rash, 1995). (In a convergent HMD with partial binocular overlap, the right monocular image is presented to the left eye, and vice versa. In a divergent binocular HMD the right monocular image is presented to the right eye and the left monocular image to the left eye.) Klymenko also reports that if the area of binocular overlap is made larger, there are fewer problems with fragmentation across the FOV.

As a rule of thumb, Kalawsky (1993) recommends using complete binocular overlap when moderate FOVs suffice, but partial overlap only if very wide FOVs are necessary.

Advantages and disadvantages of monocular, biocular, and binocular HMDs. Because a monocular HMD has only one image source and one set of optics, it is lighter, cheaper, and simpler than either the biocular or binocular HMDs. In a monocular, see-through HMD the paradisplay FOV of the unaided eye is not restricted. But in a monocular HMD, the view seen by the unaided eye may produce binocular rivalry with the image seen through the HMD; there is no stereopsis; and the intradisplay FOV is reduced compared to that achievable with either a biocular or a binocular display. For example, the monocular HMD used by the U.S. Army's AH-4 helicopter has one output from a head-steered FLIR viewed by the right eye and another output from the cockpit instruments viewed by the left eye (Wells and Haas, 1992). Some users of this HMD have difficulty switching attention from one eye to the other.

In a biocular HMD there still is only one image source, but there are two sets of optics. So, the biocular HMD is heavier than the monocular one and both eyes must be properly aligned to the system. But once the eyes are properly aligned within the biocular HMD, they both view the same image. Accordingly, the potential for binocular rivalry is reduced and monocular suppression is effectively eliminated with a biocular HMD, as compared to a monocular HMD.

Only the binocular HMDs allow stereoscopic viewing that provides 3D depth perception. Although retinal disparity allows us to perceive depth stereoscopically, sometimes distortions in binocular (*or* biocular) HMDs can cause unwanted retinal disparities that lead to misperceptions or perceptual problems. Distortions in the optical design that produce retinal disparities include (1) a magnification difference between the left and right images; (2) rotation of the left and right optics relative to each other (causing trapezoidal distortion, which is especially problematic for HMDs with only partial binocular overlap); and (3) misalignment of the left and right optic axes with respect to each other. Moreover, binocular HMDs are the heaviest, the most complex, and the costliest. They also require accurate computation of the graphics (or placement of the two sensors), careful calibration of the system's optics, and proper alignment of both eyes to avoid problems such as binocular rivalry and monocular suppression (e.g., Davis and Hodges, 1995; Hodges and Davis, 1993). Finally, not everyone possesses good stereopsis. There are amblyopes who have no stereopsis as well as some individuals who are stereoanomalous. Richards (1971) has shown that stereoanomalous people may be able to perceive stereoscopic depth for objects presented in front of the fixation point, but not behind it (or vice versa). Even some individuals with normal stereopsis may have difficulty perceiving stereoscopic depth in a visual display, but with training and feedback their stereoscopic vision can improve dramatically (e.g., Surdick, Davis, King, and Hodges, in press).

Usefulness of binocular, biocular, and monocular HMDs. So many problems have been reported with monocular HMDs that it is advisable to use either biocular or binocular HMDs instead (Blackwood et al., 1995). But in what situations and for what sorts of tasks are binocular HMDs more useful than biocular HMDs? Binocular HMDs are useful when

- A visual scene is presented in a perspective view rather than in a bird's eye view (Barfield and Rosenberg, 1992; Yeh and Silverstein, 1992).

- Monocular cues provided by a biocular display provide ambiguous or less effective information than the stereopsis provided by binocular HMDs (Yeh and Silverstein, 1992).

- Static displays are used rather than dynamic displays (Wickens, 1990; Wickens and Todd, 1990; Yeh and Silverstein, 1992).

- Ambiguous objects or complex scenes are presented (Cole, Merritt, and Lester, 1990; Spain and Holzhausen, 1991).

- Complex 3D manipulation tasks require ballistic movements or very accurate placement and positioning of tools.

Stereopsis is helpful in these situations for two primary reasons. First, it helps disambiguate elevation and distance information in providing information about the spatial layout of objects (e.g., in a perspective view). Second, it provides the user with fine depth discrimination for objects (and the shapes of objects) located within an arm's length of the user. Absolute motion parallax can provide information very similar to that provided by stereopsis. But the observer's head must move to provide depth information from absolute motion parallax, as in navigating through the environment. In some circumstances, however, it is better for the user's head to remain fixed and steady. For example, in performing a delicate surgical operation using a see-through HMD, the surgeon's hands are more likely to be steady if his or her head is not bobbing around to gain depth information provided by motion parallax.

8.2.6 Color versus monochrome

To use color or not to use color in designing HMD displays? That is the question. There are arguments for and against the use of color displays for HMDs, based on technological limitations, cost-effectiveness, and what use the HMD will have. I will consider some arguments on each side as well as some basic characteristics of human color vision and their impact on the design and use of HMDs with color displays. For a more complete review of human color vision (including issues such as opponent-process theory and C.I.E. chromaticity diagrams) see Boynton (1979), Lennie and D'Zmura (1988), and Pokorny and Smith (1986).

Some basic terminology and characteristics of human color vision. *Brightness* is an achromatic dimension. The perceived brightness of an object is directly related to the intensity or luminance of the light reflected or transmitted from that object. *Chromaticity* of an object depends on the wavelengths and purity of light reflected or transmitted from that object. The perceived *hue* depends on the wavelength of light (400 to 700 nm), whereas the *saturation* depends on how pure that light is. For example, the monochromatic light of a ruby-red laser would appear to be a very saturated red hue, whereas sunlight is a mixture of many different wavelengths and appears to be desaturated and white. In general, mixtures of different wavelengths of light result in a less saturated hue than that of a single wavelength. For observers with normal color vision there is an exception to this rule: Within the range of 550 nm to 700 nm a mixture of wavelengths can appear as saturated as the matching hue of only a single wavelength. All mixtures of lights are additive, in that combining the red, green, and blue primaries in appropriate amounts will result in white light. In principle we can independently manipulate the perceived hue, saturation, and brightness of a color, but in practice any changes along one dimension may affect another dimension. An example is the well-known Bezold-Brücke effect (e.g., Boynton, 1979), in which the perceived hue of a given wavelength can change as a function of intensity. For a dominant wavelength greater than 510 nm, increasing the intensity will cause a shift toward yellow; for one less than 510 nm, increasing the intensity will cause a shift toward blue. Another example is the Abney effect: The addition of white light not only causes desaturation, but also causes a change in perceived hue (Boynton, 1971).

Color perception arises from the operation of three different types of cone photoreceptors in the retina. The cones operate over higher luminance levels (from mesopic to photopic) and are less numerous than the rods (6 million cones per retina versus 120 million rods). Although cones are found throughout the retina, they are more densely concentrated in the fovea, where visual acuity is best. Each cone photoreceptor has a different absorption spectrum that partially overlaps those of the other two types of cones. The envelope of these three spectral sensitivity functions forms the photopic spectral sensitivity curve, whose peak sensitivity is at 555 nm. (The scotopic spectral sensitivity curve peaks at 505 nm and has lower sensitivity to the long-wavelength light or reddish hues.) Because humans have three different kinds of cones, they can metamerically match *any* perceived hue by mixing and adjusting the intensity of only three different wavelengths of light. That is, two physical objects could have *very* different spectral compositions, but once a metameric match is made, the hue, saturation, and brightness of the two objects appear to be identical. Thus, a CRT color display uses only three different phosphors (red, green, and blue) to create a palette of colors.

Arguments against the use of color displays.   Because a *color* pixel consists of at least three pixels in an RGB system, color displays usually have worse spatial resolution than do their monochrome or grayscale counterparts. (An exception is a field-sequential system, in which a monochrome display is combined with rapidly changing chromatic filters to achieve a full-color image. In this case, however, the trade-off may be with temporal resolution or frame rate rather than with spatial resolution of the color display.) Moreover, no existing color display system can produce and display the full range of chromaticities (hues and saturations) that the human visual system can perceive in the real world. (For a technical discussion of why the RGB displays fail to do this, see Boynton's (1979) discussion of imaginary primaries and the C.I.E. chromaticity diagram.)

Color displays also have less contrast, are heavier, and are more expensive than monochromatic or grayscale displays. Although human color discrimination is best between 102 and 104 cd/m^2, the CRTs and LCDs used in HMDs have much lower luminances than this. (A minimum luminance of 20 cd/m^2 for simulated images can be satisfactory for color discrimination tasks [AGARD, 1981]). Also, human color deficiencies can limit the usefulness of a color display. Approximately 8 percent of the male population (and 0.4 percent of the female population) have an *inherited* color deficiency (Pokorny and Smith, 1986) that causes some confusion about different hues (especially in the green to red region of the spectrum, corresponding to wavelengths between 550 nm and 700 nm). Moreover, both males and females will *acquire* color deficiencies as a result of normal aging processes or diseases (e.g., diabetes). Most of these acquired color deficiencies are in the blue region of the spectrum.[12] Finally, *all* users are colorblind at scotopic levels (e.g., nighttime conditions) and have poor color discrimination at low mesopic luminance levels.

For these reasons some designers of HMDs eschew color displays and, instead, vigorously pursue the development of a monochromatic system. In designing a monochromatic display a green phosphor frequently is chosen to maximize luminance. This type of phosphor is a good choice perceptually because the human photopic system is most sensitive to light in the green region of hues. HMD designers also sometimes choose to use red LEDs because they are very efficient. Visual acuity partially depends upon the retinal cone mosaic, and in the fovea, where visual acuity is best, there are twice as many red-sensitive cones as there are green-sensitive cones. Thus, the use of red LEDs is a good choice from a perceptual point of view as well as from a technical one.

---

[12]A color deficiency in the blue region of the spectrum occurs if small object images (less than 15′ of arc) are presented for a very short duration (e.g., 50 to 200 msec). This small-field tritanopia occurs both because the blue-sensitive cones are absent from the central 1° of the fovea and very sparsely distributed throughout the rest of the retina and because these cones are very sluggish in their response.

The human achromatic system also has good visual acuity, because all three types of cones feed into the system. Thus, designers of HMDs can effectively use black-and-white or grayscale displays. HMD designers should and do avoid using a blue monochrome display (especially if good visual acuity or the ability to discern fine spatial detail is required). There are no blue-sensitive cones in the very center of the fovea (Williams, MacLeod, and Hayhoe, 1981a; 1981b). Also, because blue-sensitive cones are the least numerous and the most sparsely distributed type of cone, they result in very poor visual acuity compared to that provided by the red-sensitive or green-sensitive cones. Again there is a happy coincidence between the available technology and human perceptual capabilities, because currently there are no efficient blue phosphors or LEDs available.

Some arguments for the use of HMDs with color displays. Users often prefer color displays, and color-coding in displays can sometimes be very helpful (Christ, 1975). For example, imagine playing *Doom* with a monochrome HMD. Even a WFOV immersive HMD does not draw the player into the virtual world as readily as does the addition of color. Also, in environments where both the rate and the amount of information transmission is high (e.g., HUDs or HMDs used in aviation) the use of color-coding schemes can be very appealing (Stokes, Wickens, and Kite, 1990). For example, some stable or neutral point of operation could be coded by a very desaturated white light, whereas a critical or unstable condition could be coded by a very saturated colored light (e.g., red). Why is the use of color so effective in these HMD displays? Some reasons include:

1. Memory for color is better than for shape (Cavanaugh, 1972).

2. Color can be used to unify or cluster disparate elements of a display (Christ, 1975; Christ and Corso, 1982).

3. Color is processed earlier and faster than shape information so that in many situations color seems to be processed automatically.

4. Objects are more readily identified based on color rather than on size, shape, or brightness.

5. Color-coding can significantly reduce search time.

6. Color has been used successfully in redundant coding of information.

7. Color adds to the "realism" of the display.

In designing a display with symbology to convey information (e.g., a HUD or see-through HMD used in the cockpit of a plane) rather than to simulate a real-world view, one wants to select a minimum set of colors that maximizes the visual utility and information transfer capabilities of the display (Silverstein, Lepkowski, Carter, and Carter, 1986). Visual search tasks are easier if the number of colors is limited (e.g., six to nine, according to Smallman and Boynton (1990)). Thorell (1983) recommends using red, green, blue, and yellow to represent important information because these hues are from widely separated and opposing points of the spectrum as well as being easy to learn and to remember. Yet Smallman and Boynton point out that basic colors segregate well in these tasks, not because they are universally named, but because they are well separated in color space. If hue differences and/or luminance differences are larger than some critical value, search time is approximately constant and is not affected by the number of distracters (Nagy and Sanchez, 1990; 1992). For small stimuli the critical difference has a larger value than for larger stimuli (Carter and Carter, 1988; Nagy, Sanchez, and Hughes, 1990).

In designing a display with imagery (e.g., an immersive or opaque HMD), the palette should be selected to provide the best coverage for the simulated scene. Stokes, Wickens, and Kite (1990) report that humans can recognize about nine distinct hues on an absolute basis. When hue, luminance, and saturation are all varied, however, they can discriminate many more colors (e.g., 24 or more). In this vein, Padmos and Milders (1992) suggest that displaying approximately 30 different chromaticities (combinations of hue and saturation) at each luminance level should be sufficient for performing simple tasks with this imagery. For more complicated, realistic scenes, they suggest that many more chromaticities (e.g., 300) may be needed at each luminance level. Given Stokes and colleagues' suggestions as to the perceptual capabilities of the human visual system, however, this latter suggestion seems like overkill.

Finally, whether one is designing HMD displays for symbology or for imagery, neither color boundaries nor fine detail should be depicted with deep blues or violets. Instead, boundaries and fine spatial details should be depicted in colors that will activate the red–green opponent process (Boynton, 1979; Thorell, 1985) or the achromatic system.

8.3 Perception and Performance Issues

I have now considered the question *What can we see?* in terms of the limitations and capabilities of both the human visual system and the visual displays of immersive and see-through HMDs. I have also considered some perception and performance issues of *What do we need to see?* in terms of brightness, contrast, visual acuity and spatial resolution, CFF and temporal resolution, FOV and the FOV–spatial resolution trade-off, monocular versus biocular versus binocular views, and color versus monochrome. I will now further consider *what we need to see* and why we need to see it by examining several different tasks in which the use of HMDs may be beneficial. Specifically, I will look at the use of various HMDs (1) to perform surgery, (2) to inform foot soldiers about the environment and current situation in which they are located, (3) to facilitate flight, (4) for the teleoperation of robots navigating in dangerous environments, and (5) to provide immersive entertainment.

I will consider several criteria to determine the usefulness of HMDs in performing these tasks. First, in either a real-world environment or in one created by a simulated display, some basic functions of visual perception are to detect the presence of objects, to identify or recognize those objects, and to determine the spatial layout of objects in the environment, including how far away the objects are and their spatial interrelations. Second, some tasks will emphasize immediate performance benefits, such as maneuvering a teleoperated robot through a maze, navigating through a terrain, or landing an aircraft. Other tasks will emphasize long-term comprehension, such as learning the relative locations of features or objects in 3D space for the control of air traffic, or the study of molecular structure. Third, we need to consider issues of fidelity versus the technical limitations of the HMDs. That is, we need to consider what perceptual information is crucial for performing the specific task as well as what simplifications can be made in the HMD's visual display. Related to these issues of fidelity and technical limitations, we need to consider how one effectively meshes the simulated displays of a see-through HMD with the real environment. Here, issues related to the perceived distances of simulated and real objects and their spatial interrelations make a difference. Finally, we need to consider the importance of field studies, in conjunction with relevant laboratory studies, to investigate the effects of environmental stressors and work load on task performance with HMDs.

8.3.1 What is it and where is it?

Two basic functions of the human visual system are determining *what* something is and *where* it is. Recent neuroanatomical and physiological studies suggest that there are two distinct pathways in the visual system for these two functions, namely the *parvocellular* and the *magnocellular* pathways (DeYoe and VanEssen, 1988; Livingstone and Hubel, 1988).[13] The *parvo* pathway conveys information about what

---

[13]Some earlier research on visual perception emphasized two functionally different systems, one for processing what something is and the other for where it is located (Leibowitz and Post, 1982). The first of these is a focal component operating in the central 2° of the visual field; the corresponding retinal image would fall on the fovea, where visual acuity and hue perception are optimal. This focal system is concerned with processing shape, form and color—all necessary information for determining what an object is. The second system is an ambient component that involves input from the peripheral visual

something is and the *magno* pathway conveys information about where something is. The parvo pathway conveys information about the shapes and colors of things. For example, using both shape and color information would be useful in identifying an object and distinguishing it from other objects in the environment. The magno pathway conveys information about motion, stereopsis, and depth. Using various sorts of information such as optic flow, absolute motion parallax, stereopsis, and pictorial cues would be useful in determining where an object is located and how far away it is.

Of these visual depth cues, both pictorial cues and motion parallax (or motion gradient information) could be used in either a monocular or a biocular HMD. All of these depth cues, including stereopsis, could be available in a binocular HMD. The pictorial depth cues include occlusion, relative brightness, atmospheric haze, relative size, relative height, linear perspective, foreshortening, and texture gradient. Some of them are more effective than others. For example, Surdick (Surdick et al., 1994; 1997) found that relative brightness is the least effective of these depth cues. Occlusion only provides signed depth information (i.e., one object is in front of another), but not how far apart the objects are. Texture gradient and other linear perspective cues are among the most effective and accurate, provided that symbol enhancers are used to anchor objects to the ground (e.g., Kim, Ellis, Tyler, Hannaford, and Stark, 1987). See Sec. 8.3.2 for a discussion of the ambiguity between elevation and distance in a perspective display.

The effectiveness of these depth cues varies with distance, as well. For example, stereopsis is less effective in providing fine depth information at distances beyond 10 meters, although it can still be used to distinguish an object located one mile away from one located at optical infinity (e.g., Boff and Lincoln, 1988). Surdick and his colleagues have found that the effectiveness of stereopsis significantly decreases even at a distance of two meters, compared to its effectiveness at one meter. Stereopsis is just as effective as texture gradient (with symbol enhancers) at a distance of one meter, but much less effective than texture gradient at two meters.

Both stereopsis (provided by binocular parallax in the visual display) and motion parallax can provide similar information about distance and 3D structure of objects (e.g., Rogers and Graham, 1982). Both can

---

field as well as from other senses (e.g., vestibular and auditory senses). This ambient system is concerned with position, spatial orientation, and relative motion—all necessary information for determining where an object is located. Under some circumstances or tasks, these two components may be dissociated. For example, while driving a vehicle the recognition of warning signs may depend on the focal component and steering the vehicle may depend on the ambient component (e.g., Stokes, Wickens, and Kite, 1990). This focal–ambient system has been questioned, however, because motion processes in the fovea are quite good (Koenderick, van Doorn, and van den Brind, 1985).

be used to break camouflage, or to resolve the inherent ambiguity in interpreting a perspective display. That is, stereopsis and motion parallax can disambiguate an object's height above the ground from its distance *without* the use of ground anchors or symbol enhancers. To avoid distortions in the apparent distances and spatial layouts of objects, information from either stereopsis or motion parallax of the display requires the following:

1. For video displays, the sensor(s) must be placed at the exact position(s) of the viewer's eye(s).

2. For computer graphics displays, the rendered images must be based on the appropriate geometric calculations (considering the effects of optical distortions in the display, etc.) and appropriate calibration of the HMD.

8.3.2 Immediate performance benefits versus long-term comprehension

Whether a task involves immediate performance benefits or long-term comprehension strongly influences the sort of information needed in the HMD's visual display and how it should be presented (Andre, Wickens, Moorman, and Boschelli, 1991; Wickens, Merwin, and Lin, 1994). Tasks that involve immediate performance benefits include accurate navigation through space (e.g., a teleoperated robot walking through a building or an aircraft approaching a landing runway) and complex 3D manipulation tasks (e.g., performing a surgical operation). In these sorts of tasks, an inside-out, perspective view of the environment may be most helpful. To determine where something is located in a perspective view, we need to know its azimuth (x-axis position), elevation (y-axis position), and distance (z-axis position).

In general, humans are much better at locating objects along the azimuth than in determining the absolute distance of an object (Wickens, 1990), perhaps because cues such as color that help to identify an object also help to determine its azimuth position. However, the perceived hue of an object may not help to determine its distance from the observer. (The depth cue of atmospheric haze, which has a blue cast, may be an exception to this.) Moreover, information about distance and elevation are integrated in a perspective view, so that it can be difficult to disentangle the two sources of information. Changing the angle from which a user views the perspective visual scene[14] also changes the relative compression and expansion of the y and z axes. Thus, there is an inherent ambiguity between position along the y axis (distance)

---

[14]Research shows that a viewing angle of 30° to 45° results in the best performance (reported in Wickens, 1990).

and position along the z axis (elevation). The use of visual cues such as symbol enhancers (which anchor objects to a ground perspective), stereopsis, or motion gradients can help resolve this ambiguity.

Tasks that involve long-term comprehension include scientific visualization (e.g., viewing complex 3D molecular structures) and air traffic control. Under these conditions, an outside-in (exocentric) perspective may be more useful than an inside-out (egocentric) perspective. Often, 3D (perspective) representations result in better performance than 2D (plan) views, especially for integrative information processing based on a visual image of the data rather than on verbal or prepositional knowledge (Wickens, Merwin, and Lin, 1994). In general, tasks that require holistic awareness of space may be facilitated by the use of 3D technology (Wickens, 1990). For focused attention processing, however, 3D representations may not have an advantage.

8.3.3 Fidelity versus technical limitations issues

The issues of fidelity and the image quality of HMDs matter for several reasons. First, in using HMDs for training purposes and in transferring that training to a real-environment situation, displays of poor quality may result in negative transfer of training (e.g., Padmos and Milders, 1992). Second, there may be adaptation/readaptation problems in switching between HMDs and the real environment. Third, there may be instances of simulator sickness. Fourth, there may be degraded performance in tasks undertaken with HMDs in place. Yet imitating the real world perfectly is not feasible, and not even necessary. The required quality of the display depends upon the specific simulation task.

Fidelity issues also may include the meshing of inputs between virtual and real imagery, as well as the meshing of inputs among the senses and perceptual-motor systems. Perceptual mismatches occur when the movement or position of an object viewed on the HMD's visual display differs from the movement or position of the object as perceived by other means (e.g., the real visual environment or via other senses). Perceptual inputs may lag behind motor movements because of time delays in tracking head, hand, or body movements and updating perceptual information to correspond to those movements, and in terms of the temporal resolution for updating the image (see Sec. 8.2.3). When these intersensory perceptual mismatches occur, tactile location often will dominate visual perception, which in turn dominates auditory information. (See Welch and Warren, 1986, for more information on intersensory interactions.) When perceptual mismatches occur between the real visual scene and the virtual visual imagery, it is less clear what happens. Roscoe (1984; 1993) reported that virtual objects appear farther away than real-world objects. Rolland and her colleagues have confirmed this result with a see-through HMD (Rolland, Gibson, and Ariely, 1995; Rolland, Holloway, and Fuchs, 1994). Roscoe suggests that a magnification factor of approximately 1.25 may correct for this disparity in the perceived distances of real and virtual objects.[15] Others have observed that virtual symbology or graphics in a see-through HMD almost always appear *closer* than real-world objects, perhaps because of the occlusion of real-world objects by virtual imagery (K. Moffitt, personal communication, February 1996). More research needs to be done in studying perceptual mismatches between real and virtual imagery as well as in studying intersensory perceptual mismatches.

8.3.4 Laboratory research versus field studies of perception and performance issues with HMDs

Laboratory research can disentangle various factors involved in the visual perception and performance of tasks and can help develop theoretical frameworks to apply across numerous situations. These laboratory studies also can provide requirements for the detection of objects, visual acuity, stereopsis, and other sensory tasks. However, guidelines based only on laboratory research are not enough. Tests of visual performance under actual or simulated field conditions should precede final design implementation for HMDs. The design of these field studies should be based on empirical and theoretical information from laboratory studies. Currently there are relatively few field studies of HMD-related perception and performance issues, compared to the number of laboratory studies. Most of the suggestions and recommendations made in the following examples are based on laboratory research, so the limitations of these examples should be duly noted.

Effects of environmental stressors on performance with HMDs. Future field studies should include relevant factors normally encountered in performing specific tasks in particular environments. For example, the effects of arousal or stress (such as heat stress) can restrict the processing of information in the peripheral visual field, resulting in a narrowing of the attentional field (Stokes, Wickens, and Kite, 1990). Teichner (1968) concluded from a literature review that the stress-related narrowing of attention arose from three sources: the physical environment, emotion-provoking stimuli, and information overload.

---

[15]Roscoe states that the human's eyes do not automatically focus at optical infinity (i.e., emmetropic state) when viewing collimated virtual images. Instead the eye's lens accommodates inward toward its dark focus or resting accommodation distance (on average, between one and two meters). According to Roscoe, this in turn causes misperception of the distances of virtual objects. Other researchers, however, do not agree with Roscoe's explanation for the misperception of distance in a virtual environment.

Another stressor example is that vibration causes relative motion between the line of sight and the optical axis of the HMD, so there is a decrease in reading performance and, perhaps, the performance of other visual tasks (Wells and Haas, 1992).[16] Salvendy (1987) has a more complete description of various stressors and some of their effects on human performance. Many of these stressors can influence visual performance in HMDs. I will consider some effects of these stressors as I discuss several tasks in which HMDs could be used.

8.3.5 Examples of HMD tasks involving perception and performance issues

Now that we've considered several criteria for evaluating perception and performance issues with HMDs, I will apply them to several different types of tasks that could benefit from the use of HMDs. These examples are not exhaustive; there are many other tasks and situations in which HMDs could be useful.

Aiding surgical operations with an HMD. An optical see-through HMD can be used for surgical operations in which the surgeon may see an ultrasonic, X-ray, CT, or other scanned, virtual image (e.g., of a tumor to be removed) superimposed on the appropriate portion of the patient's body. Several issues are critical here. The virtual object could be monochromatic, in a color that contrasts with that of the surrounding tissue, so that it is easy to identify and locate the virtual image. Initially, an outside-in (exocentric) view of the virtual tumor within the real body may provide a perspective on where the tumor is located (i.e., long-term comprehension of 3D space). During the surgical operation, however, an inside-out (egocentric) perspective of the virtual image may be superimposed on the see-through, real image. This maximizes immediate performance benefits and presents both virtual and real imagery in the same type of perspective display.

Proper registration of the virtual (scanned) image and the real image are necessary in the azimuth, elevation, and depth planes. To properly mesh the depths of real and virtual imagery, depth maps of both images are necessary. The previously discussed confusion between depth and elevation that occurs with an inside-out (egocentric) visual

---

[16]Moreover, rotational oscillation of the head causes vibrations in the HMD, although the eyes remain space-stable because of the vestibular ocular reflex (VOR). Thus, VOR can degrade performance in HMDs. The effects of vibration are of special concern in HMDs where the head movement is used to direct weapons, sensors, and other systems. Solutions to head vibration problems must be sought. Possible solutions include reducing the transmission of vibration to the HMD user or deflecting the HMD's image to match and offset the effects of vibration. During a high-g maneuver, moreover, the user wants an HMD that is rigid in the z axis and will not rotate on the head.

perspective can be reduced or eliminated by providing stereoscopic depth information with binocular HMDs. Stereopsis will provide depth information at close viewing distances, which is crucial for performing surgery. Stereopsis, however, may not be able to overcome conflicting occlusion cues that arise by superimposing the virtual scanned image on the patient's body. For example, the virtual image initially may have a semitransparent quality, but becomes more opaque as the surgeon's knife cuts down closer to the location of the actual target tumor. This change from a semitransparent to an opaque quality could provide an additional relevant depth cue for the surgeon. Although absolute motion parallax cues could be used, stereopsis is preferable for several reasons. First, it is easier for the surgeon's hands to remain steady if his or her head is not moving around. Second, minimal head movements will reduce temporal lag problems in updating the virtual image so that it properly meshes with the real image seen through the optics of the HMD.

In this surgical situation, fine spatial resolution and fine depth resolution are more important than a wide, intradisplay FOV. Thus, using binocular HMDs with complete overlap is preferable to using HMDs with only partial binocular overlap. The spatial resolution of the see-through optics can match the visual acuity of the human visual system, but spatial resolution of the virtual imagery probably will be of poorer quality. It is also important to mesh the real and virtual visual images with the tactile and proprioceptive feedback and motor movements of the surgeon's hands.

HMDs for foot soldiers. HMDs can serve several functions for foot soldiers:

- Navigation information (e.g., maps and current position obtained from the Global Positioning System)
- Command and control functions (e.g., messages regarding danger and troop movements)
- Output and control of a thermal weapon sight

All of these functions involve situational awareness. Accomplishing a mission and the foot soldier's survival are highly correlated with the amount and quality of information provided (Blackwood et al., 1995). The HMD's sensory fidelity, FOV, and availability of depth cues provide critical information for the foot soldier.

The choice of whether to use a monocular, biocular, or binocular HMD may depend upon whether video imagery or graphic symbology is displayed. For displays of video imagery, a binocular HMD (e.g., a see-through, binocular HMD with a combiner that integrates imagery with an outside view of the world) may be a better choice than either an opaque monocular or a see-through biocular HMD. The opaque monocular HMD has a narrow FOV and also causes binocular rivalry, suppression, nausea, and disorientation—all of which are detrimental to the foot soldier's performance. A binocular see-through HMD can avoid these problems because it has a wider FOV (especially with partial overlap), yet provides slightly different views to the two eyes so that stereoscopic depth can be perceived when the display is properly calibrated and the HMD correctly adjusted. Although the FOV may not be optimal even if the binocular display has partial overlap, the effective FOV can be increased by using head movements to scan the scene or (possibly) auditory inputs that cue the foot soldier where to look next.

For symbology displays, however, either a see-through or a look-at monocular HMD may be preferable to the other types. With either see-through or look-at monocular HMDs, the user looks up or looks to the right (or left) side to view the simulated display. Thus, there would be little or no loss of natural-viewing FOV with this type of monocular display, as well as minimal binocular rivalry or monocular suppression.

The foot soldier must wear the HMD for long periods of time. Accordingly, the bulk, weight, fragility, and comfort of the HMD become critical and can affect the soldier's mobility, as well as potentially causing heat stress and fatigue. Either a see-through or a look-at monocular HMD would be the lightest and most comfortable HMD. A biocular HMD is lighter and less fragile than the more complicated binocular system. Unfortunately, a biocular HMD does not provide the binocular parallax cues necessary for stereopsis, which can provide depth information and break camouflage when viewing video imagery. These disadvantages can be overcome by the use of pictorial depth cues, motion parallax, and a laser rangefinder to provide depth information in a biocular display. Moreover, both motion parallax and thermal imagery can defeat visual camouflage.

Based on these considerations, one might choose a biocular see-through HMD over a binocular one for viewing video imagery. Yet some recent experimental results indicate that binocular HMDs have an advantage over biocular or monocular HMDs for traversing off-road terrain at night (CuQlock-Knopp, Torgerson, Sipes, Bender, and Merritt, 1994). In this field test monocular, biocular, and binocular night-vision goggles (NVG) were tested using a within-subjects design. Performance and satisfaction were significantly better for the binocular NVG than for either the biocular or monocular NVG; there were no significant differences in either performance or satisfaction between the biocular and monocular NVGs. The fewest errors were made (e.g., bumping into obstacles or stumbling) and the terrain was traveled in the shortest amount of time when the binocular NVGs were used. Users also reported more visual confidence, less visual discomfort, better warning of eye-level and ground-level hazards, and a preference for using the binocular NVGs rather than either the biocular or the monocular ones.

Although the recognition of objects and spatial layout depend directly on the display parameters at the low light levels of nighttime, use of the display may be more problematic during the daytime (as previously discussed in Sec. 8.2.1). Often the virtual imagery is presented in a collimated display system so that there is no need to change accommodation when switching attention from the virtual imagery or symbology to the outside world. HMDs used by foot soldiers must be resistant to the effects of vibration caused by walking in often rugged terrain and resistant to heat stressors caused by wearing the HMD during long periods of moderate to vigorous physical activity. (How to adapt the HMD for these stressors is outside the domain of this chapter.)

HMD-assisted flight.   Navigation awareness—the pilot's dynamic representation or mental model of the aircraft's orientation in space and time—is important to pilots during flight. In the display perspective that represents the precise location of the plane, one needs to know the pitch, roll, and yaw as well as the x, y, and z dimensions. This information can be represented on either a two-dimensional or a three-dimensional format. The frame of reference can be either inside-out (egocentric view) or outside-in (exocentric view), and the display can be either monochrome or color.

In using see-through HMDs with symbology displays, Andre and Wickens (1991) reported that often fewer tracking errors are made with an inside-out attitude director indicator (ADI), but fewer incorrect control responses are made with outside-in ADI. There is a measure-dependent trade-off. In their own study, they found that a planar outside-in display both resulted in better main flight performance and improved the pilot's ability to recover from a disorienting event. The use of color, however, improved performance and recovery obtained with an inside-out display format. Wells and Haas (1992) suggest a hybrid display comprised of a combination of outside-in and inside-out formats. This hybrid display could avoid some confusions that occur when a pilot wears an HMD and changes his or her head position (e.g., confusions between pitch and roll information when the pilot looks either forward or to the side).

In using binocular HMDs with video imagery, the findings of Bui, Vollmerhausen, and Tsou (1994) may be relevant. They report that a binocular HMD with only partial overlap resulted in constricted eye movements, frequent head movements, higher workload, and lower confidence than one with complete binocular overlap and the same FOV. They suggest that partial overlap produced poor results because the area of overlap is very small (only 18°). Other research also suggests that the region of binocular overlap should be at least 40° so that the luning border will be moved to a region outside the normal binocular fusion and ocular tracking regions of the eye. If a larger region of binocular overlap is used (e.g., more than 40°), then perhaps an optimal HMD design would be a see-through, binocular HMD with partial overlap. A combiner would integrate the imagery with the WFOV optics view of the real environment.

Teleoperation of robots in hostile environments. In some situations the task involves operating in a dangerous or hostile environment, one that could put human life in danger. For example, suppose a bomb threat has been received and a building cleared of human inhabitants so that a bomb squad can search the building, find the bomb, and dismantle it. In this scenario a teleoperated robot is sent in to search the building. Sensors located on the remotely operated robot could provide video input to an opaque, binocular HMD with a moderate FOV. At photopic luminance levels, objects can be identified by color and shape information provided by the video and matched to templates of the supposed bomb. Other sources of information may be used under dim ambient illumination (e.g., shape information provided by infrared video, or providing the teleoperated robot with a very bright head beam to illuminate its path). The visual sensor devices could be placed in appropriate positions on the teleoperated robot to correspond to the positions of the human user's eyes (in terms of height above the ground and interpupillary distance). This would provide accurate binocular parallax, motion parallax, and ground perspective depth information to the human.

Optic flow, absolute motion parallax, and stereopsis are helpful in slowly navigating through the terrain and avoiding obstacles, as well as in reaching for and manipulating objects. Sonar or laser rangefinder information could provide additional information about the distances and locations of objects. If the spatial layout of the building is known, then the user could initially be shown an exocentric view of the building's floor plan (including the location of the teleoperated robot) to develop a strategy for searching and navigating through the building. Whether the building's spatial layout is known or not, an egocentric perspective then could be used to guide the robot's navigation through the building to identify and dismantle the bomb. Because the human user is not in physical danger while the robot performs its tasks, there is no stress-induced narrowing of attention during the search. We also do not need to worry about the effects of physical vibration on the

HMD since the human user is relatively stationary during the entire operation. We must consider, however, that the stationary human is viewing a video obtained from a perambulating robot, which means that the human's vestibular and proprioceptive cues will not mesh with the received visual information.

Immersive entertainment. The entertainment industry may be expanding its use of HMDs in the future. Some examples are interactive video arcade-style games, movies observed by passengers during long air flights, and virtual vacations to the location of one's choice (sans air flights). The user population will be diverse in terms of age, abilities, and physical characteristics. For many it may be the first time they've used an HMD, but they all will prefer an immersive and realistic experience for the short period of use. Although they can tolerate the weight of a complex HMD to provide more realistic imagery, the weight should not be so heavy that it induces headaches. Moreover, the necessary adjustments of the HMD should be fairly simple, and easy to perform. An opaque, biocular HMD with color video or computer graphics display could be used here. The biocular display could have only partial overlap, creating a wide FOV for a greater sense of immersion, but maintaining the relatively good spatial resolution necessary for a sense of realism. The use of color also would add to the sense of realism.

The biocular display's adjustments (e.g., IPD) would be simpler than those required for a binocular HMD. To add to the sense of immersion and realism, an egocentric perspective should be used. Moreover, if other depth cues are present (e.g., pictorial cues and optic flow), they can be so compelling that one does not realize stereoscopic cues are missing (e.g., King, Elliott, and Davis, 1996). For binocular video inputs, the sensors or cameras would need to be laterally displaced to match the IPD of each individual. Otherwise, depth distortions occur that detract from the sense of realism. Two scenes must be calculated for binocular computer graphics input with each object having the correct binocular parallax for each individual user. Otherwise, depth distortions may occur. By eliminating binocular parallax, the savings in computational complexity can perhaps be used to produce faster updates for rapidly changing scenery and action, again enhancing the sense of immersion and realism.

8.4 Summary and Conclusions

Head-mounted displays, like other displays, are designed to present structured information to the human senses. Although in many ways HMDs are different from more traditional displays, some of the design principles that apply to more traditional displays (Tufte, 1983, 1990) also apply to HMDs.[17] In designing displays one needs to consider the practical, technical limitations of available display hardware and software, as well as the human visual system's capabilities and the specific tasks in which those HMDs will be used.

In the first part of this chapter I examined the capabilities of the human visual system and of HMDs in terms of brightness, contrast, visual acuity and spatial resolution, CFF, temporal sensitivity and temporal resolution, the field of view, monoscopic versus biocular versus binocular views, and color versus monochrome views. In doing this I addressed the question *What can we see?* as well as *What do we need to see?* These questions are important because the human visual system has been developed to process information about objects and events in the everyday, real-world environment. Displays that are designed to capitalize on human visual processing capabilities and limitations should be more efficient and cost-effective than those that are not.

HMDs also can access and make available to the human visual system information that usually is not available to it (e.g., distance information from laser rangefinders and infrared or thermal images for the identification of objects). Thus, HMDs can overcome some of the limitations of the human visual system and add to the store of information normally available. Displays that access far more information than is available to direct observation and that overload the system, however, are not optimal. Conversely, displays that severely degrade or restrict the available information are not optimal either. The former results in resource-limited performance, and the latter results in data-limited performance (Norman and Bobrow, 1975; Stokes, Wickens, and Kite, 1990). These relate to performance issues.

In the last part of this chapter I considered some criteria for evaluating perception and performance with HMDs in various tasks and situations. Here I focused on the question *What do we need to see?* In particular, I considered the two basic functions of vision (determining what something is and where it is located in 3D space); the perspective needed for immediate performance benefits versus that for long-term comprehension; issues of fidelity versus technical limitations (including meshing virtual and real-world views); as well as laboratory versus field research (the latter including the effects of environmental stressors on perception and performance with HMDs). I then applied these criteria to several tasks and situations (e.g., surgical operations, foot soldiering, and immersive entertainment), trying to determine the necessary and/or optimal characteristics of HMDs for these various scenarios.

---

[17]These include variation in the size and color or objects and text, layering and separation, and use of blank space.

Finally, once an HMD has been systematically designed to match the human visual capabilities and the display capabilities for a particular situation or task, it is necessary to test that HMD in the field or, at least, a simulation of the field situation. This means that once a prototype is developed, it is necessary to determine what works, what does not work, and why. With careful testing of the prototype, the HMD design can be modified to improve perception and performance in those specific tasks for which it is designed.

8.5   References

Alfano, P. L., and Michel, G. E. (1990). Restricting the field of view: Perceptual and performance effects. *Perceptual and Motor Skills, 70,* 35–45.

Andre, A. D., Wickens, C. D., Moorman, L., and Boschelli, M. M. (1991). Display formatting techniques for improving situation awareness in the aircraft cockpit. *International Journal of Aviation Psychology, 1,* 205–218.

Anonymous (1995). Technology review: Headmounted displays. *VR News,* pp. 20–29.

Arditi, A. (1986). Binocular vision. In K. R. Boff, L. Kaufman, and J. P. Thomas (Eds.), *Handbook of perception and human performance.* Vol. 1. New York: Wiley.

Barfield, W., and Rosenberg, C. (1995). Judgments of azimuth and elevation as a function of monoscopic and binocular depth cues using a perspective display. *Human Factors, 37,* 173–181.

Blackwood, W. O., Anderson, T. R., Barfield, W., Bennett, C. T., Corson, J. R., Endsley, M. R., Hancock, P. A., Hochberg, J., Hoffman, J. E., and Kruk, R. V. (1995, April). *Human factors in the design of tactical display systems for the individual soldier: Phase I report.* National Research Council.

Blake, R. (1977). Threshold conditions for binocular rivalry. *Journal of Experimental Psychology: Human Perception and Peformance, 3,* 251–257.

Boff, K. R., and Lincoln, J. E. (Eds.). (1988). *Engineering data compendium: Human perception and performance.* Wright-Patterson Air Force Base, OH: AAMRL.

Boynton, R. M. (1971). Color Vision. In J. W. Kling and L. A. Riggs (Eds.), *Experimental psychology.* New York: Holt, Rinehart, and Winston.

Boynton, R. M. (1979). *Human color vision.* New York: Holt, Rinehart and Winston.

Braddick, O. (1974). A short-range process in apparent motion. *Vision Research, 14,* 519–527.

Brooks, F. (1995). Invited address in the Distinguished Lecture Series. The Graphics, Visualization, and Usability Center, Georgia Institute of Technology, Atlanta, GA.

Bui, T. H., Vollmerhausen, R. H., and Tsou, B. H. (1994). Overlap binocular field-of-view flight experiment. *Society for Information Display International Symposium Digest of Technical Papers, 25,* 306–308. Santa Ana, CA: Society for Information Display.

Carter, R. C., and Carter, E. C. (1988). Color coding for rapid location of small symbols. *Color Research and Application, 13,* 226–234.

Cavanaugh, J. P. (1972). Relation between immediate memory span and the memory search rate. *Psychological Review, 79,* 525–530.

Christ, R. E. (1975). Review and analysis of color coding research for visual displays. *Human Factors, 17,* 542–570.

Christ, R. E., and Corso, G. (1982). The effects of extended practice on the evaluation of visual display code. *Human Factors, 25,* 71–84.

Cole, R. E., Merritt, J. O., and Lester, P. (1990). Remote manipulator tasks impossible without stereo TV. *Stereoscopic Displays and Applications I, Proceedings of the SPIE, 1256* (pp. 255–265).

CuQlock-Knopp, V. G., Torgerson, W., Sipes, D. E., Bender, E., and Merritt, J. O. (1994). A comparison of monocular, biocular, and binocular night-vision goggles for traversing off-road terrain. *Society for Information Display International Symposium Digest of Technical Papers,* (pp. 309–312). Santa Ana, CA: Society for Information Display.

Davis, E. T., and Hodges, L. F. (1995). Human stereopsis, fusion, and stereoscopic virtual environments. In W. Barfield and T. Furness (Eds.), *Virtual environments and advanced interface design*. New York: Oxford.

deLange, H. (1958). Research into the dynamic nature of the human fovea-cortex systems with intermittent and modulated light. I. Attenuation characteristics with white and colored light. *Journal of the Optical Society of America, 48*, 777–784.

DeYoe, E. A., and VanEssen, D. C. (1988). Concurrent processing streams in monkey visual cortex. *Trends in Neural Science, 11*, 219–226.

Foley, J. D. (1987). Interfaces for advanced computing. *Scientific American*, pp. 82–90.

Gibson, J. J. (1979). *The ecological approach to visual perception*. Boston: Houghton Mifflin.

Graham, C. H. (1965). Visual space perception. In C. H. Graham (Ed.), *Vision and visual perception*. New York: Wiley.

Graham, N. V. S. (1989). *Visual pattern analyzers*. New York: Oxford.

Hodges, L. F., and Davis, E. T. (1993). Geometric considerations for stereoscopic virtual environments. *PRESENCE: Teleoperators and Virtual Environments, 2*, 34–43.

Holmgren, D. E., and Robinett, W. (1993). Scanned laser displays for virtual reality: A feasibility study. *PRESENCE: Teleoperators and Virtual Environments, 2*, 171–184.

Hood, D. C., and Finkelstein, M. A. (1986). Sensitivity to Light. In K. R. Boff, L. Kaufman, and J. P. Thomas (Eds.), *Handbook of perception and human performance*. (Vol. 1.) New York: Wiley.

Kalawsky, R. S. (1993). *The science of virtual reality and virtual environments*. Reading, MA: Addison-Wesley.

Kim, W. S., Ellis, S. R., Tyler, M. E., Hannaford, B., and Stark, L. W. (1987). Quantitative evaluation of perspective and stereoscopic displays in a three-axis manual tracking task. *IEEE Transactions on Systems, Man and Cybernetics, 17*, 61–71.

King, R. A., Elliott, K., and Davis, E. T. (1996). A comparison of size constancy between matched scenes with many and few depth cues. *Investigative Ophthalmology and Visual Science Supplement, 37*, p. S519.

Klein, S. A., and Levi, D. M. (1985). Hyperacuity thresholds of 1 sec: Theoretical predictions and empirical validation. *Journal of the Optical Society of America, A, 2*, 1170–1190.

Klymenko, V., and Rash, C. E. (1995). Human performance with new helmet-mounted display designs. *CSERIAC Gateway, 6*(4), pp. 1–5.

Koenderick, J. J., van Doorn, A. J., and van den Brind, W. A. (1985). Spatial-temporal parameters of motion detection in the peripheral visual field. *Journal of the Optical Society of America, A, 2*, 252–259.

Leibowitz, H. W., and Post, R. B. (1982). The two modes of processing concept and some implications. In J. Beck (Ed.), *Organization and representation in perception*. Hillsdale, NJ: Erlbaum.

Lennie, P., and D'Zmura, M. (1988). Mechanisms of colour vision. *CRC Critical Reviews in Neurobiology, 3*, 333–400.

Levelt, W. J. M. (1968). *On binocular rivalry*. Hague: Mouton.

Levi, D. M., Klein, S. A., and Aitsebaomo, A. P. (1985). Vernier acuity, crowding, and cortical magnification. *Vision Research, 25*, 963–977.

Livingstone, M., and Hubel, D. (1988). Segregation of form, color, movement, and depth: Anatomy, physiology, and perception. *Science, 240*, 740–749.

Mandelbaum, J., and Sloan, L. L. (1947). Peripheral visual acuity. *American Journal of Ophthalmology, 30*, 581–588.

Melzer, J. E., and Moffitt, K. (1989). Partial binocular-overlap in helmet-mounted displays. *Display System Optics II, Proceedings of the SPIE, 1117*, 56–62.

Melzer, J. E., and Moffitt, K. (1991). An ecological approach to partial binocular overlap. *Large Screen Projection, Avionics, and Helmet Mounted Displays, Proceedings of the SPIE, 1456*, 175–191.

Miller, J. W., and Ludvigh, E. (1962). The effects of relative motion on visual acuity. *Survey of Ophthalmology, 7*, 83–116.

Nagy, A. L., and Sanchez, R. R. (1990). Critical color differences determined with a visual search task. *Journal of the Optical Society of America, A, 7*, 1209–1217.

Nagy, A. L., and Sanchez, R. R. (1992). Chromaticity and luminance as coding dimensions in visual search. *Human Factors, 34,* 601–614.

Nagy, A. L., Sanchez, R. R., and Hughes, T. C. (1990). Visual search for color differences with foveal and peripheral vision. *Journal of the Optical Society of America, A, 7,* 1995–2001.

Norman, D., and Bobrow, D. (1975). On data-limited and resource-limited processing. *Journal of Cognitive Psychology, 7,* 44–60.

Osgood, R. K., and Wells, M. J. (1991). The effect of field-of-view size on performance of a simulated air-to-ground night attack. In *Proceedings AGARD Symposium on Helmet Mounted Displays and Night Vision Goggles, 517,* 10.1–10.7.

Padmos, P., and Milders, M. V. (1992). Quality criteria for simulator images: A literature review. *Human Factors, 34,* 727–748.

Piantanida, T., Boman, D.K., Larimer, J., Gille, J. and Reed, C. (1992). Studies of the field-of-view/resolution trade-off in virtual-reality systems. *Human Vision, Visual Processing, and Digital Display III, Proceedings of the SPIE, 1666,* 448–456.

Piantanida, T., Boman, D. K., and Gille, J. (1995). *Perceptual issues in virtual reality.* Unpublished manuscript.

Pokorny, J., and Smith, V. C. (1986). Colorimetry and color discrimination. In K. R. Boff, L. Kaufman, and J. P. Thomas (Eds.), *Handbook of Perception and Human Performance.* New York: Wiley.

Proffitt, D. R., and Kaiser, M. K. (1986). The use of computer graphics animation in motion perception research. *Behavior Research Methods, Instruments, and Computers, 18,* 487–492.

Richards, W. (1971). Anomalous stereoscopic depth perception. *Journal of the Optical Society of America, 61,* 410–414.

Robson, J. G. (1966). Spatial and temporal contrast sensitivity functions of the visual system. *Journal of the Optical Society of America, 56,* 1141–1142.

Rogers, B., and Graham, M. (1982). Similarities between motion parallax and stereopsis in human depth perception. *Vision Research, 22,* 261–270.

Rolland, J. P., Gibson, W., and Ariely, D. (1995). Towards quantifying depth and size perception in virtual environments. *PRESENCE: Teleoperators and Virtual Environments, 4,* 24–49.

Rolland, J. P., Holloway, R. L., and Fuchs, H. (1994). A comparison of optical and video see-through head-mounted displays. *Proceedings of the SPIE, 2351* (pp. 293–307). Bellingham, WA: SPIE—The International Society for Optical Engineering.

Roscoe, S. N. (1993). The eyes prefer real images. In S. R. Ellis, M. K. Kaiser, and A. J. Grunwald (Eds.), *Pictorial communication: Virtual and real environments.* Washington, D.C.: Taylor and Francis.

Roscoe, S. N. (1984). Judgments of size and distance in imaging displays. *Human Factors, 26,* 617–629.

Salvendy, G. (1987). *Handbook of human factors.* New York: Wiley.

Sandor, P. B., and Leger, A. (1991). Tracking with a restricted field of view: Performance and eye-head coordination aspects. *Aviation, Space, and Environmental Medicine, 62,* 1026–1031.

Schor, C. M. (1987). Spatial factors limiting stereopsis and fusion. *Optic News, 13,* 14–17.

Silverstein, L. D., Lepkowski, J. S., Carter, R. C., and Carter, E. C. (1986). Modeling of display color parameters and algorithmic color selection. *Advances in Display Technology VI, Proceedings of the SPIE, 624,* 26–35.

Smallman, H. S., and Boynton, R. M. (1990). Segregation of basic colors in an information display. *Journal of the Optical Society of America, A, 7,* 1985–1994.

Smets, G. J. F., and Overbeeke, K. J. (1995, September). Trade-off between resolution and interactivity in spatial task performance. *IEEE Computer Graphics and Applications,* 46–51.

Spain, E. H., and Holzhausen, K. P. (1991). Stereoscopic versus orthogonal view displays for performance of a remote manipulation task. *Stereoscopic Displays and Applications II, Proceedings of the SPIE, 1457,* 103–110.

Stokes, A., Wickens, C., and Kite, K. (1990). *Display technology: Human factors concepts.* Warrendale, PA: Society of Automotive Engineers.

Surdick, R. T., Davis, E. T., King, R. A., and Hodges, L. F. (in press). The perception of distance in simulated visual displays: A comparison of the effectiveness and accuracy of multiple depth cues across viewing distances. *PRESENCE: Teleoperators and Verbal Environments.*

Surdick, R. T., Davis, E. T., King, R. A., Corso, G. M., Shapiro, A., Hodges, L. F., and Elliot, K. (1994). Relevant cues for the visual perception of depth: Is where you see it where it is? *Proceedings of the Human Factors and Ergonomics Society 38th Annual Meeting* (pp. 1305–1309). Santa Monica, CA: Human Factors and Ergonomics Society.

Swanson, W. H., and Wilson, H. R. (1985). Eccentricity dependence of contrast matching and oblique masking. *Vision Research, 25,* 1285–1295.

Task, H. L. (1991). Optical and visual considerations in the specification and design of helmet-mounted displays. *Society for Information Display International Symposium Digest of Technical Papers, 22,* 297–300.

Teichner, W. H. (1968). Interaction of behavioral and physiological stress reactions. *Psychological Review, 75,* 271.

Thorell, L. (1983). Using color on displays: A biological perspective. In *1983 SID Seminar lecture notes,* (pp. 2–5). Santa Ana, CA: Society for Information Display.

Travis, D. S. (1990). Applying visual psychophysics to user interface design. *Behavior and Information Technology, 9,* 425–438.

Tufte, E. R. (1983). *The visual display of quantitative information.* Cheshire, CT: Graphics Press.

Tufte, E. R. (1990). *Envisioning information.* Cheshire, CT: Graphics Press.

Tyler, C. W. (1983). Sensory processing of binocular disparity. In C. M. Schor and K. J. Ciuffreda (Eds.), *Vergence eye movements: Basic and clinical aspects.* Boston: Butterworths.

Wallach, H., and Bacon, J. (1976). The constancy of the orientation of the visual field. *Perception and Psychophysics, 19,* 492–498.

Wells, M. J., and Haas, M. (1992). The human factors of helmet-mounted displays and sights. In I. Marcel Dekker (Ed.), *Electro-optical displays.* New York: Marcel Dekker.

Wenzel, E. M. (1991). Localization in virtual acoustic displays. *PRESENCE: Teleoperators and Virtual Environments, 1,* 80–107.

Wenzel, E. M., Wightman, F. L., and Foster, S. H. (1988). A virtual display system for conveying three-dimensional acoustic information. *Proceedings of the Human Factors Society 32nd Annual Meeting* (pp. 86–90). Santa Monica, CA: Human Factors Society.

Westheimer, G., and McKee, S. (1975). Visual acuity in the presence of retinal-image motion. *Journal of the Optical Society of America, 65,* 847–850.

Wickens, C. D. (1990). Three-dimensional stereoscopic display implemention: Guidelines derived from human visual capabilities. *Stereoscopic Displays and Applications I, Proceedings of the SPIE, 1256,* 2–11.

Wickens, C. D., Merwin, D. H., and Lin, E. L. (1994). Implications of graphics enhancements for the visualization of scientific data: Dimensional integrality, stereopsis, motion, and mesh. *Human Factors, 36,* 44–61.

Wickens, C. D., and Todd, S. (1990). Three-dimensional display technology for aerospace and visualization. *Proceedings of the Human Factors Society 34th Annual Meeting* (pp. 1479–1483). Santa Monica, CA: Human Factors Society.

Williams, D. R. (1985). Aliasing in human foveal vision. *Vision Research, 25,* 195–205.

Williams, D. R., MacLeod, D. I. A., and Hayhoe, M. M. (1981a). Foveal tritanopia. *Vision Research, 21,* 1341–1356.

Williams, D. R., MacLeod, D. I. A., and Hayhoe, M. M. (1981b). Punctate sensitivity of the blue-sensitive mechanism. *Vision Research, 21,* 1357–1376.

Yeh, Y. Y., and Silverstein, L. D. (1992). Spatial judgments with monoscopic and stereoscopic presentation of perspective displays. *Human Factors, 34,* 583–600.

Designing HMD Systems for Stereoscopic Vision

David A. Southard

Stereoscopic HMD systems offer depth perception, which is required for applications such as teleoperation, augmented reality, and virtual environments. Tight coupling of the display device to the human visual system for these applications requires special attention to the human factors of binocular vision. This chapter reviews significant factors of stereoscopic vision and recommends a mathematical model for computer image generators that promotes comfortable and correct viewing conditions.

9.1   Introduction

The defining feature of a head-mounted display (HMD) is that the display stays with the user as he or she moves about performing a task. The advantage is that information is always immediately available. The converse to this is that the display is *always* there—potentially used for long periods of time, and perhaps the user's *primary* interface with the system. In such cases, it becomes important to have a human-centered design. Vision is an extremely important and personal sense. Anyone who has experimented with the poorly done stereoscopic effects of some 3D comics or grade-B movies can probably attest to the real possibility of fatigue, or even outright discomfort. As system designers, we have a responsibility to be sensitive to the user's comfort, or else the systems simply will not be used. On the other hand, a truly well-executed stereoscopic system will attract an enthusiastic following.

9.1.1   The problem

Although the possibilities offered by stereoscopic computer graphics are exciting, initial experiences with stereoscopic images are sometimes disappointing. The viewer may experience discomfort, or may be unable to fuse the left/right image pairs. After some experimentation, I concluded that some recommendations found in the stereoscopic literature, which for the most part were modeled after stereophotography, are inappropriate for producing computer-generated stereograms. Moreover, the methods can be applied successfully to only a limited class of scenes. Much of the literature has been devoted to reviewing the many particular types of stereoscopic viewing apparatus (Lane, 1982; Hodges & McAllister, 1985; Hodges, 1992). Although the central ideas of stereoscopy are clear, the results can fall short of expectations, not only because of inadequate viewing devices, but also because of inappropriate production methods.

9.1.2   Design goals

In virtual environment applications, one attempts to place the user in a computer-generated synthetic environment to perform some task that would be too difficult, too hazardous, or too expensive to perform in a real setting. From a display system designer's point of view, this can be done using either stationary head-tracking displays (Deering, 1992), movable displays mounted on counterbalanced articulated armatures (McDowall, Bolas, Pieper, Fisher, & Humphries, 1990), large-screen projection displays (Cruz-Neira, Sandin, & DeFanti, 1993), or HMDs (Teitel, 1990). Although much of the discussion applies equally well to all types of stereoscopic displays, this chapter will concentrate on HMDs.

An accurate viewing model is especially important for use in stereoscopic HMDs because in many applications the display almost completely replaces the accustomed natural visual cues. These displays, and the viewing models used to calculate stereoscopic image pairs, must be matched to the human visual system. If such considerations are not taken into account, there is a potential for disorientation and fatigue after using the display for a short period of time. The goal of this chapter is to explore the issues surrounding stereoscopic HMDs, to examine how to identify and measure the HMD parameters critical to stereoscopic vision, and to specify a geometric projection algorithm that provides correct stereoscopic vision in an immersive graphical environment.

9.2 Background

Lipton (1982) reports in his study of stereoscopic cinema that stereopsis was discovered by Wheatstone, who developed the mirror stereoscope in 1833 and published his findings in 1838. Although it had been known since ancient times that each eye saw a slightly different view, stereopsis was not recognized as a distinct visual perception until Wheatstone. Brewster invented the lenticular stereoscope in 1844, and the familiar View-Master stereoscope is essentially his design. Anderton's 1891 patent suggested the use of polarized light for stereoscopic image selection, and this method became practical with Land's invention of the sheet polarizer in 1934. Both Lipton and Valyus (1966) provide many examples of variations and improvements in photographic and optical techniques, and the concepts of stereoscopy have changed little since those original developments.

9.2.1 Computers and stereoscopy

Computer stereograms are appropriate wherever a regular 3D perspective can be used, and when understanding the spatial relationships in the picture is critical. Grotch (1983) describes the uses of stereoscopy in scientific data analysis and provides references for applications in aerial photogrammetry, medical imaging, crystallography, molecular modeling, and atmospheric studies. Roese and McCleary (1979) suggest its use in flight simulation for aircraft carrier landing approaches, and Bridges and Reising (1987) proposed 3D and stereoscopic displays for use in aircraft cockpits. Smedley, Haines, Van Vactor, and Jordan (1989) report on stereoscopic displays for ocean bottom and coastal terrain visualization. Veron, Southard, Leger, and Conway (1990) used visual simulations of views from low-flying aircraft for applications in mission planning, command, and control workstations. Gorski (1992) evaluated a prototype visual system, developed for NASA's Johnson Space Center, for training astronauts to use the space shuttle's remote manipulator arm and for space station docking and construction maneuvers. Williams, Parrish, and Nold (1994) describe experiments in tracking objects in the presence of a cluttered visual space. Fisher (1995) and Hoberman (1995) used stereoscopic displays in public installations designed for high-traffic public relations presentations and interactive museum experiences.

9.2.2 Applications for stereoscopic HMDs

Simple information displays using text, line graphics, and video are appropriate for monoscopic and biocular HMDs. Stereoscopic displays should be used whenever the application calls for enhanced 3D spatial understanding. Applications for stereoscopic HMDs can be classified into three categories: teleoperation, augmented reality, and virtual environments.

In *teleoperation,* the HMD is the master device for a stereoscopic camera setup. The cameras can be mounted on an articulated slave platform that mimics the motion of the user's head. Alternatively, the cameras are fixed to a viewing platform, and the user adjusts the platform's position with a separate hand controller. A stereoscopic display is beneficial for enhancing eye-hand coordination when the user is grasping and moving objects with mechanical manipulator arms. Examples that require a fine degree of spatial awareness include:

Hazardous and inaccessible environments.   Human-controlled tele-robots can safely operate in hostile environments, such as in a toxic waste cleanup site. They can also provide "hands-on" access to remote places, for such activities as deep-ocean or space-based maintenance and construction (Woods, Docherty, & Koch, 1994).

Surgical procedures.   An increasing number of medical procedures are performed using laparoscopic or endoscopic instruments. Optical fibers have been incorporated into these instruments to transmit images from inside the body to the surgeon. This is a form of teleoperation, and there are plans to extend this idea to the full-fledged concept of telesurgery, wherein the surgeon may be far removed from the patient. Although stereoscopic imagery is not usually employed in the current generation of these instruments, there is strong potential for aiding the surgeon during delicate procedures. Previous generations of bulky HMDs were unsuitable for use in surgery, but new lightweight, high-resolution designs are generating considerable interest (Jones & McLaurin, 1994; McLaurin & Jones, 1994).

In *augmented reality,* computer-generated information displayed on the HMD is registered and overlaid on the real world (Sharma & Molineros, 1995). These applications use see-though HMDs and sensing devices that track the operator's viewpoint to coordinate the computer-generated view with the external world. The benefit here is that information is presented spatially in concert with the current task. This allows the operator to move quickly to the correct location in space to perform a procedure or verify a configuration.

One application for augmented reality being explored at Boeing Aircraft Corporation involves overlaying plans for wire harnesses on layout boards, and stepping technicians through construction procedures, one wire at a time. Modern aircraft require many wire harnesses, and their manufacture currently requires many special, full-sized jigs. Replacing the costly physical jigs with electronic versions offers cost savings and productivity gains.

Another augmented reality application overlays medical sensor data, such as ultrasound or computed tomography (CT) imagery, directly on a patient. This technique literally gives the physician "x-ray vision," which can aid diagnosis as well as planning, training, and performing surgical procedures.

In a *virtual environment,* we replace the video cameras used in teleoperation with a computer image generator (CIG). We then provide tools and controllers that allow the user to interact with the computer-simulated environment. The CIG provides simulated visuals that reflect the current state of the simulated environment. An HMD provides the critical interface that transforms the simulation into an *immersive* environment. That is, from the user's point of view, he or she is interacting directly with a simulated environment.

Virtual environments have the benefit of being able to simulate situations that are too costly, dangerous, or unreal to reproduce in any other way. At the same time, they provide the kind of spatial awareness that ordinarily would come only with experience in a real environment. Example applications include:

Training and education. Using a virtual environment, we can train for any teleoperation task. In this respect, virtual environments have existed for years in the form of aircraft flight simulators.

However, HMDs posit the elimination of simulator hardware in favor of an entirely simulated environment. This situation saves both time and money; a virtual environment can be reconfigured by software for many different training applications, and the need for costly hardware mock-ups is reduced or eliminated.

Entertainment. A rapidly growing application of stereoscopic HMDs currently is entertainment. Location-based entertainment centers provide patrons with role-playing experiences. Stereoscopic HMDs provide total immersion in the environment, which promotes interactive participation and enhances the user's experience.

9.3 Characteristics and Limitations of Stereoscopic HMDs

HMDs can be classified by the type of interface they present to the human visual system:

Monocular. The display is designed for one eye to view the information display while the other eye continues to look at the user's surroundings.

Biocular. The user views the same display image with both eyes. Each eye may have separate optics, but the image is essentially identical for both eyes. This is logically equivalent to looking with both eyes at a regular display monitor, but in this case it is head-mounted.

Binocular. Each eye has its own distinct perspective view of the same 3D scene. A binocular display is required for *stereopsis* (stereoscopic vision), the visual ability to sense depth in a scene. A binocular display can be achieved through separate optics for each eye, or by means of a shared light path using an image-selection mechanism to separate the views for each eye.

Each of these types is useful in various applications. For example, a monocular display might be appropriate for applications in which the user is primarily engaged in tasks within normal work surroundings. The display can be positioned where it will not interfere with the user's activities, but where it can be glanced at occasionally as needed. A monocular display can effectively present video, still pictures, and text.

A biocular display might be suited for a more restricted environment. Since both eyes can see the image, biocular displays can be more comfortable to look at and potentially less fatiguing. They also command more of the user's field of view and attention, so they are appropriate for tasks where information is critical. Biocular displays are still 2D devices that can present only a flat image floating in space. Like monocular displays, they can present both video and text.

Binocular displays are best used when the application requires close interaction between the user and a 3D environment. With real environments, a binocular display can supplement the real world by superimposing 3D imagery and data. In virtual environments, a binocular display can immerse the user in computer-generated worlds. Such worlds can be used to present new architectural or product designs, molecular models, medical models, or abstract data spaces.

This chapter focuses on binocular stereoscopic HMDs. Stereoscopic displays can be driven by video cameras or a computer image generator (CIG). Today, CIG technology that was formerly associated only with expensive flight simulators is becoming more widely available for deskside graphics workstations, desktop computers, arcade games, and home video games. Much of our discussion will focus on techniques developed for CIG applications, but the principles used are still applicable to camera-based systems.

9.3.1 Advantages

It seems evident that stereoscopic vision confers an enormous advantage on creatures that must survive in a complex 3D environment. Nature has seen fit to endow most vertebrates (mammals, birds, fish, and reptiles) with binocular vision. If users of HMDs are to work in complex 3D environments, they would surely be handicapped without stereoscopic vision.

Stereoscopic vision itself is something of a mystery. Research psychologists and physiologists are still discovering the biological mechanisms that enable us to infer depth. Geometrically, however, it is clear that depth information can be obtained by comparing two perspective views of the same scene. The small relative shift in corresponding points in a left/right image pair—called *homologous* points—is proportional to the depth of a particular point relative to the viewer. Our eye-brain visual system is able to compare the left and right visual channels to extract this depth information and produce the sensation known as stereopsis. However, stereopsis is not the only contributor to depth perception, and probably not even the most important contributor (Lipton, 1982; Davis & Hodges, 1995). What is important to realize is that stereopsis is coordinated and consistent with all the other depth cues available to us. In order to simulate stereoscopic vision in an HMD, we must carefully design all the visual simulation components to work together, producing a synergistic visual effect.

Another aspect to depth perception is judgments about weight and volume. Stereopsis enhances our ability to estimate physical properties of the objects that surround us, and thus facilitates our work as we interact with and manipulate our environment. With stereoscopic vision, textures assume an almost tactile quality. We can actually see minute surface variations and judge whether an object will be dull or sharp, smooth or rough, slippery or sticky, hard or soft, sleek or fuzzy. Reflections and specular highlights provide additional clues to the environment. Even the small change in position between our eyes is enough to create a big difference in these reflective effects. That difference is sufficient to provide a wealth of additional information: surface materials, positions of lights, background and obscured objects, and sense of surround space.

Stereopsis plays a critical role in eye-hand coordination. Studies have shown that tasks requiring eye-hand coordination benefit from stereoscopic vision (Sollenberger & Milgram, 1993). There is evidence that complex abstract objects such as node-and-arc network graphs, which cannot be unambiguously represented on a 2D surface, are better understood when presented in a stereoscopic environment (Arthur, Booth, & Ware, 1993; Ware & Franck, 1994). Stereoscopic displays have also been shown to be effective for tasks that require tracking of targets in the midst of clutter (Williams et al., 1994).

When all the perceptual cues work together harmoniously, we have a feeling of immersion in the synthetic environment. This immersive quality can enhance our sense of contact with the environment, and thus improve productivity within the environment. Stereoscopic display can contribute significantly to the sense of immersion, but it is a dual-edged sword. If mishandled, the stereoscopic display can *detract* from the task at hand.

9.3.2 Pitfalls

It is clearly a challenge for display systems to live up to the standards set by the capabilities of natural vision. In a stereoscopic system, any discrepancies in the display system are only exacerbated and brought to the fore. Beyond the basic questions of resolution, brightness, and contrast, however, stereoscopic displays create some unique issues that must be addressed.

Anyone who has attended a 3D motion picture has probably come away with a certain degree of eyestrain. Generally, the closer one sits to the front of the theater, the more likely one is to get a headache. The reasons for this are discussed extensively by Lipton (1982). A 3D movie is designed to be watched from the center seat of a certain theater. Any other seat, or any other theater, is likely to depart from the ideal viewing conditions. Sitting closer to the screen is usually harder on the eyes, and sitting farther from the screen is easier.

The culprit is *parallax,* the distance measured between homologous points on the display screen. As one sits closer to the display screen, the parallax remains the same, but our eyes must converge to increasing angles. This additional effort can cause fatigue, and ultimately pain. It is easy, quite by accident, to create conditions that cause our eyes to diverge or cross at excessive, uncomfortable angles.

Fortunately, as HMD system designers, we can design an optimal "theater" for our patrons. Most causes of viewer fatigue are avoidable. The geometric parameters of the display can be designed to create perspective views that are comfortable to viewers.

Our vision is very sensitive to vertical misalignments. We have a tolerance of only 0.16 degree of vertical disparity between views for each eye (Lipton, 1982). There is a much greater tolerance for horizontal disparities. Vertical disparities can have several causes: misalignment of the image pairs; misalignments internal to the imaging optics; distortions such as barrel, pincushion, or keystoning effects in the display device; distortion inherent to the optics (especially wide-angle optics); mismatches between camera and HMD optics; or, equivalently, incorrect perspective algorithms used in CIGs. With careful design and attention to mechanical and optical tolerances, virtually all of these distortions can be designed out of the system.

Parallax greater than the interpupillary distance causes divergence of the eyes. Lipton (1982) recommends that in stereoscopic cinema the divergence should be limited to a total of 1 degree. However, in a CIG-HMD system it is unnecessary to tolerate *any* divergence. Assuming that the optical system is properly aligned, the most likely cause of divergence is magnification of the stereopair images, which also magnifies the parallax. If the user sets the magnification, this information must be relayed to the software so that the perspective transformations can account for new viewing parameters. For a camera-based system, the cameras must be carefully aligned and matched for focal length, exposure, and depth of field.

Stereoscopic displays require that we exercise our eyes in a manner to which they are not accustomed. Normally, we automatically adjust the focus of our eyes as we look at various depths. In stereoscopy, although we turn our eyes to converge at different depths, we must remain focused on the display screen at all times. Some users have difficulty with this task, and this difficulty is termed an *accommodation/convergence conflict* (A/C conflict). Children typically have less difficulty than adults. Most people have no difficulty within certain established limits, but when these limits are exceeded a total breakdown of stereopsis can occur.

A/C conflicts place a practical limit on the range of depth in a scene. Even though certain depth-convergence combinations are geometrically correct, they may still be a source of fatigue in a stereoscopic display (Miyashita & Uchida, 1990). The fatiguing effect can often be reduced over time as the user gains experience with stereoscopic displays. One remedy is to focus the display at a distance that maximizes the comfortable depth range. Another method is to clip off elements of the scene that are too close or too far for comfortable viewing.

A/C conflicts are probably the only source of fatigue that currently cannot be completely avoided. They are inherent to stereoscopic displays. To eliminate A/C conflicts altogether would require a holographic scene, or dynamically refocusing the optics to the appropriate depth as the user's eyes roam the 3D scene. These approaches, which are current research topics, are likely to remain infeasible for practical HMDs in the foreseeable future.

Sometimes stereoscopic displays are presented with exaggerated depth effects, termed *hyper-stereoscopic*. A familiar example is provided by an ordinary pair of prism binoculars. The prisms not only fold the optical path into a compact package, but they also expand the distance between the primary lenses (the *interaxial* distance). This arrangement provides depth discrimination at distances beyond the normal viability of stereopsis. Other examples include a pair of stereoscopic cameras mounted on the sides of a helmet, or teleoperation of a robot having a wider-than-normal camera separation. A baseline larger than the average interpupillary distance of 63–65 mm extends the range, and thus the utility, of stereoscopic perception.

Hyperstereoscopic setups are commonly used in map making. Stereopair aerial or satellite photographs are viewed in a special plotting instrument. The hyperstereoscopic effect affords additional accuracy when plotting isoelevation contour lines.

Another way to exaggerate depth is to use excess perspective. An example of this is using wide-angle camera lenses with a display having a normal field of view. The perceived effect is an expanded sense of depth, and this technique can be pleasurable when used in moderation.

Exaggerated depth is not always beneficial. There is a natural desire to "spice up" the display, but unless there is a specific reason to do so it is best not to yield to temptation. That's because the effects do not always work out as well as one might hope. The visual mind is accustomed to operating in the real world, and it attempts to interpret everything we see in that experiential context. Therefore, if presented with unusual depth effects, the mind may be forced to reinterpret the scene under new and sometimes unexpected assumptions. Depending on the subject matter, the mind may decide that the objects in the scene are much smaller than intended. Instead of, say, flying through a full-sized city, we may experience a toy model of a city. Researchers have found that users exhibit errors in spatial estimation tasks when first exposed to hyperstereoscopic displays, but they adapt to the new situation with feedback and experience (Milgram & Krüger, 1992).

The converse condition, too little depth, can be a result of small interaxial distances, excessive image magnification, or weak perspective. This condition is easily caused by using zoom or telephoto camera lenses with a normal display field of view. If the zoom is not accompanied by a corresponding increase in the camera baseline, as it is in the case of prism binoculars, the depth information is lost. People and objects can appear as cardboard cutouts—flat surfaces with relative depth, but no thickness of their own. This condition is termed *cardboarding*.

Keep in mind that the three basic characteristics—interaxial distance, image size, and perspective (field of view)—are inextricably linked by our visual sense in a synergistic relationship. A change to any individual parameter will introduce a distorted depth perception that cannot be compensated for by adjusting the other values.

9.3.3 Performance issues

A CIG-based system can be very sensitive to the increased demands of stereo. If dual-image generators are used, their video signal and their frame-buffer update must be synchronized. If separate computers are used for each channel, this synchronization can be problematic because software, operating systems, and networks do not always support the required real-time response. It is best to have image generators that have hardware support for multichannel synchronization.

The CIG must be capable of updating the scene at interactive rates. Update rates as low as 10 frames per second can be usable, but may contribute to *simulator sickness* if the user is immersed for a significant period. Simulator sickness is similar to motion sickness, and it can be induced by discrepancies between a user's head movements and corresponding movements in the visual scene. An update rate of 15 frames per second should be considered a minimum, although experts recommend real-time update rates of 30 or 60 frames per second to minimize the possibility of simulator sickness.

One of the most obvious requirements of stereoscopy is the need for two displays. In many cases designers will select separate display systems for each eye. This results in higher costs for dual displays, optics, and driving electronics. It can also add to the weight and bulk of the HMD. A benefit of dual systems is that each eye receives the full resolution and brightness of its display.

An alternative that reduces costs is to share the display and optics for both eyes. In this case, the left- or right-image selection occurs through other mechanisms. For example, the left and right images can be time-multiplexed using mechanical or electronic shutters to route the correct display to each eye. This configuration has the disadvantage that each eye gets only a part of the resolution or brightness of the display. In both dual and shared displays, the size, geometry, brightness, contrast, and color must be carefully matched. Minor variations in displays can become apparent when used in a stereoscopic setup (Lipton, 1984).

9.3.4 User acceptance

In an appropriate application, a stereoscopic display can boost user acceptance of an HMD. After all, an HMD often restricts the movements of the user. An immersive HMD limits the user's ability to interact with the normal environment. To gain acceptance, there must be a clear advantage to using the HMD. Stereoscopic capability can be part of the advantage. When combined with the visual immersion, support for viewpoint control using natural head movements, and a task that requires enhanced spatial understanding of a 3D environment, a stereoscopic HMD provides a unique experience.

Approximately 15 percent of people have some degree of stereoscopic impairment (Lipton, 1982). Approximately 2 percent are stereo-blind and simply do not possess the sense of stereopsis. Perhaps another 2 to 5 percent have physiological conditions such as hyperphoria, strabismus, or diplopia. The remainder have varying degrees of stereoscopic perception, and they experience some fatigue or discomfort with even the best displays. Since most of this group nevertheless have functional binocular vision, these conditions do not preclude them from working effectively with a stereoscopic display. They may experience a reduction of the potential benefits, however. Since stereoscopic displays are primarily beneficial for tasks requiring enhanced 3D understanding, the question is whether people with stereoscopic impairment are suitable for such tasks in the first place. They would be as impaired in the real world as with an HMD.

It is possible that users may experience eye fatigue, blurring, or disorientation after using a stereoscopic HMD for long periods. Sources of fatigue such as misregistration, vertical parallax, relative brightness, and other binocular asymmetries are attributable to the HMD design. Any discrepancy in these areas should be considered inexcusable. Divergent parallax and excessive parallax are completely avoidable with careful software design. In a properly designed HMD system, most fatigue is traceable to A/C conflicts. This leads us to believe that if the display hardware and software are conscientiously designed to match normal vision as closely as possible, there should be no impediments to using stereoscopic displays for periods up to an hour or more. Although I know of no physiological studies that confirm this, experience indicates that once the major viewing parameters are correct, one becomes more adept at using stereoscopic displays and afterward adapting back to real-world vision, with practice. However, industrial health guidelines advise that we should rest our eyes frequently when using any display, not just HMDs.

9.3.5 A systems approach to HMD design

The purpose of this section is to suggest some ideas and promote discussion about how HMDs fit into a comprehensive human-computer interface system. These suggestions are intended to bring out HMD design issues that impact the usability of the system.

An HMD with excellent optical and mechanical design can be rendered useless by CIG software that is mismatched to the HMD's characteristics, or that cannot adapt to different users' adjustments. In addition to basic adjustments associated with fitting the HMD to the user's head, some HMD designs offer optical adjustments that affect the stereoscopic viewing parameters. If these adjustments are not made known to the CIG system, the display system has failed from the user's point of view. Ideally, sensors should be provided so that application software can determine the parameters without resorting to manual measurements and keyboard input. In this way, the associated CIG software can adjust the perspective view generation algorithms automatically.

An adjustment for interpupillary distance (eye separation) should be considered a must, because this varies significantly among users. Some optical systems have a wide exit pupil, which makes them less sensitive to variations among users. In this case, only a software calibration procedure (Min & Jense, 1994) is required. A knob or other mechanism mounted directly on the HMD could be provided for this purpose, along with a serial port so that parameters can be communicated directly to the CIG software. Since for immersive applications the CIG tracks the position of the HMD anyway, it might make sense to integrate these two inputs into the same I/O stream.

Many people who are not stereoscopically deficient wear corrective lenses, so naturally HMDs should be designed to accommodate eyeglasses. If the optics can be focused for each eye separately, many people who wear simple corrective lenses may prefer to use the HMD without eyeglasses. This option may be more comfortable for users who do not have severe astigmatism. There is no need for sensors to report fine focus adjustments, but if the focal length of the optical system is drastically altered, as with a zoom control, that should be sensed and reported because it alters the field of view and apparent distance of the display screen.

In some designs, the display images can be optically or mechanically moved to provide various binocular field-of-view configurations. As before, a sensor should be provided to supply the CIG with calibrated information on location and orientation of the display images. This could be accomplished with sensor switches and corresponding preset display modes.

Even without a sensor capability, it is important for HMD manufacturers to provide accurate display configuration measurements with their products, so that system integrators and users can derive accurate stereoscopic parameters.

9.4 Challenges for Successful Design

Successful application of computer-generated stereograms requires an appreciation for the depth cues our visual system uses to interpret the sensory input provided by our eyes. If the light stimulus to our eyes can be sufficiently duplicated by computer simulation, a stereoscopic perception will result. Given a reasonable image selection mechanism (a stereoscopic viewing device), the success of the visual simulation depends primarily on the production of the stereogram, and only secondarily on the properties of the selection mechanism.

9.4.1 Matching human vision

An important viewing consideration is maintaining the normal image size, perspective, and depth relationships between the objects in the stereoscopic scene. This is called *orthoscopy,* or in our application, *orthostereoscopy.* Perfect orthostereoscopy can be achieved only under highly controlled conditions, if at all. For this reason, Lipton (1990) views orthostereoscopy as rarely of practical value. HMDs in particular offer much more control than other stereoscopic display technologies. Our goal is not only to present pleasing images but also to faithfully communicate the size and depth relationships in a simulated scene. Regard for the viewer's comfort and everyday visual experience suggests that we use orthostereoscopy as a guide, from which we judiciously depart only by necessity.

Many visual depth cues are learned through experience with our environment. These are the psychological depth cues listed in Table 9.1. In contrast, the physiological depth cues listed in Table 9.2 correspond to our sensory capabilities. If we are careless it is possible to construct stereograms in which the depth cues contradict each other. In these cases the perception of the image may be unpredictable or bizarre. The mind will often choose one of these interpretations:

- One cue will be ignored in favor of another.
- The depth of the image will be exaggerated or compressed.
- The viewer will experience discomfort.
- The binocular images will not fuse, causing a double image.

TABLE 9.1   Psychological Depth Cues

Name	Example
Image size	Large → close, small → distant
Linear perspective	Parallel lines converge in distance.
Interposition	Close objects overlap distant objects.
Aerial perspective	Sharp → close, hazy → distant
Light and shade	Bright → close, dim → distant
Textural gradient	Detail → close, no detail → distant
Motion parallax	Fast → close, slow → distant

Source: Lipton, 1982, pp. 53–38

TABLE 9.2   Physiological Depth Cues

Name	Description
Accommodation	Focusing for a certain distance
Convergence	Rotation of eyes to point of interest
Retinal disparity	Difference between left & right views
Stereopsis	Literally, "solid-seeing"

Source: Lipton, 1982, pp. 58–61

MacAdam (1954) emphasizes the synergistic relationship of linear perspective, familiarity with the objects represented, and disparity in a stereoscopic image. In particular, he notes that if either perspective or disparity deviates from normal human experience, there is no way to compensate adequately for one by adjusting the other. However, he demonstrates that interpretation of perspective and disparity are not absolute, but relative to each other and to the familiarity of the image. For example, an image of a human face displayed twice as large as normal will nearly always be perceived as a normal-sized face at a distance consistent with the viewer's normal experience. The relative disparity in the stereoscopic image will then be interpreted proportionally. MacAdam also refutes the popular fallacy that distance in a stereoscopic image is sensed by "triangulating" the convergence of the eyes to the point of interest. On the contrary, the primary means of determining distance from the viewer is image size and knowledge of the object; any depth cue from convergence must be strictly relative to other points in the scene.

Besides coordinating the proper use of visual depth cues, the programmer must be aware of several constraints that have a direct bearing on the viewability of the stereogram. If these constraints are not considered, it is very likely that the viewers will be uncomfortable, or will not experience stereopsis. Tolin (1987) discusses these considerations as "pitfalls" for the display designer to avoid. Lipton (1984) characterizes them in terms of binocular symmetries:

Illumination.   Both left and right fields should have equal patterns of luminance across the display surface. Note that *uniform* luminance is not necessarily required, only *symmetrical* luminance: homologous points on each display have the same brightness.

Aberration.   The optical systems for both fields need not be perfect, but at least they must be symmetrical in terms of optical measures such as focus and distortions across the image field.

Chromaticity.   The color balance and uniformity of the images must be symmetrical. If the color differences are too great, the image will appear to shimmer due to retinal rivalry.

Geometry.   This criterion states that the size and shape of the image space must match across stereoscopic fields. For example, misaligned optics or poor designs can cause distortions, such as keystoning, which shift homologous points and create vertical parallax.

Registration.   The image fields for the left and right views must overlap properly to avoid excess or divergent parallax and, in a time-multiplexed system, to provide a steady image. A subclass of this is *raster* registration. Raster lines in a spatially multiplexed or interlaced transmission system may be displaced, producing effects such as vertical parallax and jitter.

Temporal.   Asymmetries may be caused by a device that displays unsynchronized stereoscopic fields. They may also be caused by a sampling device, such as a camera or CIG, which captures stereoscopic image pairs that are temporally displaced. This second type of temporal asymmetry is more insidious. It may not be noticeable at first, but it becomes evident with movement of the viewpoint or as the speed of objects increases.

The only desirable departure from symmetry is parallax, which is necessary for stereopsis. Parallax is the distance between homologous (corresponding) image points in a stereo-pair, measured at the display screen. Table 9.3 summarizes the various types of parallax. A point imaged behind the display screen is defined as having *positive* parallax; on the screen, *zero* parallax; and in front of the display screen, *negative* parallax. Parallax that exceeds the interpupillary distance

TABLE 9.3 Parallax

Horizontal	Horizontal (with respect to viewer) distance between homologous points, measured at the display screen
Zero	Lines of sight converge at screen
Positive	Lines of sight converge behind screen
Negative	Lines of sight converge in front of screen
Divergent	Positive parallax greater than interpupillary distance
Vertical	Vertical (with respect to viewer) distance between homologous points, measured at the display screen

is termed *divergent*. A properly designed stereoscopic display and projection algorithm should never allow divergent parallax, which will cause discomfort to the user. *Vertical* parallax represents an unnatural situation, and is likely to be painful for the viewer. When mapped to the appropriate eye by the image selection mechanism, the other types of parallax (negative, zero, and positive) correspond to the retinal disparity, which contains the stereoscopic depth information.

9.4.2 Perspective

Perspective is not a property of vision; rather, it is a property of three-dimensional space. It is a geometric property that causes foreshortening of objects and makes parallel lines seem to converge in the distance. These phenomena are encompassed by the field of *projective geometry* (Penna & Patterson, 1986).

Projective geometry is the technology, established during the Renaissance, that allows us to record a scene accurately as seen from any position in space. CIGs use the rules of projective geometry to simulate the normal optics of a camera and produce interactive pictures of 3D scenes. When connected to a stereoscopic display, however, difficulties may arise. If the geometrical viewing conditions in the HMD do not exactly match the projective geometry used to create the stereoscopic images, an image distortion will result.

As mentioned in a previous section, excess perspective in a stereoscopic display causes exaggerated depth, and too little perspective causes cardboarding. Intuitively, we can understand this effect by imagining the scene as a deformable solid. Excess perspective attempts to squeeze a wide-angle view into a narrow-angle display, which forces the scene to expand in the dimension of depth. Conversely, stretching a narrow-angle view onto a wide-angle display causes the scene to thin out.

Altering the perspective may be desirable in some applications, but in most situations normal perspective is desired. For example, abnormal perspective can affect eye-hand coordination (Milgram & Krüger, 1992). The user might be able to learn to compensate, but such training would be counter to ordinary experience and could even cause momentary confusion when the user returns to the real world.

Ideally, perspective calculations should match the subtense of the display field of view. To do this, we need an accurate representation for all the relevant parameters. We now begin to define terms and to construct a mathematical viewing model, which can be used to form correct perspective in HMDs.

9.4.3 Field of view

Perspective projections are sometimes specified by the desired field of view (FOV). This approach usually assumes that the center of perspective is centered exactly on the display surface, but in HMDs this is not always the case. It is easier to get correct perspective by direct measurements of the viewing geometry. However, the FOV is still a useful parameter for characterizing and comparing various types of displays. Referring to Fig. 9.1, there are three quantities of interest:

Monocular field of view. The FOV is the angle subtended horizontally by the display surface for each eye. I use the notation α_l and α_r for the left and right FOV, respectively. The FOV for each eye can be calculated (approximately) by

$$\alpha = 2\tan^{-1}\left(\frac{w}{2l}\right) \qquad (9.1)$$

where w is the width of the image and l is the focal length of the optical system. (If the image plane is rotated or the eyepoint is offset from center, this formula is approximate. However, these factors do

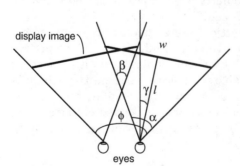

Figure 9.1. Field of view.

not affect the projection algorithm.) We can similarly define a vertical field of view for each eye.

Binocular overlap. This is an angular measurement of the region visible to both eyes, within which stereopsis can occur. I define it as

$$\beta = \frac{1}{2}(\alpha_1 - 2\gamma_1 + \alpha_r - 2\gamma_r) \tag{9.2}$$

where γ is the angle of each FOV's centerline from the forward coordinate axis. The overlap can be zero or negative when the central shared portion of the view is obstructed. In this case, stereopsis is impossible.

Binocular field of view. This is the total field of view with both eyes viewing simultaneously. Note that stereopsis is not possible in peripheral regions because the binocular overlap is strictly less than or equal to the binocular field of view. Even when the binocular overlap is equal to the binocular field of view, the peripheral regions are exclusive to each eye, due to the displacement between the eyes. This situation exists even in normal vision; a stereoscopic display is just somewhat more restricted than normal vision. The binocular field of view can be calculated as

$$\phi = \alpha_1 + \alpha_r - \beta \tag{9.3}$$

9.4.4 Range of depth

The angular measurement of eye movement is termed *vergence.* Using the small angle approximation $\omega = \tan \omega$, and referring to Fig. 9.2, we can derive the relation (Southard, 1992)

$$\theta \equiv \alpha - \beta \approx \frac{d_{ip}}{d_{focus}} - \frac{d_{ip}}{d_p} \tag{9.4}$$

Valyus (1966) found that most people can comfortably tolerate a vergence θ up to $\pm 1.6°$ (± 0.0279 radian). More recently, Yeh and Silverstein (1990) measured a vergence tolerance ranging from $\theta_{min} = -4.93°$ (-0.0860 radian) to $\theta_{max} = 1.57°$ (0.0274 radian).

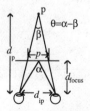

Figure 9.2. Parallax and convergence.

The vergence tolerance has implications for the depth range recommended for display by a particular device. Based on Valyus's guideline, Southard (1992) calculates that for a stereoscopic display focused at infinity, the closest recommended distance from the viewer to an object is 2.3 m. The greatest comfortable depth range within the vergence limits will be achieved, however, when objects at infinity are near the outer vergence limit instead of at zero vergence. When the display is focused at 2.4 m, this arrangement provides a comfortable stereoscopic depth range from about 1.2 m to infinity. This consideration leads to a recommendation that HMDs should be designed to be focused at approximately 2.4 m from the viewer, not at infinity. The comfortable viewing range can be brought closer to the viewer by decreasing the focal distance, but this implies that viewing distant objects may cause eyestrain.

The possibility of excessive parallax can be eliminated by setting the near and far clipping plane distances, n and f respectively, to limits specified by the vergence tolerance, θ_{min} and θ_{max}. The designer may prefer to select limits deemed appropriate for the application, as measured by Valyus, Yeh, and Silverstein, or another authority. By substituting θ_{min} and θ_{max} in Eq. (9.4) and rearranging, we can get expressions for the near and far clipping plane distances,

$$n = \frac{d_{ip}}{d_{ip}/d_{focus} - \theta_{min}} \tag{9.5}$$

and

$$f = \frac{d_{ip}}{d_{ip}/d_{focus} - \theta_{max}} \tag{9.6}$$

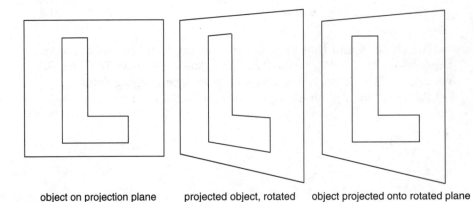

object on projection plane   projected object, rotated   object projected onto rotated plane

Figure 9.3. Keystoning leads to vertical parallax.

9.4.5 Rotated perspective

The operators for rotation and perspective projection are not commutative. If the projective planes are rotated, keystoning of the stereoscopic image pairs can introduce *vertical parallax* (Fig. 9.3). The reason for this is that one edge of the image is now closer to the eye than the other, and so has a longer projected length. Vertical parallax should be eliminated to remove this source of eye fatigue. If the viewing apparatus is properly designed and maintained, vertical parallax can be eliminated by compensating for rotated perspective in the projection algorithm.

9.5 Simulating Stereoscopic Vision

Early workers proposed using a simple rotation of the object model to create a stereoscopic pair of images. Most researchers understand that this is an incorrect, albeit expedient, model. Hodges and McAllister (1989, 1990) describe the problems caused by the rotation algorithm, including vertical parallax and depth distortions toward the periphery of the stereo image. The rotation method can be effective for isolated objects floating in space, but it is inappropriate for extended scenes such as those encountered in virtual environments.

Baker (1987) articulated the parallel projection algorithm, in which the image planes are modeled as rectangular areas centered on each eyepoint, with parallel optical axes from the eyes to the projective planes. During viewing, the projected images can be shifted horizontally to control the relative positioning of the scene with respect to the image surround. When the parameters are carefully selected, this method can produce correct perspective. Most implementations provide the user with a horizontal shift and interaxial distance controls. These allow the user to adjust the scene to various preferences, but the perspective is likely to be incorrect in relation to the dimensions of the scene. The parallel projection algorithm has a serious shortcoming, in that the left and right vertical edges of the scene cannot be made to match properly without an additional masking step. Due to this deficiency, the parallel projection algorithm is suited to isolated objects floating against a neutral background, but not to displays of extended scenes.

It is now widely recognized that some form of off-axis perspective projection should be used for stereoscopic visual simulation, including virtual environments. Previously, I presented a method termed the "stereoscopic window" algorithm (Southard, 1992) that is appropriate for stationary planar displays. The algorithm models the display screen as a window, through which we view a scene. Objects that extend beyond the display surface are properly clipped at the display edges, so that correct perspective and occlusion by the display surround are maintained.

The stereoscopic window algorithm correctly models displays for which the left and right images are coplanar and fully overlapped on the display surface. For HMDs, however, neither of these assumptions may hold, as illustrated in Fig. 9.1. The design constraints associated with head-mounted displays often dictate that each eye has its own display device. Furthermore, it may not be physically feasible to mount the display devices and their associated optics so that the perceived images are coplanar and parallel to the face. Thus, each image plane has its own position and orientation. This work extends the stereoscopic window model to account for various HMD designs.

Robinett and Rolland (1991, 1992) provide a viewing model for the VPL EyePhone™, Models 1 and 2. This HMD uses backlit flat-panel LCD displays. The displays are offset and do not overlap, but they are coplanar. The EyePhone uses LEEP wide-angle optics (Howlett, 1990) having a radial distortion that emphasizes detail in the central region of the image, with less detail in peripheral regions. These optics were originally designed for telepresence camera systems, which are fitted with corresponding compensation lenses. In their original use, LEEP optics provide a wide-angle, orthostereoscopic view for telepresence systems. Unfortunately, they introduce distortion in the computer-generated images of virtual reality systems. Since the wide-angle image field exhibits a nonlinear radial distortion, it is problematic for computer image generators to compensate for this effect. Fisher at NASA Ames accomplished this by aiming the LEEP telepresence cameras at conventional computer monitors.

Robinett and Holloway (1994, 1995) describe a viewing model for HMDs in terms of *quaternions*. Quaternions offer advantages including compact representation and intuitive interpretation, and they are computationally efficient and robust. However, some operations such as shearing, nonuniform scaling, and perspective projection cannot be represented by quaternions, whereas they can be represented by homogeneous matrix operations. Most graphics hardware and software available today support matrix transformations. Therefore, we prefer a matrix formulation for a complete description of our stereoscopic viewing model.

9.5.1 Recommended method

The perspective model presented here is generalized to any HMD with planar (or very nearly planar) display surfaces. This includes most HMDs that use LCD or CRT displays, but it excludes displays using special optics that do not preserve normal perspective. In overview, the model has the following characteristics:

Viewing position. The viewing position may be located anywhere in front of the display screen. The eyepoint does not have to be centered on its respective display.

Scene scaling. The scene will be properly scaled to actual measurements. This property allows it to be used for interactive augmented reality or virtual environment applications.

Display dimensions. The transformations account for independent placement of rectangular display screens of any size.

Display orientation. The transformations account for independent orientation of the display screens.

Partial overlap. The left- and right-eye displays may overlap fully or partially, either coplanar or at an angle. Stereopsis will be possible in the overlapped section. The remaining portions will still serve to provide monoscopic "peripheral" vision.

Head tracking. The algorithm supports input from a head-tracking device. The algorithm accounts for the offset between the tracking sensor and the user's head.

9.5.2 Matrix operators

We now describe the viewing algorithm in detail. The essential parameters for calculating a perspective projection are the size, position, and orientation of the image plane, and the position of the center of projection (the eyepoint). For HMDs, we must assume that these parameters are different for each eye. The algorithm is stated in terms of parameterized matrix transformations. Positions are denoted by homogeneous coordinates (Foley, van Dam, Feiner, & Hughes, 1990) written as a row vector $p = [x\ y\ z\ 1]$. A subscript x, y, or z indicates selection of a component of p; otherwise, subscripts are descriptive. Orientation coordinates are specified by a 3-tuple of Euler angles $a = \left(h\ p\ r \right)$, where h is the heading (azimuth), p is the pitch (elevation), and r is the roll (twist). Subscripts h, p, and r are used to denote component selection. Transformations are 4×4 matrix operators, written as A. Operations are concatenated by matrix multiplication, $N = AB$. Vectors are transformed by multiplication with the operators, written as $p' = pA$.

We use a right-handed world coordinate system for object modeling and positioning (Fig. 9.4). By convention the positive x axis extends to the right (east), the positive y axis extends forward (north), and the positive z axis extends up. For historical reasons, viewing projections are specified in a left-handed coordinate system, with the x axis to the right, the y axis up, and the z axis extending into the display screen

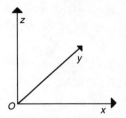

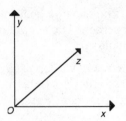

Figure 9.4. Right-handed world coordinate system.

Figure 9.5. Left-handed viewing coordinate system.

(Fig. 9.5). Other systems may follow different conventions, which will require minor modifications to the coordinate axes transformations.

The complete algorithm is defined in terms of the following basic operators. It is possible to implement each of these operators as a procedure call that initializes a matrix. The matrices can then be multiplied together to build composite operators. Alternatively, the subroutines can be thought of as operators that apply their transformation to an input matrix. In addition, graphics libraries such as GL (McLendon, 1991) support many of these operations directly. Here, we follow the notation conventions used by GL and IRIS Performer (Hartman & Creek, 1994). PHIGS (ISO, 1989), PEX (Gaskins, 1992), and OpenGL (Neider, Davis, & Woo, 1994) use column vectors, for which the matrix operators are the transpose of those presented here.

Translation.

$$T(p) = \begin{bmatrix} 1 & 0 & 0 & 0 \\ 0 & 1 & 0 & 0 \\ 0 & 0 & 1 & 0 \\ p_x & p_y & p_z & 1 \end{bmatrix} \tag{9.7}$$

Rotation.

$$R(a) = \begin{bmatrix} c_h c_r - s_h s_p s_r & s_h c_r + c_h s_p s_r & -c_p s_r & 0 \\ -s_h c_p & c_h c_p & s_p & 0 \\ c_h s_r + s_h s_p c_r & s_h s_r - c_h s_p c_r & c_p c_r & 0 \\ 0 & 0 & 0 & 1 \end{bmatrix} \tag{9.8}$$

where $c_h = \cos a_h$ $s_h = \sin a_h$
$c_p = \cos a_p$ $s_p = \sin a_p$
$c_r = \cos a_r$ $s_r = \sin a_r$

R is derived from component Euler angle rotations around the modeling coordinate axes (Foley et al., 1990) and is equivalent to $R_y(a_r)R_x(a_p)R_z(a_h)$.

Scaling.

$$S(s_x, s_y, s_z) = \begin{bmatrix} s_x & 0 & 0 & 0 \\ 0 & s_y & 0 & 0 \\ 0 & 0 & s_z & 0 \\ 0 & 0 & 0 & 1 \end{bmatrix} \tag{9.9}$$

Shearing to z axis.

$$Sh_z(a, b) = \begin{bmatrix} 1 & 0 & 0 & 0 \\ 0 & 1 & 0 & 0 \\ a & b & 1 & 0 \\ 0 & 0 & 0 & 1 \end{bmatrix} \tag{9.10}$$

Perspective.

$$P_{-1,1}(d) = \begin{bmatrix} 1 & 0 & 0 & 0 \\ 0 & 1 & 0 & 0 \\ 0 & 0 & (1+d)/(1-d) & 1 \\ 0 & 0 & -2d/(1-d) & 0 \end{bmatrix} \tag{9.11}$$

$$P_{0,1}(d) = \begin{bmatrix} 1 & 0 & 0 & 0 \\ 0 & 1 & 0 & 0 \\ 0 & 0 & 1/(1-d) & 1 \\ 0 & 0 & -d/(1-d) & 0 \end{bmatrix} \tag{9.12}$$

The $P_{-1,1}$ form of the perspective transform is used with graphics systems such as GL that normalize to the range $[-1, 1]$ for clipping. The $P_{0,1}$ form is used with systems that normalize to the range $[0, 1]$.

Transformation of coordinate axes.

$$N_{rl} = \begin{bmatrix} 1 & 0 & 0 & 0 \\ 0 & 0 & 1 & 0 \\ 0 & 1 & 0 & 0 \\ 0 & 0 & 0 & 1 \end{bmatrix} \tag{9.13}$$

N_{rl} is a transformation of coordinate axes from the right-handed modeling coordinate system, to the left-handed graphics coordinate system. This has the effect of exchanging the y and z axes.

9.5.3 Viewing algorithm

Matrix transformations can be constructed from primitive matrix operators as described above. The viewing algorithm can be expressed as a composite transform, which places the viewer at the correct location in the model space, and then projects an image onto the viewing plane for each eye. Note that the projection transformation will be different for each eye, because we assume that the viewing geometry is different for each eye. We can express this as

$$N = N_{\text{viewpoint}}^{-1} N_{\text{proj}} \qquad (9.14)$$

where N is the composite transform, $N_{\text{viewpoint}}$ describes the position of the eyepoint for each eye, and N_{proj} is the perspective transformation for each eye. The inverse of the modeling transformation matrix for any object is the viewing transformation from that object's position and orientation. However, it is not always necessary to explicitly calculate the inverse of $N_{\text{viewpoint}}$ to get the viewing matrix. Software packages such as IRIS Performer (Hartman & Creek, 1994) accept $N_{\text{viewpoint}}$ directly. For GL, we can obtain the inverse by negating or inverting the parameters and reversing the order of the matrix operators (recognizing that R is a composite transformation). This procedure systematically "undoes" the transformations in $N_{\text{viewpoint}}$, and thus is equivalent to the inverse. Alternatively, there is a quick way to compute the inverse of a matrix N composed entirely of rotations and translations (Wu, 1991):

$$N^{-1} = \begin{bmatrix} & & & 0 \\ & N_{\text{rot}}^{\text{T}} & & 0 \\ & & & 0 \\ -r_0 r_3 & -r_1 r_3 & -r_2 r_3 & 1 \end{bmatrix} \qquad (9.15)$$

where $N_{\text{rot}}^{\text{T}}$ is the transpose of the 3×3 rotation submatrix, comprising the first three rows and columns of N; r_i is the ith row of N; and $r_i r_j$ is the inner (dot) product of r_i and r_j. This procedure works because $N_{\text{rot}}^{\text{T}}$ is orthonormal, and the inverse of an orthonormal matrix is just its transpose.

We now proceed by describing how to construct $N_{\text{viewpoint}}$ and N_{proj}. The situation for a head-mounted display is illustrated in Fig. 9.6, and the required parameters are summarized in Table 9.4. To characterize a head tracker, we have two points, p_{tracker} and p_{sensor}. The coordinates p_{tracker} and a_{tracker} are the location and orientation on the viewing platform where the tracker's emitter is mounted. The tracker measures positions relative to that point. p_{sensor} is the position measured by the tracker's sensor, and a_{sensor} is the measured orientation. The sensor can be mounted anywhere on the HMD. Various sensor mounting points are accounted for by p_{head} and a_{head}, which are measured relative to the sensor's position and orientation. We assume that the user has some kind of control that allows for movement through the virtual environment. This could be accomplished, for example, by a joystick controller, by hand gestures using a glove input device, or by voice command. The head tracker information is relative to the platform's

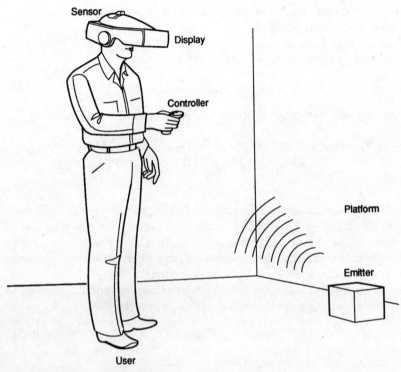

Figure 9.6.   Head-mounted display.

TABLE 9.4   Stereoscopic Viewing Model Parameters

Parameter	Description	Source
h	height of displayed image	display specifications
w	width of displayed image	display specifications
n	distance from eyepoint to near clipping plane	Eq. (9.5)
f	distance from eyepoint to far clipping plane	Eq. (9.6)
p_{platform}	position of motion platform in world coordinates	user input
a_{platform}	orientation of motion platform in world coordinate	user input
p_{tracker}	position of tracker origin relative to platform	measurement
a_{tracker}	orientation of tracker origin relative to platform	measurement
p_{sensor}	position of sensor	measured by tracker
a_{sensor}	orientation of sensor	measured by tracker
p_{screen}	position of center of displayed image relative to the user's head	measurement or specifications
a_{screen}	orientation of displayed image relative to the user's head	measurement or specifications
p_{head}	position of the center of baseline of the eyes, relative to position sensed by tracker	measurement or specifications
a_{head}	orientation of the center of baseline of the eyes, relative to orientation sensed by tracker	measurement or specifications
p_{ipd}	$[\pm d_{ip}/2 \ 0 \ 0 \ 1]$, + for right eye, − for left eye	measured from user

coordinate system. The position of the platform is given by $\boldsymbol{p}_{\text{platform}}$ and the orientation by $\boldsymbol{a}_{\text{platform}}$.

The first step is to calculate the position of each eyepoint relative to the position of its corresponding display screen:

$$\boldsymbol{p}_{\text{eye}} = \boldsymbol{p}_{\text{ipd}} - \boldsymbol{p}_{\text{screen}} \tag{9.16}$$

Next, we calculate the viewpoint transformation:

$$
\begin{aligned}
\boldsymbol{N}_{\text{viewpoint}} = {} & \boldsymbol{T}(\boldsymbol{p}_{\text{eye}})\boldsymbol{R}(\boldsymbol{a}_{\text{head}})\boldsymbol{T}(\boldsymbol{p}_{\text{head}})\boldsymbol{R}(\boldsymbol{a}_{\text{sensor}})\boldsymbol{T}(\boldsymbol{p}_{\text{sensor}}) \\
& \boldsymbol{R}(\boldsymbol{a}_{\text{tracker}})\boldsymbol{T}(\boldsymbol{p}_{\text{tracker}})\boldsymbol{R}(\boldsymbol{a}_{\text{platform}})\boldsymbol{T}(\boldsymbol{p}_{\text{platform}})
\end{aligned} \tag{9.17}
$$

This HMD viewpoint transformation is illustrated in Fig. 9.7 as a nested hierarchy of coordinate transformations. The eye position is transformed by the head tracker's orientation and position, which in turn is transformed by the orientation and position of the platform (i.e., the room the user is in) in the world model.

Equation (9.18) calculates the position of each eyepoint relative to the frame of reference of the rotated display screen:

$$\boldsymbol{p} = \boldsymbol{p}_{\text{eye}}\boldsymbol{R}(\boldsymbol{a}_{\text{screen}}) \tag{9.18}$$

We can now write the projection transformation:

$$\boldsymbol{N}_{\text{proj}} = \boldsymbol{R}(\boldsymbol{a}_{\text{screen}})\boldsymbol{N}_{\text{rl}}\boldsymbol{Sh}_z\left(\frac{-\boldsymbol{p}_x}{\boldsymbol{p}_y}, \frac{-\boldsymbol{p}_z}{\boldsymbol{p}_y}\right)\boldsymbol{S}\left(\frac{-2\boldsymbol{p}_y}{wf}, \frac{-2\boldsymbol{p}_y}{hf}, \frac{1}{f}\right)\boldsymbol{P}\left(\frac{n}{f}\right) \tag{9.19}$$

The first term rotates the world model to compensate for the display screen's orientation. The second term changes from the right-handed world coordinate system to the left-handed graphics coordinate system. The third term shears the scene in depth to account for the off-axis

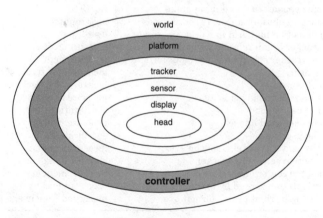

Figure 9.7. HMD nested transformations.

position of the center of projection. The shear parameters are just the ratio of the perpendicular off-axis distance to the perpendicular distance from the image plane to the center of projection, for both the horizontal and vertical directions. The fourth term scales the image dimensions to fit a canonical viewing frustum. The uniform scaling factor of $1/f$ normalizes the far clip distance to 1. The x and y scaling factors normalize for the screen dimensions and viewing distance. The fifth term is the perspective transformation. The parameter n/f is the position of the near clipping plane in the canonical viewing frustum.

9.6 Conclusion

We have presented a viewing model for virtual environment head-mounted displays that applies to the class of head-mounted stereoscopic displays with planar images. The model accounts for the rotated and offset image planes present in some HMD designs. The algorithm eliminates the possibility of vertical parallax due to image plane rotations, and parameters can be selected to eliminate accommodation/convergence fatigue for most viewers by means of clipping planes. We have applied this model to HMDs in our laboratory. The algorithm successfully compensates for partially overlapping noncoplanar image planes on an HMD. This viewing model has immediate application for immersive displays and visual simulation, such as those encountered in virtual environments and training applications. Quantitative evaluations of the viewing model are left for future study.

9.7 References

Arthur, K., Booth, K. S., & Ware, C. (1993). Evaluating performance in fish tank virtual reality. *ACM Trans. Information Systems, 11* (3), 239–265.

Baker, J. (1987). Generating images for a time-multiplexed stereoscopic computer graphics system. *True 3D Imaging Techniques and Display Technologies, Proceedings of the SPIE, 761,* 44–52.

Bridges, A. L., & Reising, J. M. (1987). Three-dimensional stereographic pictorial visual interfaces and display systems in flight simulation. *True 3D Imaging Techniques and Display Technologies, Proceedings of the SPIE, 761,* 102–111.

Cruz-Neira, C., Sandin, D. J., & DeFanti, T. A. (1993). Surround-screen, projection-based virtual reality: The design and implementation of the CAVE. *Computer Graphics (Proceedings SIGGRAPH '93), 27,* 135–142.

Davis, E. T., & Hodges, L. F. (1995). Human stereopsis, fusion, and stereoscopic virtual environments. In W. Barfield & T. A. Furness III (Eds.), *Virtual environments and advanced interface design* (pp. 145–174). New York: Oxford University Press.

Deering, M. (1992). High-resolution virtual reality. *Computer Graphics (Proceedings SIGGRAPH '92), 26* (2), 195–202.

Fisher, S. S. (1995). Recent developments in virtual experience design and production. *Stereoscopic Displays and Virtual Environments II, Proceedings of the SPIE, 2409,* 296–302.

Foley, J. D., van Dam, A., Feiner, S. K., & Hughes, J. F. (1990). *Computer graphics: Principles and practice* (pp. 229–281). Reading, MA: Addison-Wesley.

Gaskins, T. (1992). *PEXlib Programming Manual.* (pp. 1105–1012). Sebastopol, CA: O'Reilly & Associates.

Gorski, A. M. (1992). User evaluation of a stereoscopic display for space training applications. *Stereoscopic Displays and Applications III, Proceedings of the SPIE, 1669,* 236–243.

Grotch, S. L. (1983). Three-dimensional and stereoscopic graphics for scientific data display and analysis. *IEEE Computer Graphics and Applications, 3 (8),* 31–43.

Hartman, J., & Creek, P. (1994). *IRIS Performer programming guide* (Doc. No. 007-1680-020). Mountain View, CA: Silicon Graphics Inc.

Hoberman, P. (1995). Bar Code Hotel: Diverse interactions of semi-autonomous entities under the partial control of multiple operators. *Stereoscopic Displays and Applications II, Proceedings of the SPIE, 2409,* 303–310.

Hodges, L. F., & McAllister, D. F. (1985). Stereo and alternating-pair techniques for display of computer generated images. *IEEE Computer Graphics and Applications, 5 (9),* 38–45.

Hodges, L. F., & McAllister, D. F. (1989, July/August). Computing stereographic views. *Course Notes #24: Stereographics.* ACM SIGGRAPH '89, Boston, MA.

Hodges, L. F., & McAllister, D. F. (1990). Rotation algorithm artifacts in stereoscopic images. *Optical Engineering, 29,* 973–976.

Hodges, L. F. (1992). Time-multiplexed stereoscopic computer graphics. *IEEE Computer Graphics and Applications, 12 (2),* 20–30.

Howlett, E. M. (1990). Wide angle orthostereo. *Stereoscopic Displays and Applications, Proceedings of the SPIE, 1256,* 210–223.

ISO. (1989). *Information processing systems—Computer graphics—Programmer's hierarchical interactive graphics system (PHIGS)—Part 1: Functional Description* (ISO/IEC 9592-1). New York: American National Standards Institute.

Jones, E. R., & McLaurin, A. P. (1994). Single camera three-dimensional laparoscopic system. *Stereoscopic Displays and Virtual Reality Systems, Proceedings of the SPIE, 2177,* 156–160.

Lane, B. (1982). Stereoscopic displays. *Processing and Display of Three-Dimensional Data, Proceedings of the SPIE, 367,* 20–32.

Lipton, L. (1982). *Foundations of the stereoscopic cinema: A study in depth.* New York: Van Nostrand Reinhold.

Lipton, L. (1984). Binocular symmetries as criteria for the successful transmission of images. *Processing and Display of Three-Dimensional Data II, Proceedings of the SPIE, 507,* 108–113.

Lipton, L. (1990, August). Some aspects of stereoscopic composition. *Course Notes #6: Stereographics.* ACM SIGGRAPH '90, Dallas, TX.

MacAdam, D. L. (1954). Stereoscopic perceptions of size, shape, distance, and direction. *Journal of the Society of Motion Picture and Television Engineers, 62,* 271–293.

McDowall, I. E., Bolas, M., Pieper, S., Fisher, S. S., & Humphries, J. (1990). Implementation and integration of a counterbalanced CRT-based stereoscopic display for interactive viewpoint control in virtual environment applications. *Stereoscopic Displays and Applications, Proceedings of the SPIE, 1256,* 136–146.

McLaurin, A. P., & Jones, E. R. (1994). Virtual endoscope. *Stereoscopic Displays and Virtual Reality Systems, Proceedings of the SPIE, 2177,* 269–273.

McLendon, P. (1991). *Graphics library programming guide* (Doc. No. 007-1210-040). Mountain View, CA: Silicon Graphics Inc.

Milgram, P., & Krüger, M. (1992). Adaptation effects in stereo due to on-line changes in camera configuration. *Stereoscopic Displays and Applications II, Proceedings of the SPIE, 1669,* 122–134.

Min, P., & Jense, H. (1994). Interactive stereoscopy optimization for head-mounted displays. *Stereoscopic Displays and Virtual Reality Systems, Proceedings of the SPIE, 2177,* 306–316.

Miyashita, T., & Uchida, T. (1990). Cause of fatigue and its improvements in stereoscopic displays. *Proceedings of the Society for Information Display, 31,* 249–254.

Neider, J., Davis, T., & Woo, M. (1994). *OpenGL programming guide (appendix G).* Reading, MA: Addison-Wesley.

Penna, M. A., & Patterson, R. R. (1986). *Projective geometry and its applications to computer graphics* (pp. 220–321). Englewood Cliffs, NJ: Prentice Hall.

Robinett, W., & Rolland, J. P. (1991). A computational model for the stereoscopic optics of a head- mounted display. *Stereoscopic Displays and Applications II, Proceedings of the SPIE, 1457,* 140–160.

Robinett, W., & Rolland, J. P. (1992). A computational model for the stereoscopic optics of a head-mounted display. *Presence, 1,* 45–62.

Robinett, W., & Holloway, R. (1994). *The visual display transformation for virtual reality* (Technical Rep. TR94-031). Chapel Hill, NC: Dept. of Computer Science, University of North Carolina.

Robinett, W., & Holloway, R. (1995). The visual display transformation for virtual reality. *Presence, 4,* 1–23.

Roese, J. A., & McCleary, L. E. (1979). Stereoscopic computer graphics for simulation and modeling. *Computer Graphics (Proceedings SIGGRAPH 79), 13,* 41–47.

Sharma, R., & Molineros, J. (1995). Role of computer vision in augmented virtual reality. *Stereoscopic Displays and Virtual Reality Systems II, Proceedings of the SPIE, 2409,* 220–231.

Smedley, K. G., Haines, B. K., Van Vactor, D., & Jordan, M. (1989). Digital perspective generation and stereo display of composite ocean bottom and coastal terrain images. *Proceedings of the SPIE, 1083,* 246–250.

Sollenberger, R. L., & Milgram, P. (1993). The effects of stereoscopic and rotational displays in a three-dimensional path-tracing task. *Human Factors, 35,* 483–500.

Southard, D. A. (1992). Transformations for stereoscopic visual simulation. *Computers & Graphics, 16,* 401–410.

Teitel, M. A. (1990). The Eyephone: A head-mounted stereo display. *Stereoscopic Displays and Applications, Proceedings of the SPIE, 1256,* 168–171.

Tolin, P. (1987). Maintaining the three-dimensional illusion. *Information Display, 3 (11),* 10–12.

Valyus, N. (1966). *Stereoscopy.* London: Focal Press.

Veron, H., Southard, D. A., Leger, J. R., & Conway, J. L. (1990). Stereoscopic displays for terrain database visualization. *Stereoscopic Displays and Applications, Proceedings of the SPIE, 1256,* 124–135.

Ware, C., & Franck, G. (1994). Viewing a graph in a virtual reality display is three times as good as a 2D diagram. In *Proceedings IEEE Symposium on Visual Languages* (pp. 182–183) Los Alamitos, CA: IEEE Computer Society.

Williams, S. P., Parrish, R. V., & Nold, D. E. (1994). Effective use of stereoptic 3D cueing to declutter complex flight displays. *Stereoscopic Displays and Virtual Reality Systems, Proceedings of the SPIE, 2177,* 223–231.

Woods, A. J., Docherty, T., & Koch, R. (1994). Field trials of stereoscopic video with an underwater remotely operated vehicle. *Stereoscopic Displays and Virtual Reality Systems, Proceedings of the SPIE, 2177,* 203–210.

Wu, K. (1991). Fast matrix inversion. In J. Arvo (Ed.), *Graphics gems II* (pp. 342–350). San Diego, CA: Academic Press.

Yeh, Y. Y., & Silverstein, L. D. (1990). Limits of fusion and depth judgment in stereoscopic color displays. *Human Factors, 32,* 45–60.

Brain-Actuated Control and HMDs

Victoria Tepe Nasman

Gloria L. Calhoun

Grant R. McMillan

Brain-actuated control, or BAC, is introduced and proposed as an alternative control technology that might be integrated with HMDs. Several BAC methodologies are considered. In each case, a particular brain-wave response or pattern is used to control a device or computer task.

A variety of potential applications are considered, including assistive and rehabilitative control, video and entertainment, and aviation and military uses. The basic hardware and software requirements for BAC are reviewed to demonstrate that BAC/HMD integration will probably require very little, if any, modification to basic HMD design or configuration.

10.1   HMDs: The Need for Control Alternatives

Helmet/head-mounted displays represent an engineering success story. The multitude of HMD-related technologies necessary to meet today's performance requirements are already mature and available, as shown in preceding chapters. Head-worn information displays have been tested and improved over many years of research and development. As a current and well-established technology, HMDs offer much potential for a wide variety of present and future applications in aviation, business, medicine, education, and entertainment.

The development of HMD technology has, to date, focused heavily on generating displays at optical infinity and on the integration of optional capabilities, such as projection as an overlay on the real world, stereo presentation, and display update with head position. Such advances are worthwhile objectives when considered against the tremendous demands placed upon users by modern work environments. However, the research and development mission thus far has emphasized the goal of transmitting information *from the system to the user*. The goal of transmitting information or commands *from the user to the system* has received far less attention.

In order to exploit fully the potential advantages of the HMD-based interface, a departure from conventional control is needed. Keyboards, joysticks, and head-down panel-mounted controls fail to optimize—and may even compromise—the potential advantages of head-up, hands-off technology. Further, manual controls are commonly overloaded and tend to restrict information and command flow between the user and system. Control technologies that are presently considered desirable for HMD operation include virtual hand controllers, head-aimed control, eye line-of-sight control, gesture recognition, voice-activated control, and touch-sensitive panels with switch information superimposed from the HMD. Each of these control technologies requires physical manipulation (of the hand, fingers, voice, eyes, and/or head) to control one or more aspects of the task or scene presented on the HMD (see Furness, 1986). Although each of these options may support communication from the user to the system, each presents a variety of possible cost and implementation problems. For example, voice recognition systems may be ineffective in high-noise environments. During high acceleration in aviation, precise head or hand positioning may be difficult or impossible. It may be difficult to integrate an eye-tracking interface with some HMD optical path designs (see Borah, 1989). Display flicker (even when the flicker is not perceptible) and vergence at infinity may also affect eye movement (Neary, 1994; Neary and Wilkins, 1989). Although one or more of these existing forms of control may be achieved and be effective in various application environments, their limitations highlight the need to consider alternatives that may be more resilient to specific constraints associated with either the HMD or the context in which it is operated.

10.2 Brain-Actuated Control: A Unique Control Technology

"Mind over matter" is a popular theme in science fiction and has been the focus of countless psychic investigations. The human imagination has long been captivated by the notion that its own organ, the brain, might achieve direct control over objects or events in the physical world. Only recently has this idea received serious attention as a matter of scientific investigation and systems engineering. Research has shown that brain electrical activity can be applied in real time as a virtually direct control interface to physical devices and computers. "Brain-actuated control" (or BAC) is one of several alternative control technologies now under investigation in our own laboratory and in others.

For HMD applications, control based on brain electrical (electroencephalographic, or EEG) activity could be critical for harnessing the combined power of the human mind and computer technology. A direct brain interface can facilitate the head-up, hands-free advantage of the HMD and so may increase the effective bandwidth of control over a wide range of applications and environments (Table 10.1 and Fig. 10.1). The potential benefits of BAC extend well beyond its relatively unobtrusive and natural character as a nearly direct link between mind and machine. Indeed, many basic conceptions of physical control (see Keele, 1986) may not apply to EEG-based control. For instance, physical movement involves delays associated with arousal, cognition, response processing and planning, and response execution itself. Because a BAC interface operates close to the source of control initiation, it may offer a much faster response output than that normally associated with physical movement.

There are obvious advantages to integrating BAC in head-mounted systems. For example, physical disability, injury, or situational constraints may make it difficult to use traditional manual controls or to control head position or eye line-of-sight with respect to an external

TABLE 10.1   Candidate Applications of Brain-Actuated Control and HMDs

Application	Example operation	Advantage
Aviation	Change radio frequency	Hands remain free
Chemical/biological agent environment	Access environmental readings and control vehicle/subsystems	Alternative when conventional controls are inaccessible
Entertainment	Control video game	Novelty, fun
High-noise environment	Access task-related information	Alternative when voice systems are unusable
Maintenance	Access technical reference information	Hands remain dedicated to other tasks
Rehabilitation	Control home systems	Alternative when conventional controls are inaccessible
Weightless environment	Access display formats	Alternative when conventional controls are inaccessible

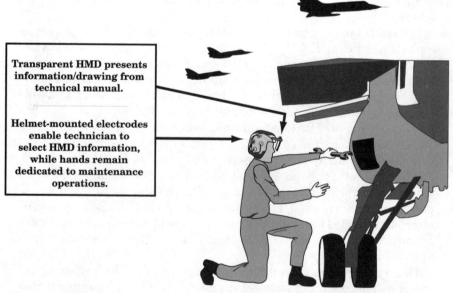

Transparent HMD presents information/drawing from technical manual.

Helmet-mounted electrodes enable technician to select HMD information, while hands remain dedicated to maintenance operations.

Figure 10.1.   Schematic illustrates the use of brain-actuated control with HMD for flight line maintenance to provide head-up, hands-free display and control of technical information in a high-noise setting.

display. Among the most obvious applications of BAC/HMD are those which address the need for mobile assistive or rehabilitative technology. Integrated as a primary control technology, BAC might be employed to operate and interact directly with HMD displays and programs. For example, BAC might be used to operate home control programs that enable selection and on/off switching of home facilities and appliances. In this application the disabled user would, by modulating a specific brain response, control the position of a selection box or cursor on a display that contains several control functions (e.g., telephone, television, lights). When the selection box is positioned on a function for a criterion time period, the display would change to provide subfunctions for that particular home system (preprogrammed phone numbers, a keypad, and return key).

Many tasks in military environments would be enhanced by the convenience, portability, and flexibility beyond that provided by traditional controls. This is frequently the case in modern fighter aircraft cockpits, where the effects of high-acceleration flight and dozens of traditional switches, displays, and devices conspire to produce extremely high cognitive and physical workload. Although it is difficult now to imagine a day when pilots might be able to use BAC as a reliable primary control, it is easier to foresee the implementation of BAC as a secondary control by which pilots or navigators might perform tasks such as multifunction display operation, weapons selection, radio frequency switching, or target selection. There are a multitude of in-flight secondary control tasks to which BAC might be safely and easily applied, using a head-mounted system to record and monitor brain electrical activity and to deliver the necessary task information and feedback to the user. To the extent that BAC/HMD may help reduce the effects of overall workload on pilot cognitive and physical abilities, it may ultimately prove to be a necessary cockpit control option. HMDs are also being considered for use as technical reference displays in maintenance and space environments, where BAC might offer a "hands free" approach to HMD operation.

All human control is ultimately dependent on neural activity. Brain-actuated control simply provides a more direct connection to this fundamental control source. Brain electrical activity can be monitored and recorded noninvasively from the surface of the scalp. Recorded EEG may be used to monitor sensory and cognitive function, to identify changes in physiological status, and to measure human operator mental workload. Any of these strategies may be used to support advanced human monitoring systems, including adaptive HMD formats. For the purposes of this chapter, however, EEG recording is considered primarily as necessary to the implementation of an alternative control technology. Here, the purpose of BAC is to enable the human operator to exercise voluntary, direct control over systems such as HMDs.

Controlled modulation of the scalp-recorded EEG provides the essential basis for this exciting new candidate control. Sometimes referred to as a "direct brain interface," BAC represents an entirely new class of alternative control technology that may substitute directly for muscle effector systems and thereby replace traditional control devices, such as buttons and joysticks. To the extent that BAC can satisfy both of these requirements, it will offer an alternative control source that is truly independent of physical capability.

In the following sections, we will review current approaches to BAC, its present limitations, and the significant progress that has been made in this field over a five-year period of research. We describe the essential hardware and software components of BAC system technology, as well as a variety of human factors issues that must be considered for its application. We discuss primary considerations for the use of BAC with HMDs. Finally, we provide recommendations for future research and technology development. It is our desire to propose a direction for subsequent efforts, and thereby promote BAC/HMD integration.

Most of the BAC methodologies and systems considered here attempt to exclude all non-EEG biopotential sources. However, we will include in our discussion so-called hybrid systems that employ EEG in combination with other bioelectric signals (e.g., forehead muscle activity (EMG), eye movements (EOG)). Although hybrid systems require limited facial and/or eye muscle control capacity, this requirement may be met by many severely impaired individuals. Hybrid systems may be well suited to tasks that require multiaxis control or frequent and flexible control activation/deactivation. Thus, although hybrid systems do not rely exclusively on EEG-based control, their development and testing may play a role in the long-term development of true BAC.

10.2.1 History and background

Research spanning three decades has shown that human beings can be trained, by operant conditioning and biofeedback, to regulate various aspects of their own brain electrical activity (Elbert, Rockstroh, Lutzenberger, and Birbaumer, 1984; Fehmi, 1978; Finley and Johnson, 1983; Kuhlman, 1978; Miltner, Larbig, and Braun, 1986; Rockstroh, Birbaumer, Elbert, and Lutzenberger, 1984; Roger and Galand, 1981; Travis, Kondo, and Knott, 1975). Biofeedback and operant conditioning are procedures that use timely feedback and reinforcement to support the acquisition of new abilities such as those involving the self-regulation of biological rhythms. Specifically, information may be presented to enable the human subject to observe changes that occur as a result of his or her own brain processes. When changes are observed as reliably related to a particular thought or behavior, that thought or behavior can be replicated voluntarily to reproduce the intended effect. Cognitive strategies may vary. Once trained, individuals often find it difficult to describe their strategies. Perhaps this is due to the development of an automatized response, or more simply due to the fact that there is no direct sensation or perception associated with neural activity. Whatever the case, both animal and human studies provide convincing evidence that learned and voluntary self-regulation of brain electrical activity can be achieved as a real and nontrivial phenomenon, rather than as a secondary effect of bioelectric artifact or peripheral muscle activity (Elbert et al., 1984; Elbert, Rockstroh, Lutzenberger, and Birbaumer, 1980; Finley and Johnson, 1983; Nasman, Ingle, and Schnurer, 1994; Roger and Galand, 1981; Rosenfeld and Hetzler, 1978).

The potential value of EEG-based control was recognized as early as 1964, by Edmond Dewan, who suggested that the ability to "manipulate a switch simply by altering thoughts in some way" might be applied to the field of prosthetics. Dewan applied the EEG alpha rhythm as an on/off "switch" to communicate messages in Morse code (Dewan, 1964, 1966). Later investigators have pursued similar efforts, but with the additional assistance of biofeedback training to facilitate EEG self-regulation (Elder, Lashley, Kedouri, Regenbogen, Martyn, Roundtree, and Grenier, 1986). More recent applied efforts have considered a variety of different methodologies in the attempt to identify specific brain electrophysiological responses, components, and/or patterns that might provide the basis for effective control. In general, the various paradigms proposed to date can be classified as employing either "endogenous" (cognitive event–related) or "exogenous" (sensory evoked) brain electrophysiological indices of information processing.

10.2.2 Endogenous versus exogenous control signals

Endogenous brain electrical activities are those that correspond in predictable fashion to specific perceptual and cognitive operations such as stimulus categorization (Farwell and Donchin, 1988), verbal information processing (Fujimaki, Takeuchi, Kobayashi, Kuriki, and Hasuo, 1994; Hiraiwa, Shimohara, and Tokunaga, 1990; Keirn and Aunon, 1990), and the control of body movement (Mason and Birch, 1995; McFarland, Neat, Read, and Wolpaw, 1993; Pfurtscheller, Flotzinger, Mohl, and Peltoranta, 1992; Pfurtscheller, Flotzinger, and Neuper, 1994; Wolpaw and McFarland, 1994; Wolpaw, McFarland, Neat, and Forneris, 1991). Endogenous activities may be adapted as control signals with or without the use of operant conditioning or biofeedback. For example, Farwell and Donchin (1988) have employed the "P300" brain response, an endogenous potential whose signal characteristics vary as a function of stimulus probability and task relevance. The P300 response was used to identify words or characters selected by the user from among a stimulus series on a visual display. Provided that task design is conducive to the production of an endogenous response, the natural variance of that response may be sufficient for task control (see also Polikoff, Bunnel, and Borkowski, 1995). Alternatively, the human user may be trained either to modify or to produce an endogenous response at will. Wolpaw and McFarland (1994) have trained human subjects to control cursor activity on a computer monitor by modifying their EEG "mu" rhythm, a frequency pattern normally associated with physical movement.

For the purpose of direct brain interface design, the use of an endogenous control signal presents a number of challenges related to response production, identification, monitoring, and contamination. Under even well-controlled experimental conditions, endogenous components may be difficult to identify and verify as independent of other effects of mental, emotional, or physical processes. In some cases, endogenous control may require advanced pattern recognition technology (see Hiraiwa et al., 1990; Pfurtscheller et al., 1992, 1994).

By contrast, *exogenous* brain responses are those that reflect more basic operations associated with the transmission of sensory (e.g., visual, auditory, tactile) information to primary sensory areas of the brain. *Exogenous* refers not to the neural source of the response, but rather to the fact that a response is evoked by, and thus is immediately and strictly related to, a specific external and causal stimulus. Its implementation thus requires a system design that can harness the natural and effectively "automatic" variation of the exogenous response as a source of task-relevant control, or that employs a biofeedback training protocol that supports the acquisition of voluntary control over the exogenous response.

Sutter (1984, 1992) describes an exogenous brain response interface whereby the natural variation in cortical visual evoked potentials are monitored to determine direction of gaze on a matrix of flickering stimuli. One area of our own research uses biofeedback and operant conditioning to bring an exogenous visual evoked response under voluntary control. This approach emerged from earlier efforts by Junker and his colleagues to define and develop a "closed-loop" control system using a steady-state visual evoked response (SSVER) generated by a sinusoidally modulated light stimulus (Junker, Schnurer, Ingle, and Downey, 1988; Schnurer, Ingle, Downey, and Junker, 1988). The SSVER is well defined and is readily identifiable as a frequency-specific response to a controlled stimulus. The use of an external evoking stimulus presents an obvious engineering challenge concerning the need for unobtrusive stimulus presentation in applied settings. However, there are numerous advantages to this approach, including tight experimental control and a high level of resistance to potentially confounding effects of distraction and artifact (e.g., eye and muscle movement). Because the SSVER is time-locked to an externally driven stimulus, it permits synchronous signal processing, which in turn improves noise tolerance and reduces or eliminates potentially confounding effects of other activities and rhythms, such as broad-band alpha. In applied settings these advantages may relieve a variety of physical and psychological constraints.

10.2.3 Current status of BAC: Research and performance

The possibility of an applied direct brain interface presents a variety of theoretical and applied questions that have yet to be answered. Ultimately, these questions must be addressed to support informed judgments concerning possible applications and to achieve system designs that are best suited to specific applications and to users' needs. For example, what are the fundamental neurocognitive mechanisms by which EEG self-regulation is accomplished? What might be the ultimate effects of user distraction and/or increased workload? How might BAC best be coordinated with other tasks and other forms of control? Under what conditions is BAC performance optimized, and under what conditions is it made more difficult? How is training best accomplished? Must BAC be practiced on a regular basis, or is this acquired skill sufficiently robust to endure periods of nonuse?

Current efforts to design and develop direct brain interface technology also highlight the need for fundamental improvements in signal acquisition, signal processing quality and speed, and system control logic. Independently and in combination, these components of the interface exert a direct influence on the precision, reliability, and overall effectiveness of a user's control. Although EEG-based control remains limited to rudimentary actions, impressive progress has been demonstrated over the past five years of research and there already exists a fairly impressive record of achieved control in various laboratories using a variety of tasks and methodologies. Current laboratory BAC tasks and demonstrations include target interception, roll angle tracking, maze navigation, and icon or "button" selection. A limited amount of work has already been done to demonstrate dual-axis control. In the remainder of this section, we summarize four of the leading methodologies currently under investigation.

As we noted earlier, the BAC system designed in our laboratory employs an EEG-based control signal derived from the user's visual evoked response to a sinusoidally modulated light source (McMillan, Calhoun, Middendorf, Schnurer, Ingle, and Nasman, 1995). Two fluorescent lamps mounted behind a diffusing screen are modulated in intensity at a rate of 13.25 Hz to evoke an SSVER in the user's visual cortex (Fig. 10.2). The SSVER is recorded as a bipolar (differential) response over occipital scalp sites O1 and O2 (left and right occipital scalp, respectively, in accordance with the 10-20 International System for scalp surface electrode placement). The response is then amplified and synchronously processed by a lock-in amplifier system. This amplifier system provides a continuous measure of the magnitude of the (13.25-Hz SSVER) control signal, which in turn controls the roll position of a flight simulator. This information is simultaneously provided to the user's feedback display and to the simulator's control algorithm. The algorithm requires the user to hold the response magnitude above or below individually calibrated thresholds for a set duration before a discrete output is generated. Typical parameters produce an output when the SSVER remains above or below threshold for 75 percent of the samples in a one-half second interval. These settings ensure that brief, unintentional fluctuations in magnitude will not interrupt system control, and so require that control be achieved by sustained, intentional changes to the magnitude of the SSVER. Parameter settings are also adjustable for individual users, levels of training, and different device applications.

By attending to near real-time feedback of SSVER magnitude, users learn to self-regulate their response, which in turn enables them to control the simulator's roll position. A task display inside the simulator provides a random series of commands that instruct the user to roll right or left by controlling SSVER magnitude. The user modulates SSVER magnitude in response to each command, and the

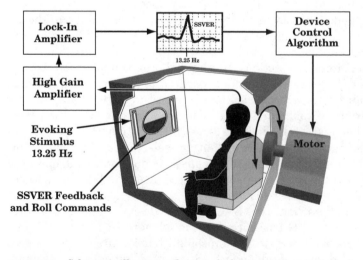

Figure 10.2. Schematic illustrates the use of a brain interface to control the roll position of a flight simulator. Roll position is determined by self-regulation of the steady-state visual evoked response (SSVER).

simulator rolls in accordance with the output of the control algorithm. As the user increases SSVER magnitude above a given threshold, a feedback display cursor shifts toward the right, followed by simulator roll motion to the right. When the SSVER magnitude is reduced below a lower threshold value, the feedback cursor moves to the left, and the simulator then rolls to the left. The operation of this control-duration-based (or "pulsatile") algorithm can be observed as simulator motion steps separated by one-half second intervals (Fig. 10.3).

The control logic described and illustrated here represents an improvement to methodologies employed previously in our laboratory. Pulsatile logic has served as an effective means to reduce the effects of noise due to brief, task-unrelated variation in SSVER magnitude. Most new subjects are able to achieve some level and corresponding sense of control after a single 30-minute training session, and they typically demonstrate fairly dramatic learning curves over the course of their first few training sessions (Fig. 10.4). Trained users typically are able to acquire 80–95 percent of the roll angle targets presented in each trial.

Erich Sutter (1992) of the Smith-Kettlewell Eye Research Institute has developed a system that permits visual stimulus selection by

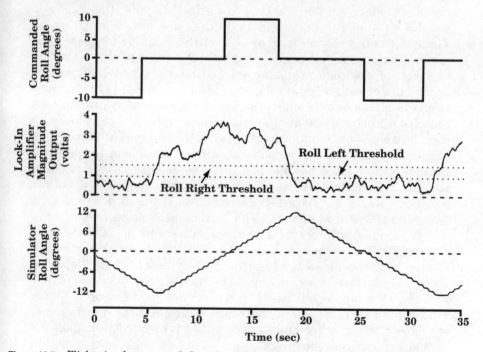

Figure 10.3. Flight simulator control. Sample trial data demonstrate control after approximately 12 hours of subject training. (Positive designates roll right, negative roll left.)

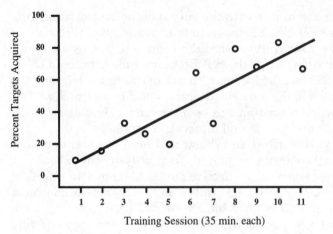

Figure 10.4. Individual learning curve for simulator control task. Linear regression (solid line) based on means (16 trials/session). Subject had no prior biofeedback or simulator training.

monitoring the visual evoked response. This approach is based on the fact that a modulated stimulus in the center of the visual field evokes a much larger cortical response than one presented in the visual periphery. An effect known as cortical magnification makes it possible for a user to select stimuli (in this case, letters, numbers, or words) by fixating foveal gaze directly at the stimulus of choice. The visual evoked potential as recorded in this manner is very sensitive to effects of electrode misplacement and muscle artifact. Successful routine use of such a system therefore requires precise individualized electrode placement. A reference evoked response template must be collected during a 10–20 minute preliminary evaluation session. However, because the elicitation of the visual cortical evoked response is involuntary, and because computer algorithms are able to discriminate fixation points on the basis of response size, no special user training is needed to support this type of system. Sutter reports that character selection rates of 40 characters per minute have been achieved with better than 90 percent accuracy. This level of success clearly represents a viable communication rate for persons with severe disabilities.

Jonathan Wolpaw and his colleagues (1991, 1994) at the Wadsworth Center for Laboratories and Research in New York have developed an EEG-based system for controlling the position of a cursor on a computer monitor. This approach is based on self-regulation of sensorimotor (or 8–12 Hz mu rhythm) amplitude, which determines the direction and step size of cursor movement. Cursor position is updated every 0.05–0.3 second; skilled users can direct the cursor to a stationary target within 2–6 seconds. Single- and dual-axis cursor control tasks have been designed and tested, producing accuracy ranges of 80–95 percent and

40–70 percent, respectively. Dual-axis control is supported by a control algorithm that generates a control signal as either the sum (vertical cursor movement) or the difference (horizontal cursor movement) of activity at two scalp recording sites. New users may achieve some degree of control as early as the first training session. Between 5 and 20 hours of training and practice are generally necessary to achieve reliable control at high levels of accuracy. Additional training is necessary to achieve dual-axis control with mu rhythm self-regulation.

A very different approach to BAC is represented by the work of Gert Pfurtscheller and his colleagues (1992, 1994) at the Graz University of Technology in Austria. Their system employs a neural network to recognize alpha (10–12 Hz) and gamma (30–40 Hz) frequency band EEG patterns that precede specific body movements (e.g., finger, tongue, or toe). No user training is required, but the neural network must be trained to recognize relevant characteristics of the user's EEG; this requires 100–200 repetitions of each associated movement. Once trained, the neural network requires only one second of EEG data to identify the intended movement. The speed of the neural network analysis allows near real-time operation and control. The off-line system has achieved 89 percent accuracy in predicting finger movements.

It is clear that BAC performance will evolve as a result of ongoing and future research in several laboratories. Application of BAC will require laboratory demonstration, testing, and evaluation with respect to safety issues. Successful use in the real world will require supporting system technology that is portable, flexible, convenient, and relatively unobtrusive. However, we believe that the existing technology and the already impressive level of control demonstrated by laboratory experimentation are sufficient to support and warrant first-generation application of BAC. It is also clear that most BAC systems require an investment of time in training; however, this requirement is certainly no more demanding than that required to learn other new skills. Observations of researchers in the field suggest that no unique skills or individual characteristics are required to learn how to perform BAC. As with any skill, capability may vary among individuals as a function of motivation, commitment, and interest in the use of this new technology.

10.3 Potential Applications of BAC

Despite all that is not yet understood about BAC, the majority of investigators in this field are conducting near-term efforts to demonstrate possible applications of BAC as an assistive technology, and it is in this area that BAC will likely achieve its earliest and most significant impact. It is clear that BAC presents a valuable alternative control modality for individuals who are physically impaired or unable to make effective use of traditional manual or physical devices and computer controls. Other alternative control and communication devices (e.g., sip-puff tubes, chin-operated joysticks, mouth-held wands) are typically awkward in their appearance and require spared physical capabilities. These devices are of little help to individuals whose profound physical disabilities leave them "locked in," with no remaining motor control or capability. For persons with spinal cord injuries, BAC also offers a means by which to bring back direct involvement of the brain as an essential element of control. Most users report a strong sense of accomplishment while performing BAC; it is reasonable to expect that this new technology may provide yet more meaningful satisfaction to individuals who are unable to make use of traditional controls.

Efforts are already underway to implement BAC as an assistive/rehabilitative control technology. One example of this objective has already been demonstrated by the integration of BAC and a neuromuscular stimulator designed to exercise paralyzed muscles (Calhoun et al., 1995). Other laboratories are considering BAC to control limb prostheses (Humphrey, 1995) and as a communication tool for persons otherwise disabled by injury or disease (McCane, Vaughan, McFarland, and Wolpaw, 1995). Incorporating BAC in an HMD system would enhance its applicability as a rehabilitative/assistive technology. A head-stabilized display system may be essential for many users with spasticity or limited movement control.

Early applications of BAC are also likely to emerge in the area of interactive video entertainment. Game applications are motivated primarily by the novelty of a "no-hands" technology. Hybrid systems that combine EEG/EMG or EEG/EOG bioelectric signals have already been used to operate simple video games, such as Pong, and maze tasks (Junker, Berg, Schneider, and McMillan, 1995; Patmore and Knapp, 1995). Once again, such applications may provide satisfaction beyond simple entertainment value to individuals who are unable to use standard video game controls.

Vehicle and process control represent the most demanding and long-term applications of BAC. Before BAC can be used for this purpose, it must demonstrate performance superior to traditional controls, reduced workload, and/or special capabilities or advantages to serve specific or extreme environmental or user demands. Distinct advantages may be identified in environments where movement is constrained (e.g., in high-acceleration flight, during extra-vehicular activity in space flight) or where physical control workload is extremely high. Our laboratory has demonstrated control of flight simulator roll-axis motion, although operator performance continues to fall short of normal joystick operation. At present, we are focusing our research efforts on testing BAC as a means to control secondary cockpit functions, such as radar operation, weapons selection, and other discrete switching tasks. Anticipated scenarios might involve offering BAC as an optional control interface to be activated or deactivated at the user's convenience or preference. Thus, a user might choose to engage a direct brain interface to operate a specific secondary device, such as a radio, under difficult flight conditions in which manual resources are fully dedicated to controlling the aircraft itself.

Applications in the military domain are not strictly limited to vehicular control. One intriguing application now under consideration is that BAC might be used as an alternative controller for users who must operate in environments where there are dangerous chemical and biological agents. The protective gear and gloves that are currently necessary for safety make it difficult, if not impossible, to operate small manual switches and controls.

Whatever the intended application, portable BAC systems can be produced at relatively low cost. Hardware and software can be hosted on a laptop computer, with dedicated amplifier and microcontroller chips configured to provide inputs through keyboard, mouse, or joystick ports. Peripheral components (e.g., electrodes, preamplifiers) are also reasonably priced. Integrated HMDs would certainly enhance the commercial appeal of portable BAC technology, making it easier to use. Because BAC display requirements are modest, integration could probably be achieved very affordably.

10.4   BAC: Systems Engineering and Technology

Successful integration of BAC and HMD technologies will require an understanding of the principles and performance capabilities of each. This section reviews the key elements of BAC systems design. Although a wide variety of specific components and operations have been tested, most BAC systems combine essentially the same basic elements (see Fig. 10.5). Hardware and software are integrated to accomplish signal acquisition, signal processing, control, and performance feedback. BAC systems that rely upon exogenous response control signals also incorporate an evoking stimulus to generate the EEG signal of interest (see Sutter, 1992; McMillan et al., 1995).

10.4.1   Signal acquisition and processing

The EEG signals of interest in BAC fall within the 1–40 Hz frequency range, with amplitudes ranging between 1 and 50 microvolts. Although these signals are very small, and great care must be taken to prevent contamination by other electric and bioelectric signals, commercially

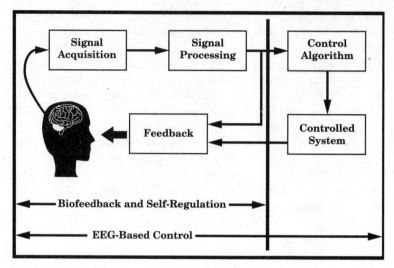

Figure 10.5   Basic elements of EEG-based control.

available biological signal amplifiers are acceptable and are commonly used. The most commonly used scalp electrodes are standard small (approximately 1-cm diameter) gold or silver/silver-chloride disks developed for EEG recording. In most cases accurate electrode placement is easily learned, although specific methodologies (see Sutter, 1992) may require a high degree of precision. BAC systems that use EEG recorded from many areas around the scalp (e.g., Pfurtscheller et al., 1994) may employ commercially available electrode caps that provide standardized placements and relatively faster application.

Standard EEG electrode application is a relatively straightforward and easily learned skill, but it remains somewhat inconvenient, and occasionally messy; it typically requires some amount of assistance; and it needs to follow safety guidelines (see Pivik, Broughton, Coppola, Davidson, Fox, and Nuwer, 1993). The application area of the skin is cleaned with a mild abrasive in order to reduce impedance at the skin-electrode interface. Low impedance (typically less than 5 kΩ) can be verified by an inexpensive meter. The electrical contact is maintained with a conductive paste; electrodes are affixed with small gauze pads, adhesive rings, or tape. It is rarely necessary to shave hair from the scalp, although this procedure may be used to ensure reliable placement and contact. Gel electrodes may also be used to record from the scalp. However, most gel electrodes are manufactured with adhesive collars that do not adhere well to hairy areas of the skin; it is usually necessary to secure gel electrodes with pressure, e.g., using an elastic headband or hat. Although not yet commercially available, dry electrode recording systems are highly desirable for real-world applications.

In most BAC systems, analog brain electrical activity recorded at the level of the scalp is amplified and immediately converted to digital format using a sampling rate of at least 100 Hz. The EEG power within specific frequency bands is then estimated by Fast Fourier Transform (see Wolpaw et al., 1991; Wolpaw and McFarland, 1994), digital bandpass filters (see Pfurtscheller et al., 1992, 1994), or some other proprietary technique. Personal computer systems of the 80486 class are sufficient to support these computations, although digital signal processing boards may also be employed. Theoretical considerations or empirical results typically drive the choice of which frequency band(s) to apply as control signal.

Signal processing may also be performed within the analog domain, as illustrated by the system we have used to control the roll motion of a simple flight simulator (McMillan et al., 1995) and to operate a neuromuscular stimulator (Calhoun, McMillan, Morton, Middendorf, Schnurer, Ingle, Glaser, and Figoni, 1995). Our system uses a lock-in amplifier system synchronized with the sinusoidally modulated evoking stimulus. The lock-in amplifier continuously measures the magnitude of the SSVER produced by the light. First developed in 1987 (Schnurer et al., 1988), this analog system provided very fast processing that was then difficult to obtain with digital techniques. However, a digital version of the same system is presently under development and is expected to provide superior performance and reliability.

Other BAC systems have been proposed that employ neural networks to identify and harness EEG patterns (e.g., selected frequency bands or response potentials) that precede and predict intended hand movements, joystick movements, or simple utterances (Hiraiwa et al., 1990; Pfurtscheller et al., 1992, 1994). Although no real-time control has yet been demonstrated, the objective of the research is to develop control algorithms and devices to execute the desired types of behavior based on their preparatory EEG activity.

10.4.2 Control algorithms and feedback displays

Direct brain interface system design has had to overcome numerous challenges with respect to the development of effective software-based control algorithms. Special problems are posed by the use of self-regulated (versus spontaneous) EEG activity, in which intended response variance may be as brief as a fraction of a second within the context of ongoing, unintended signal variability. Low-pass filtering, signal averaging, and logic based on threshold and duration requirements have been employed to deal successfully with this problem (Junker et al., 1995; McMillan et al., 1995; Wolpaw et al., 1991; Wolpaw and McFarland, 1994). However, the development of yet more effective control logic represents a continuing challenge for researchers in this field. Engineers and programmers must work to achieve a balance between reliable control and system responsiveness to individual user signal characteristics and capabilities.

Systems that employ self-regulated EEG rely heavily on the use of biofeedback to enable the user to control otherwise spontaneous or unintended changes in EEG. User feedback may be implemented either as an integral element of the BAC task itself or as an independent display element. For example, cursor movement may provide the necessary feedback to determine response magnitude; cursor motion relates directly to (presumably intended) changes in the EEG-derived control signal (see Wolpaw et al., 1991, Wolpaw and McFarland, 1994). However, other tasks may require additional displays to provide the timely feedback that is necessary to support near real-time control. For instance, in our own laboratory the EEG-controlled roll-axis motion of a flight simulator is delayed by 0.5–1.5 seconds due to signal processing and control algorithm requirements. A separate feedback display in the form of a simple bar graph provides more timely feedback to the user concerning his or her SSVER response variation.

10.5   BAC and HMDs

All current approaches to BAC can be described as rudimentary. Our consideration of BAC for specific HMD applications is thus limited to examining the interface and to identifying primary research concerns. In the sections to follow, we address the issues of HMD design and utilization that are especially relevant to future integration of BAC and HMDs. Finally, we make specific recommendations with respect to research and development strategies, concept demonstration and testing, and technology transfer.

Special considerations regarding the integration of EEG-based control are related primarily to information exchange between system and user. An HMD system that integrates BAC (see Fig. 10.6) should support (a) the communication of EEG-based commands from the user to the system, and (b) the display of task requirements and performance feedback from the system to the user. We consider each of these interface requirements separately, below.

10.5.1   EEG-based command and control

The communication of EEG-based commands from the user to the BAC or BAC/HMD system requires the integration of hardware and software technology necessary to support signal acquisition and conditioning, task-relevant control logic, control activation, and command format.

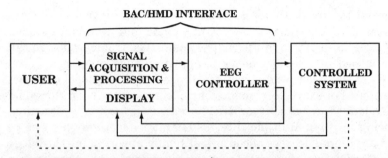

Figure 10.6 Information exchange in HMD/BAC. BAC interface supports transmission of EEG-based commands from user to system, as well as presentation of task information and performance feedback from the system to its user.

In its minimum configuration, a BAC system requires head-mounted components to support *EEG signal acquisition*; the basis for control input is the voltage of the EEG response or pattern of interest. Although scalpwide recordings are often used for the purpose of research, applied control is more commonly derived from just a few electrodes placed at selected sites. A smaller number of electrodes makes the conventional application procedure feasible for HMD applications. However, electrode application is certainly not as simple, quick, and unobtrusive as donning an HMD. Moreover, perspiration and head movement may tend to degrade the conductive properties of electrode pastes and gels. Many researchers anticipate the development and refinement of dry electrode systems to minimize these problems. Head-mounted preamplifiers and filters are also desirable to reduce the possible effects of high electrode impedance and resultant susceptibility to electrical noise. Whatever BAC methodology might be selected, consideration must be given to the degree to which recording instrumentation can be comfortably integrated within HMD headgear. Components must be as small and lightweight as possible and should be distributed in such a way as not to alter significantly the weight or center of gravity of the assembly. Generally, the assembly must be comfortable to wear and use in its intended environment.

Overall BAC/HMD system performance may be affected by several critical procedures and parameters that relate to signal conditioning. These include signal filtering and averaging, response threshold criteria, and time delay control. Additional signal conditioning may also be required if eye movements introduce noise in the EEG data. In each case research and testing should be performed independently, and with respect to specific EEG methodology, prior to BAC/HMD integration. There is no reason to expect that HMD hardware and electronics will impose additional constraints or limitations on selected BAC parameters or algorithm specifications.

Essential to any BAC methodology is a control algorithm that (a) translates EEG responses into control signals and (b) applies *system control logic* to variation in the control signal. Control logic design can be approached in a variety of ways, and it should be determined on the basis of the needs and objectives of particular methodologies and applications. For example, a single control signal criterion or threshold may be adequate to support an EEG-based on/off "switch" or function-sequencing routine. Multiple thresholds and criteria may be used to support additional degrees of control, although any given methodology may impose inherent limitations on the number of control states that can be effectively achieved under various conditions of mental or attentional workload.

Additional bandwidth may be achieved with control algorithms that perform multiple transformations on EEG responses from only two or three recording sites. Wolpaw et al. (1991, 1994) have employed an algorithm that derives control signal data based on the sum and/or difference of response amplitudes at recording sites over two hemispheres. Other possibilities include the use of additional scalp recording sites (see Ingle and Nasman, 1994), additional EEG responses, or multiple parameters (e.g., frequency, amplitude, latency) that describe the variance of one or more specific EEG components. However, additional laboratory experimentation, testing, and evaluation are needed to determine whether any of these strategies can be successfully employed and, if so, to which EEG-based methodologies they may be best suited.

Control algorithms may also be written to produce a system response that is directly proportional to EEG-based control signal variance. In general, proportional system response will tend to be less stable by itself than in combination with response criterion thresholds that control against undesired effects of natural variance in EEG. Variance beyond criterion threshold is assumed to be the result of intentional self-regulation, and it may also be applied proportionally to system control as a function of distance from threshold or as a function of rate of change.

With signal acquisition and control logic in place, the system and its resident task(s) must be designed to provide an effective means for *control activation*. Traditional control devices provide a natural basis by which to initiate or disengage control; for instance, one may grasp or release a mouse or joystick. However, EEG-based control is effectively "on" at all times. This requires that there be some procedure whereby the user can activate or deactivate the control interface (Table 10.2).

Relatedly, multiple-task environments may present the need for differentiation between intended and unintended control signal changes. One method is to avoid the problem entirely by requiring a consent response to designate or confirm the user's intent that the EEG control system act upon the most recent EEG control signal input. However,

TABLE 10.2   Options for EEG-Based Control Interface Activation/Deactivation

Activate interface	Control task	Consent response	Deactivate interface	EEG recording
	EEG signal 1	Non-EEG response		Continuous
	EEG signal 1	EEG signal 2		Continuous
Non-EEG response	EEG signal 1		Non-EEG response	During task
EEG signal 1	EEG signal 2		EEG signal 1	Continuous

this may be distracting to the user and could impose delays that may interfere with task performance. Another method is to require that the user issue a unique "switch" response via manual, voice, or eye line-of-sight command to activate and deactivate the EEG control interface. Of course, this solution assumes at least limited physical capability on the part of the user. Absent any such capability, an independently selected EEG-based response may be reserved for the purpose of control interface activation. An EEG-based response would in this case require continuous recording, and it should be selected to represent a deliberate response that is easily achieved. It is reasonable to expect that voluntary self-regulation of more than a single EEG response, component, or pattern should be feasible. At present, however, there is no research to demonstrate successful self-regulation of multiple and independent EEG components or patterns.

Whatever method is employed to activate/deactivate EEG control, the user must be informed that the EEG-based control is "on" or "off." This information could be presented as a continuous element of the display to which the user may refer at his or her own convenience. If an additional display element is undesirable, operational status may instead be designated as the presence ("on") or absence ("off") of EEG feedback.

Once EEG-based control is activated, there are two general modes of control, or *command formats*, to which it might be applied. EEG-based control might be applied to situations in which the user needs to initiate or enter a command without prompting; this situation is most closely analogous to the use of a keyboard or a joystick (see Heetderks and Schwartz, 1995). Alternatively, BAC might be used to select one of several display or menu features, icons, toggle "switches," or "buttons." Although future research and development may improve EEG-based control so that it may become sufficiently flexible to accomplish both "keyboard" and "menu" type operations, its present state seems best suited to the latter. Our own laboratory tests suggest that users may find it easier to produce a change in their brain response (SSVER) magnitude than to sustain that change over long periods of time (longer than 2 seconds). It should be noted, however, that repetitive on/off switching may also be difficult to achieve and maintain as a form of task control; it may be useful to combine discrete and sustained control as complementary features or options within a single task.

10.5.2 System operations feedback

There are at least two fundamental sources of information to which the BAC user must have frequent if not constant access. First, the user must receive reliable *biofeedback* to monitor real-time changes in the EEG-based control signal. Because there is no physical control feedback, and because brain electrical activity provides no direct sensory experience of its own, the information delivered by the BAC/HMD system will constitute the sole source of control information to the user. Second, the user should be given timely *performance feedback* and/or information concerning task status. Although it is not necessary in principle to deliver biofeedback and performance feedback strictly as visual information, this is often the most convenient and effective option.

Several BAC methodologies employ visual displays that could be presented within a head-mounted display platform. Visual displays may include EEG biofeedback and activation status, alphanumeric displays for character selection, and/or light-source stimulation for the purpose of evoking a specific brain response. In general, the use of an EEG-based controller will impose no new performance requirements with respect to HMD image quality. Moreover, EEG-based control will not necessarily involve the coordination of computer-generated imagery with real-world scenery. For these reasons many HMD design features and parameters will be unaffected by EEG control integration; these features include update rate, resolution, focus, color, field of view, exit pupil size, eye relief distance, interpupillary distance, and in the case of transparent systems, percentage reflection and transmission figures (see Kocian and Task, 1995).

However, there are other aspects of HMD design whose impact on EEG-based control must be considered. Depending on the specific EEG-based control methodology selected, HMD display design must take into consideration possible effects of (a) monocular versus binocular viewing, and (b) luminance. To our knowledge, every BAC methodology tested to date has involved binocular viewing. Although monocular systems have not been tested, there is little reason to expect that monocular viewing should interfere with BAC itself. However, the question of ocularity may be of special concern for EEG-based systems that employ an exogenous brain-evoked response. For example, monocular stimulation may lead to reduced visual evoked response magnitude (Regan, 1989). Although luminance reportedly has only a small effect on visual evoked potential latency (see Halliday and Callaway, 1980), it is necessary to confirm that an adequate response is evoked in dynamic environments and applications in which highly variable luminance may produce eye strain or changes in visual attention. This should be a special concern in the design and application of transparent displays.

Also important is the fact that the user needs current information concerning the status of the task, and confirmation of each interactive event. This information may also be sufficient as EEG biofeedback, provided that brain response variation corresponds to switch sequencing or cursor movement. If this is not the case, then dedicated feedback is necessary and the format must be determined with respect to overall display design. Considerations will include location within the visual field, source (raw versus processed signal information), form (e.g., absolute value, relative trend, deviation from threshold), sensory modality, and period (continuous, occasional, on demand, or automatic). For continuous visual feedback, an HMD system update rate on the order of 60 Hz is probably sufficient. In general, any delay in the feedback of the control signal will tend to reflect the speed of the EEG signal processing system, rather than the HMD system's performance.

10.6   Conclusions and Recommendations

The burden of new research and development toward BAC/HMD integration rests heavily with investigators in the area of BAC. There is much yet to be learned about the manner in which BAC may be most effectively trained, implemented, and applied. To our knowledge there is no effort yet underway by HMD investigators to apply BAC as a control system. However, efforts toward integrating BAC and HMD technologies can and should begin immediately, for the purpose of early testing and to determine what, if any, additional integrative design issues should be considered throughout the course of research and development. To the extent that HMD designers may be interested in applying BAC technology to their work, that interest will help to sustain, inform, and encourage BAC-related research in general.

Exploratory research will tend to dictate its own evolution and course of progress. However, we would like to offer specific recommendations by which to frame the objectives of research and development as these relate specifically to the goal of BAC/HMD integration.

10.6.1   Research and development strategies

Research in BAC has progressed quickly thus far, supported by basic scientific questions whose answers are necessary to the development of successful systems. Although initial test applications are possible without much basic information about the means and mechanisms of neural self-regulation, reliable real-world application will ultimately depend on a clear understanding of how the user achieves task control. To this end it is necessary to establish BAC as a nontrivial (EEG-related) phenomenon, that is, to conduct controlled research and testing to better understand by what mechanisms BAC is achieved. This requires that the neurophysiological basis for BAC be defined as it relates to observed, task-related control. For example, we have reported preliminary evidence that SSVER self-regulation may be accomplished by the generation of interhemispheric differences in signal timing (phase) (Nasman et al., 1994). It is also important to address the extent to which task-related control might vary as a function of feedback parameters (e.g., format, modality, and timing with respect to control signal variation). Functional and neuroanatomical hypotheses might also be derived from scalpwide recording studies in which EEG components or dominant frequencies can be observed as a dual function of time and spatial distribution (topography) (e.g., see Nasman, Ingle, and Calhoun, 1995). Topographic studies are especially useful in the attempt to disentangle primary sensory and higher-order perceptual, associative, or attentional phenomena.

Basic human factors research should proceed in parallel, addressing the need to refine and test a variety of task and display designs, stimulus and feedback parameters (e.g., sensory modality), control logic requirements, and training protocols. Probably the greatest challenge in this area is the need to define, for any particular BAC methodology and task or application, the critical measures by which to evaluate user performance. Relatedly, training protocol design should address the question of what performance measures should be provided as feedback to the user. These might include multiple or composite scores that reflect task performance or target acquisition time, number of targets acquired, control signal response latency, proportion of correct responses, and/or errors of omission and commission. It is useful to conduct training studies that involve a variety of performance measures to determine how well they reflect skill acquisition and improvement.

"Concept demonstrations" have become increasingly popular, and increasingly necessary, as a form of applied scientific communication. Demonstration is also helpful as an objective by which to guide coordinated efforts among multiple laboratories and specialists in various disciplines. By implementing comparable tasks and simulations, test conditions, variables, and parameters, it is possible to produce an overview from which may emerge a better understanding of specific methodological advantages and difficulties. Early pilot demonstrations are also useful to explore the landscape of challenges and possibilities that need be considered with respect to integrated technologies. We favor early concept demonstration as an initial step toward BAC/HMD integration, and we recommend emphasis on the broad applicability of BAC/HMD to a variety of situations and persons.

Regarding applied research specifically, we support a model that conceives of BAC not strictly as a substitute technology, but rather as one among several interrelated and integrated technologies (McMillan, Eggleston, and Anderson, in press). Futuristic, so-called self-adjusting systems are likely to involve a variety of on-line adaptive and intelligent technologies, such as physiological workload monitors, specialized control interfaces, and sophisticated display platforms. It is presumed, but not yet known, that sophisticated systems such as these will enable the human user to function more effectively. Successful application may depend heavily on flexibility and adaptation to user preferences, needs, choices, and capabilities.

BAC requires a diverse assortment of specialists to address requirements for sound scientific design, neurophysiological recording and monitoring, psychology, human factors engineering and analysis, electrical engineering, and software design. Specific development projects might also require the skills of experts in aviation, developmental disabilities, spinal cord injury, and rehabilitative medicine. Finally, the integration of BAC and HMD technologies will require open communication and cooperation among researchers in each field.

In its present state of development, BAC lacks the resolution and reliability afforded by many conventional controls. It may be elegantly appropriate for some tasks, while clearly inappropriate for others. Its suitability should be considered in advance and, if possible, evaluated with careful attention to the character and demands of each application. Again, we emphasize that special attention should be paid to the needs and capabilities of intended users.

10.6.2 Technology transfer priorities and issues

Head-mounted displays were once a highly specialized tool of interest only to scientists and engineers working in aviation and cockpit interface design. Today, HMD research targets a broader market of specialized technology (e.g., virtual reality) for civilian use. In general, federal laboratories no longer focus exclusively upon military applications, but now consider that it is also important to transfer new technologies to the civilian sector. Consistent with this, we emphasize multiple BAC applications across a variety of intended users and environments. Specifically, we consider that BAC is a potentially valuable assistive technology for persons whose performance is limited by physical disability, environmental constraint, or task demands.

Experts often find it difficult to conceive of or identify applications outside their own areas of expertise. This makes it difficult to "push" a specific technology into new domains of application, or to anticipate and allocate resources in a manner appropriate to successful transfer. We agree with strategists who recommend instead that new technologies be evaluated by the method of "technology pull." Proponents of this strategy recommend advance consideration of specific needs that can be met by subsequently identified technologies. User needs should be identified as specifically as possible, in terms that relate to engineering specifications for task-related performance (see Assistive Technologies Group, 1994). Translation from biological or behavioral needs to engineering specifications can reveal commonalities and broader markets that otherwise might not be apparent. Technology-relevant needs analysis of this type helps to support the identification of potentially broad markets for technology transfer.

Certainly, the potential market for BAC is much larger than any particular subset or group of people with a particular disability or limitation. Rather, it may be adapted to meet a variety of needs and address a variety of limitations, whether these are due to physical or environmental constraints. A new control technology of this type should be considered in its functional context, which might involve a variety of life activities and many other devices, tools, or systems that can be used in conjunction with BAC. It is our hope that BAC and HMD technologies may be integrated to stimulate a broad range of new applications.

10.7   References

Assistive Technologies Group. (1994, March). *Toward a new methodology for the development of assistive technologies. Forum on technology transfer and people with disabilities.* Washington, DC: National Technology Transfer Center.

Borah, J. (1989). *Helmet mounted eye tracking for virtual panoramic display systems. Volume II: Eye tracker specification and design approach* (AAMRL-TR-89-019). Wright-Patterson Air Force Base, OH: Armstrong Aerospace Medical Research Laboratory.

Calhoun, G. L., McMillan, G. R., Morton, P. E., Middendorf, M. S., Schnurer, J. H., Ingle, D. F., Glaser, R. M., and Figoni, S. F. (1995). Functional electrical stimulator control with a direct brain interface. *Proceedings of the RESNA 18th Annual Conference,* 696–698.

Dewan, E. M. (1964). *Communication by electroencephalography* (AFCRL-64-910). Bedford, MA: Air Force Cambridge Research Laboratories.

Dewan, E. M. (1966). Communication by voluntary control of the electroencephalogram. *Proceedings of the Symposium on Biomedical Engineering, 1,* 349–351.

Elbert, T., Rockstroh, B., Lutzenberger, W., and Birbaumer, N. (1980). Biofeedback of slow cortical potentials. *Electroencephalography and Clinical Neurophysiology, 48,* 293–301.

Elbert, T., Rockstroh, B., Lutzenberger, W., and Birbaumer, N. (1984). *Self-regulation of the brain and behavior.* New York: Springer-Verlag.

Elder, S. T., Lashley, J. K., Kedouri, N., Regenbogen, D., Martyn, S., Roundtree, G., and Grenier, C. (1986). Can subjects be trained to communicate through the use of EEG biofeedback? *Clinical Biofeedback and Health, 9,* 42–47.

Farwell, L. A., and Donchin, E. (1988). Talking off the top of your head: Toward a mental prosthesis utilizing event-related brain potentials. *Electroencephalography and Clinical Neurophysiology, 70,* 510–523.

Fehmi, L. G. (1978). EEG biofeedback, multichannel synchrony training, and attention. In A. Sugarman and R.E. Targer (Eds.), *Expanding dimensions of consciousness.* New York: Springer.

Finley, W. W., and Johnson, G. (1983). Operant control of auditory brainstem potentials in man. *International Journal of Neuroscience, 21,* 161–170.

Fujimaki, N., Takeuchi, F., Kobayashi, T., Kuriki, S., and Hasuo, S. (1994). Event-related potentials in silent speech. *Brain Topography, 6,* 259–267.

Furness, T. A. (1986). The super cockpit and its human factors challenges. *Proceedings of the Human Factors Society 30th Annual Meeting, 1,* 48–52.

Halliday, R., and Callaway, E. (1980). Time shift evoked potentials (TSEPs). In C. Barber (Ed.), *Evoked potentials* (pp. 325–327). Lancaster, England: MTP Press.

Heetderks, W. J., and Schwartz, A. B. (1995). Command-control signals from the neural activity of motor cortical cells: Joy-stick control. *Proceedings of the RESNA 18th Annual Conference,* 664–666.

Hiraiwa, A., Shimohara, K., and Tokunaga, Y. (1990, September). EEG topography recognition by neural networks. *IEEE Engineering in Medicine and Biology, 9,* 39–42.

Humphrey, D. R. (1995). Intracortical recording of brain activity for control of limb prostheses. *Proceedings of the RESNA 18th Annual Conference,* 650–658.

Ingle, D. F., and Nasman, V. T. (1994). Self-regulation of the steady-state visual evoked response as recorded at multiple scalp sites [Abstract]. *Psychophysiology, 31,* S58.

Junker, A., Berg, C., Schneider, P., and McMillan, G. R. (1995). *Evaluation of the Cyberlink interface as an alternative human operator controller* (AL/CF-TR-1995-0011). Wright-Patterson Air Force Base, OH: Armstrong Laboratory.

Junker, A. M., Schnurer, J. H., Ingle, D. F., and Downey, C. W. (1988). *Loop-closure of the visual cortical response* (AAMRL-TR-88-014). Wright-Patterson Air Force Base, OH: Armstrong Aerospace Medical Research Laboratory.

Keele, S. W. (1986). Motor control. In K. Boff, L. Kaufman, and J. Thomas (Eds.), *Handbook of perception and human performance* (pp. 30.1–30.60). New York: John Wiley and Sons.

Keirn, Z. A., and Aunon, J. I. (1990, March). Man-machine communications through brain-wave processing. *IEEE Engineering in Medicine and Biology Magazine, 9,* 55–57.

Kocian, D. F., and Task, H. L. (1995). Visually-coupled systems: Hardware and human interface. In W. Barfield, and T. A. Furness III (Eds.), *Virtual environments and advanced interface design* (pp. 175–257). New York: Oxford University Press.

Kuhlman, W. N. (1978). EEG feedback training: Enhancement of somatosensory cortical activity. *Electroencephalography and Clinical Neurophysiology, 45,* 290–294.

Mason, S. G., and Birch, G. E. (1995). Processing of single-trial, movement-related ERPs for human-machine interface applications. *Proceedings of the RESNA 18th Annual Conference,* 673–675.

McCane, L., Vaughan, T. M., McFarland, D. J., and Wolpaw, J. R. (1995). Development of direct brain-computer communication in a person with amyotrophic lateral sclerosis. *Proceedings of the RESNA 18th Annual Conference,* 684–686.

McFarland, D. J., Neat, G. W., Read, R. F., and Wolpaw, J. R. (1993). An EEG-based method for graded cursor control. *Psychobiology, 21,* 77–81.

McMillan, G. R., Eggleston, R. G., and Anderson, T. R., (in press). Nonconventional controls. In G. Salvendy (Ed.), *Handbook of human factors and ergonomics (2nd ed.).* New York: John Wiley and Sons.

McMillan, G. R., Calhoun, G. L., Middendorf, M. S., Schnurer, J. H., Ingle, D. F., and Nasman, V. T. (1995). Direct brain interface utilizing self-regulation of the steady state visual evoked response. *Proceedings of the RESNA 18th Annual Conference,* 693–695.

Miltner, W., Larbig, W., and Braun, C. (1986). Biofeedback of visual evoked potentials. *International Journal of Neuroscience, 29,* 291–303.

Nasman, V. T., Ingle, D. F., and Calhoun, G. (1995). Topographic and task-related effects of SSVER self-regulation [Abstract]. *Psychophysiology, 32,* S55.

Nasman, V. T., Ingle, D., and Schnurer, J. (1994). Differential hemispheric activation as a possible mechanism for SSVER self-regulation [Abstract]. *Psychophysiology, 31,* S71.

Neary, C. (1994, October). Eye pointing and visual image displays. *Proceedings of the Workshop on Human Factors / Future Combat Aircraft, Vol. 1. Four Power Air Senior National Representatives Technical Group on the Supermaneuverability of Combat Aircraft* (IABG-WTF). Ottobrunn, Germany.

Neary, C., and Wilkins, A. J. (1989). Effect of phosphor persistence on perception and eye movements. *Perception, 18,* 257–264.

Patmore, D. W., and Knapp, R. B. (1995). A cursor controller using evoked potentials and EOG. *Proceedings of the RESNA 18th Annual Conference,* 702–704.

Pfurtscheller, G., Flotzinger, D., Mohl, W., and Peltoranta, M. (1992). Prediction of the side of hand movements from single-trial multi-channel EEG data using neural networks. *Electroencephalography and Clinical Neurophysiology, 82,* 313–315.

Pfurtscheller, G., Flotzinger, D., and Neuper, C. (1994). Differentiation between finger, toe and tongue movement in man based on 40 Hz EEG. *Electroencephalography and Clinical Neurophysiology, 90,* 456–460.

Pivik, R. T., Broughton, R. J., Coppola, R., Davidson, R. J., Fox, N., and Nuwer, M. R. (1993). Guidelines for the recording and quantitative analysis of electroencephalographic activity in research contexts. *Psychophysiology, 30,* 547–558.

Polikoff, J. B., Bunnel, H. T., and Borkowski, W. J. (1995). Toward a P300-based computer interface. *Proceedings of the RESNA 18th Annual Conference,* 678–680.

Regan, D. (1989). *Human brain electrophysiology.* New York: Elsevier.

Rockstroh, B., Birbaumer, N., Elbert, T., and Lutzenberger, W. (1984). Operant control of EEG and event-related and slow potentials. *Biofeedback and Self-Regulation, 9,* 139–160.

Roger, M., and Galand, G. (1981). Operant conditioning of visual evoked potentials in man. *Psychophysiology, 18,* 477–482.

Rosenfeld, J. P., and Hetzler, B. E. (1978). Significance and mediation of neural and other feedback. *International Journal of Neuroscience, 8,* 233–250.

Schnurer, J. H., Ingle, D. F., Downey, C. W., and Junker, A. M. (1988). A real-time frequency analysis methodology for evoked potential loop-closure. *Proceedings of the 1988 NAECON, 4,* 1530–1535.

Sutter, E. E. (1984). The visual evoked response as a communication channel. *IEEE / NSF Symposium on Biosensors,* 95–100.

Sutter, E. E. (1992). The brain response interface: Communication through visually-induced electrical brain responses. *Journal of Microcomputer Applications, 15,* 31–45.

Travis, T. A., Kondo, C. Y., and Knott, J. R. (1975). Alpha enhancement: A review. *Biological Psychiatry, 10,* 69–89.

Wolpaw, J. R., and McFarland, D. J. (1994). Multichannel EEG-based brain-computer communication. *Electroencephalography and Clinical Neurophysiology, 90,* 444–449.

Wolpaw, J. R., McFarland, D. J., Neat, G. W., and Forneris, C. A. (1991). An EEG-based brain-computer interface for cursor control. *Electroencephalography and Clinical Neurophysiology, 78,* 252–259.

Design Issues in Human Performance-Based Test and Evaluation of HMDs

Jeff Gerth

The human performance test and evaluation (HPT&E) is a system-oriented approach that is user-centered. HPT&Es provide HMD designers and evaluators with a systematic and holistic tool to ensure that the desired performance level is met. The result is the design or selection of a user-friendly HMD that supports the user in his or her tasks.

11.1 Introduction to the Human Performance Test and Evaluation Process

11.1.1 What is a test and evaluation process?

Suppose you were a member of a large therapy services facility and were asked by your partners to evaluate two commercial HMDs, Products X and Y, for use as a virtual-reality display in the treatment of people with fear of heights (or some other phobia). For a patient suffering from fear of heights, the initial step might be to simulate placing that person at a great distance from the height, perhaps looking at the top of a tall building from across the street. Gradually, the scene would change to provide a closer look, climbing the stairs, looking through the open door to the roof, and finally peering over the edge. It could easily be argued that low-to-medium-fidelity representations of these scenes, progressively closer to the view from the top, would be less threatening initially to the phobic person. The initial low-fidelity exposures to the feared situation could then be followed with an actual visit to a tall building.

You are asked to evaluate the effectiveness of these two HMDs and select the best one for this application. Initially you talk to a few therapists who specialize in treating phobias about situations feared by the "typical" client. You personally try out the two HMDs. Since you don't have a computer-image generator available, you watch several vendor-supplied demonstrations varying in resolution and scene content. You recognize that there are many small differences in both Product X and Y, such as color balance, visual artifacts, and slight problems in adjustments and calibration. None of these problems seem to be significant enough by themselves to warrant acceptance of one HMD over the other, but you have no systematic tests to summarize such problems. After all, if you don't notice these problems when you view these demonstrations, they probably aren't serious problems for someone else, even if they suffer from phobias, are they? In summarizing your impressions you note that

- Both of the HMDs seem pretty comfortable to wear for the short time you wore each of them.
- Both HMDs cost around $1200.
- Both HMDs appear to be sturdy in their construction.
- The data sheets supplied by each of the vendors have similar or identical lists of features, although one HMD is considerably heavier than the other.

You conclude that either of the HMDs would be a good choice. You flip a coin to select between Products X and Y and make Product Y your recommendation to your partners in the therapy-services facility. You conclude this will probably be a good general-purpose HMD for use in a variety of clinical applications for your therapy-services facility.

You have just completed a hypothetical test and evaluation process of Products X and Y.

How good an evaluation is the example above and how does it differ from a *human performance test and evaluation (HPT&E)?* In considering the quality of this evaluation three key distinctions need to be defined for a HPT&E. First, the context of any HPT&E is that it be a *holistic* system-oriented approach centered on the intended user—not just HMD-device oriented. This means that you consider the entire environment as an interacting system. Second, the process is by necessity *systematic* and produces conclusive repeatable results; HPT&E utilizes a scientific method of testing by establishing predictions *a priori* and explicitly testing them through collection of relevant test data. Third, this example illustrates one type of HPT&E intended for establishing that the HMD satisfies the desired performance requirements or preferences. There are several other forms of HPT&E specifically devoted to repeated evaluations during the developmental process of designing an HMD and at the point of initial acceptance of an HMD. Repeated testing during the process of developing an HMD yields substantially better initial products that are easier to use. In summary, HPT&E *is a holistic system-oriented approach, is systematic, and tests* a priori *predictions with collected data.*

How could the evaluation of an HMD for the phobia therapy described in the introduction be improved?

1. Test intended users using representative imagery and tasks.

2. Establish the tasks (e.g., reduce fear at great distance from height) and the typical scenarios of use. (E.g., "(1) View feared image 1. (2) Record fear response. (3) If fear reduces or is below acceptable level, then repeat with feared image 2, etc.; if not, continue viewing image 1.")

3. *Define the imagery* that would be presented during phobia treatment rather than just imagining differences based on imagery samples you are given for a demonstration.

4. *Define a set of evaluation tasks* that capture the activities during actual sessions.

5. Define the duration and other key aspects of these tasks.

6. Specify an expected outcome and/or required performance goal *in advance of collecting any data.*

7. Don't rely on unverified manufacturers' claims as your source of data—*collect the requisite data whenever possible.*

8. Make recommendations based on specific task performance and/or your predefined requirements.

11.1.2   Basics of performance testing

The example in Section 11.1.1 illustrates one type of HPT&E, which is to establish the performance effectiveness of an operational system. However, the holistic approach to HPT&E recommended here supports a wide range of test settings. In any case, the successful HPT&E requires a disciplined testing approach that includes careful planning and execution. However, conducting a structured test is not the only test alternative. The testing methods presented in this chapter can be used separately for limited testing, recognizing that a comprehensive test will produce the best results. To conduct a performance test, two basic issues must be addressed. First, the HMD user(s) in the setting(s) and the scenarios of use must be determined. Second, the HPT&E test goal must be established. Addressing either of these issues in ignorance of the other may risk the validity of the HPT&E, either by not measuring what was intended or applying the results improperly to other situations.

11.1.3   HMD users, setting, and scenarios of use

The HMD user should be tested in the setting of intended use. Knowledge of the setting of intended use is necessary to properly select whom you should test. Diversity in HMD users and the settings of intended use is emerging as HMDs begin to find their way into the marketplace. Emerging user groups are described in Table 11.1.

Each of these user groups in Table 11.1 dictates vastly different user requirements. In the example at the beginning of the chapter, a person suffering from fear of heights was gradually desensitized by means of HMD simulations placing the person first at a great distance from the height, looking at the top of a tall building from across the street, and then gradually changing the scene until the person was finally peering over the edge. This initial low-fidelity exposure to the feared situation was then followed with an actual visit to a tall building. In this case, the HMD scene generation could be accomplished with a low resolution, monochrome, head-coupled HMD. If short training sessions were conducted, weight and comfort might also be less important user requirements.

In contrast to the requirements for people with phobias, consider the requirements for surgeons using HMD presentation of endoscopic or

TABLE 11.1 Emerging HMD Users

Users	Setting	Scenarios of use
Aerospace pilots (military, NASA, general aviation)	Piloting, simulator training, mission rehearsal	1. Wayfinding with maps and databases 2. Enhancing situation awareness in astronaut extravehicular activities (EVA) by providing range, rate, and targeting information 3. Enhancing displays with symbology overlays (e.g., FLIR, IR), gun sighting 4. Full mission rehearsal (e.g., Hubbell space telescope repair)
Medical professionals (surgeons, medical technicians)	Doctors' offices, surgical (clean) environments	1. Conducting surgical procedures 2. Training surgical procedures or reviewing/debriefing post-surgery
People with disabilities	Computer workstation, possible future use in mobile wayfinding in-home, office, and public areas	1. Visual steering and pointing activities with computers when using CAD, drawing, and mouse pointing 2. Enhancements to wayfinding, visual enhancements (e.g., magnification for the visually impaired)
Research professionals	Computer workstation in research laboratories and offices	Data visualization and manipulation
Entertainment/consumers	Multimedia-equipped computers in the home and arcade game centers	Virtual reality games, arcade games
Sports/consumers	Attending and participating in sports events	1. Watching an event in the bleachers while viewing televised game presentation 2. Monitoring HMD presentation of dive parameters (e.g., depth, remaining air) during underwater skin diving
People in therapy	Therapists' offices	1. Desensitizing and/or reducing anxiety from everyday activities (e.g., fear of heights, fear of public places) 2. Interacting with virtual characters/places

laprascopic procedures. Minimally invasive surgical procedures using endoscopes allow surgeons to work through extremely small openings. The surgeon can insert the endoscope and watch the progress of the tip of the scope on a TV monitor or through an HMD. High-fidelity color display is an important requirement, given that the surgeons' wayfinding through the human body requires them to recognize anatomical landmarks and the presence of medical anomalies (e.g., blockages, unhealthy tissue, and tumors). Low overall weight and proper center of gravity and comfort are also important, since surgical procedures can be lengthy. Surgeons may also find flip-up or see-through capabilities highly desirable. Surgery's "clean procedures" also dictate that all calibration and adjustment operations be hands-off or done for the surgeon wearing the HMD by surgical technicians.

None of these users could be easily represented without considerable understanding of their background and the intended setting of HMD use. HPT&E is a process that should cement a collaboration among designers, subject matter experts (SMEs), and the intended users of the system during the entire HMD product development. This collaboration takes place in design evaluation meetings called *focus groups,* which are described below. One note of caution: SMEs are themselves professionals for hire who are often "retired" users and may not be one and the same as the intended users who will actually buy and use the HMD. The point is that there is no substitute for the real user in the intended setting.

11.1.4   HPT&E test goal

HPT&E tests can be conducted for a variety of reasons but generally fall into one of three categories (Meister, 1985): acceptance testing of a new or evolving system, testing design modifications of a fully operational system, or establishing the performance effectiveness of an operational system. Although traditional test and evaluation involves collection and summary of test data without explicit testing of predictions against expected results, this approach does explicitly test predictions of system performance. The systems-oriented approach considers the HMD user in the setting of intended use, the HMD device itself, and the outcomes of use for the intended activities—all as part of a common dynamic system. Such systems-oriented testing employs a user-centered design approach by testing early and often and involving real users in the design process. Human performance-based testing and evaluation thus makes an important contribution to good design. Establishing the test goal explicitly as T&E of acceptance, design modification, or performance effectiveness creates clarity in the conduct of testing and openly establishes the intention of the HPT&E. A mixture of motives or overall lack of clarity can derail an otherwise effective evaluation design.

A number of benefits are realized from a well-positioned and clearly defined HPT&E:

Reduce R/D cost. As with any emerging technology, the final design can be unclear during the formative development phases. Design elements that don't contribute to a final product raise R/D expenses. With early HPT&E, mistakes, problems, and dead ends can be identified and eliminated, preventing unnecessary expenses.

Comply with proven standards and customization by design. Complying with human factors interface standards and other product-specific requirements results in a product that is based on proven design. Areas of the HMD product that are not addressed or are exceptions to such design practices can then be properly customized for the intended user. It is essential to avoid the mindset of, "That looks about right" (see Newman and Haworth, 1995).

Increase initial user acceptance and ease of training users. Early and frequent involvement by the intended users helps ensure that the HMD design will reflect user requirements upon initial product issue.

11.2 Testing HMDs

11.2.1 Essential content areas

Selecting HPT&E activities from each of the essential content areas for an HMD test ensures that your evaluation will be comprehensive. This supports the philosophy of HPT&E as holistic and systematic.

What content areas are essential for fully representing HMD technology and for revealing the strengths and weakness of the HMD(s) being tested? Findings by Brown et al. (1996) and recommendations by Benedict and Gunderman (1992) indicate five content areas of particular importance to HMD evaluations: usability, comfort, information display, maintainability, and safety. Although these evaluations considered only military applications, the comments in each of these areas represent performance issues that are common to many settings and intended uses in emerging areas of HMD applications.

Two caveats are that the HMD systems considered here represent high-fidelity systems and current technology HMDs. Some of the issues discussed in these content areas are not as important if lower fidelity is acceptable (e.g., implying less setup and recalibration time), and leading edge technology innovations in HMD design not yet in use are considered (e.g., high-resolution, low-weight LCD). Also keep in mind that the philosophy of HPT&E described in this chapter considers the expected practical limits of the performance envelope in defining content areas to ensure a rich yield in HPT&E information.

1. *Usability.* Usability is a problem for long-duration activities because of requirements for long setup times and frequent readjustments. Currently available HMDs require extensive in-helmet calibration and alignment procedures at donning, necessitating long setup times and periodic readjustment. Helmet stability also continues to be a problem. Perceptions of instability will cause the user to overtighten the helmet, which can lead to fainting. Heat accumulation and perspiration inside the helmet can cause it to slip, forcing users to stabilize it with one hand. This is a particular problem when the task requires two hands or quick head movements.

2. *Comfort.* Comfort is a problem for training. HMDs are somewhat uncomfortable immediately after donning; after an hour or so, discomfort may become a particular problem. Areas of discomfort most often noted include lack of dissipation of heat; ocular differences or mismatches in color, brightness, and focus; simulator sickness; discomfort associated with problems accommodating large variations in a number of critical anthropometric dimensions, such as head size, eye height, or interpupillary distance; and weight or center of gravity.

3. *Information Display.* HMDs are well suited for display of information that does not require fine visual acuity and for tasks with inherent restrictions on field of view. HMD resolution does not approach that of the human visual system, especially at threshold distances. Limited field of view (usually 40–80 degrees) is an inherent characteristic of HMD technology. Thus, not all information is appropriate for display on an HMD. Other obstructions or distortions of displayed images may result from CRT and LCD anomalies, such as CRT burns or LCD lattice effects, or from dirt or condensation on the optics portion of the HMD.

4. *Maintainability.* HMD technology is complex; well-trained electronics specialists must be available for maintenance of the HMD.

5. *Safety.* Safety is not an insurmountable problem in the use of HMDs. However, safety needs to be assessed in the environment of use before evaluation, training, or use of an HMD. Some areas of consideration include the danger of high voltage electronics close to a user's head, the presence of a potentially explosive environment, the possibility of impact or penetration, and the impact of sound quality on communication.

Following is a set of questions to use as a starting point to create a checklist of content area questions to consider in HPT&E testing:

- What is the profile of your user (setting and scenarios of use)?

- How long will your user be engaged in the intended activity? Will periodic readjustments of the HMD disrupt the intended activity?

- What kind of activity will your user engage in? What movements will it entail? Will a quick detachment from the HMD be needed (connecting cables or entire ensemble)?

- Have you considered calibration and setup as the entry point into usage, and have you allowed for the length of time needed for these activities?

- During typical scenarios of use, have you asked users about their perceptions of heat dissipation, ocular differences, and general physical comfort?

- Is the information you need to display appropriate for an HMD? Is color presentation of information required? Does the information display require high resolution and wide FOV?

- Are appropriate support personnel available to maintain the equipment?

- Is the environment safe for HMD use?

- Does the HMD provide adequate head protection?

- Are the sound attenuation and speech intelligibility adequate for effective communication?

These content-area-specific questions can be further refined and elaborated by considering the specific test focus for testing our HMD.

11.2.2 Establishing the test focus

Is the HMD a design concept or a reality? The developmental maturity of the HMD sets the focus for testing. Determining the proper focus of testing is crucial to the success of HPT&E because of the complex issues surrounding HMDs and the evaluation of human performance. Useful testing opportunities exist at many levels of the HMD development process. A fundamental consideration is whether the *attributes* of the system under evaluation or the *performance* of the total person-machine system is the appropriate focus of testing (Meister, 1985). Both types of evaluations can be conducted at the appropriate times. However, tied to these two types of evaluations are two types of standards. Attribute standards characterize how the system should appear or function, whereas performance standards characterize how the system should perform during use. Understanding the differences between these two forms of testing will assist you in planning your evaluation.

Attribute testing matches system characteristics against standards. Attribute-based evaluations are appealing because they can be done in less formal settings with developmental design concepts. Developmental design concepts can be featured in static evaluations of drawings, procedures, or with limited performance evaluations often employing models or mockups of the design concept. Checklists are an example of an attribute testing instrument. A checklist should include items that can aid in identifying design-induced operator errors and ensure that the breadth of design issues has been addressed. For example, checklist items for design-induced operator errors include:

- Does the display of items include events that will change rapidly?
- Is visual feedback provided for critical decisions?
- Do controls require excessive precision in their operation?
- Are long sequences of setup activities required (e.g., calibration) where multiple dependencies between steps are possible (e.g., earlier and later steps influence each other)?
- Are critical controls that could lead to device damage (e.g., CRT burning with input video loss) properly safeguarded from accidental activation (or deactivation)?
- Are controls and displays crowded together without regard to necessary clearances?
- Are similar controls grouped together and placed in logical association with the displays they influence or adjust?

In addition to checklists, if even a portion of the user interface can be simulated, you may also be able to record the users' performance in terms of

- Performance times (e.g., time to complete an activity, time between activities, and latency to start an activity). Data at this point may suggest possible problems if excessively long or short times are observed.
- Errors (e.g., number and type of "unexpected" activities, such as performing tasks in the wrong order, skipping steps, and turning knobs in the wrong direction). Consistent errors by users may suggest error-prone and/or unclear procedures.
- Failures and malfunctions (e.g., snapping off knobs and optics transports slipping off the tracks or repeatedly hitting limits). Careful attention to when the failures and malfunctions occurred and what

the user was doing are helpful for diagnosing problems and initiating troubleshooting. Problems encountered also relate to the stability of the design.

- Interviews with users on the outcome of use (e.g., best features, worst features, and what they would change). Informal interviews with users after use can deepen the designer's understanding of the HMD from the user's point of view. Users may also be able to describe their expectations about usage (e.g., "I expected clockwise turns of the knob to bring the optics closer") and recommend solutions to the problems they encountered.

Attribute-based testing can be an important contribution to early formative evaluations of HMDs, especially with the components that have evolved from proven standards and practices. For instance, several useful military standards address human performance requirements in the design of control-panel knobs and adjustment clearances and general guidelines for human interfaces such as display formats and symbology sets (e.g., MIL-STD 1472D and MIL-STD 1295A). These standards are useful for military systems but may not be as helpful with civilian applications. This illustrates one aspect of an obvious problem in any emerging technology such as HMDs. There are no established specific standards to serve as baselines in attribute-based evaluations.

Another problem with standards in general is that they do not by themselves establish the individual importance of the attributes, nor do they characterize which will be important to the performance of the system. Perhaps the major disadvantage of formative evaluations is that design for the HMD is changing and evolving. This is why users and SMEs must be repeatedly consulted to establish the importance of these attributes to the HMD under development. It must be emphasized that valuable contributions to system design are gained by attribute-based testing. However, stopping short of utilizing more formalized performance-based methods leaves unanswered an important question concerning the outcome of the design: Does the HMD under development achieve the intended objectives of the user effectively and efficiently?

Performance-based testing assesses human performance against performance standards and user-defined requirements at the system/subsystem level (Meister, 1985). Unlike matching system characteristics in a static evaluation of drawings, procedures models, or mockups of the design against standards (attribute testing), performance-based testing assumes that a working prototype of the HMD system exists. The HMD prototype must be sufficiently developed so that it can be used in an actual or simulated operational setting or settings.

The fidelity of the operational setting should be as high as possible for testing the HMD, since this will reflect on the intended setting of use. The effort to achieve this operational setting is an important consideration to ensure that the effort is well-spent and can be achieved.

As a general rule, if you are unsure of the fidelity of the setting, go back to the intended users and ask them through interviews, surveys, or focus groups (discussed below). Spending too much effort on developing high fidelity for unimportant aspects of the setting (e.g., comfortable chairs) and not enough on important aspects (e.g., working adjustments for the optics) wastes the testing opportunity. The constraints imposed on the setting also affect how much control the HPT&E designer will have over the test setting and exactly what performance can be measured. Developing a performance-based test without careful evaluation of these factors will usually lead to inconclusive and disappointing outcomes of use, where the designers are no better informed about what to improve or how the HMD performed than before the test. To adequately consider all these factors, it is recommended that a written plan be developed for the performance-based test.

11.2.3 Interviewing potential users and SMEs

Using *surveys*, potential users and SMEs can be interviewed concerning their product expectations, intended uses, and pitfalls with other similar technologies, and can also be interviewed after using the HMD. Avoid reliance on open-ended questions (e.g., "What did you think?") by developing a variety of questions of various styles with quantifiable answers (e.g., yes or no, ratings, short answer, or comparisons), possibly overlapping in content to assure coverage. Follow-up questioning should push for elaborations and probe into what the user was thinking, seeing, distracted by, and so forth. Be sure to include questions asking the user what was best and worst, and what they would change about the HMD. A variety of sources in marketing research and psychology can be consulted for assistance in the design and conduct of interviews or surveys.

Conducting written surveys can usefully extend the information or leads that were revealed by interviewing potential users and SMEs. Careful attention to the design of the survey will ensure that every respondent will answer the intended questions. Surveys can be pretested on a limited sample, with follow-up interviews as a final check. Rate of return on surveys is always a concern. A rule of thumb is that approximately one-third of well-designed surveys mailed out will be returned.

Focus groups are small groups of people with a particular background profile (i.e., similar or diverse in age, socioeconomic status, profession, etc.) who meet for an in-depth discussion with an experienced moderator. The moderator is trained to use questions, cues, probes, and stimulation material to obtain user attitudes, opinions, reactions, issues, and expectations related to a specific topic. In preparing for and conducting a focus group several important points need to be considered (Kelly, 1995), including the setting, the leader, the agenda, the discussion process, and documentation.

The setting for the focus group should provide adequate seating, conference tables for group discussions, and whiteboards and an overhead projector for organizing and expressing the group's conversation. Generally, focus group participants are paid for their time and provided a meal (and/or another form of compensation). It is a common practice to audio- and videotape the focus group activities and to allow observers to watch and take notes on closed-circuit TV or through one-way mirrors. Participants have rights. Informed consent from each participant must be obtained for such recording activities and should be elicited using the guidelines provided by the Institutional Review Board for your organization or university.

A member of the design team who is experienced with conducting HPT&E focus groups should lead the session. A dry run with another group of designers can be useful to gain experience and formulate an approach. The leader should be skilled at involving participants, preventing individuals from dominating the conversation, summarizing the deliberations, and keeping to the timeline and goals of the focus group.

The agenda should contain specific design features and options to discuss. Avoid specifying too many design options, but never raise a design issue without at least one design option. Set the tone by establishing that the result of the focus group is to specify the design features and to select or define the best design options to achieve those features.

Group consensus should be reached for each agenda item. The consensus should be to accept, reject, modify, or table (for further consideration) each design option. If additional issues are uncovered in the process of the discussions, the moderator can choose to add additional items to the agenda or can table the issues for further elaboration and inclusion in a later focus group. Each subsequent focus-group session should briefly review the prior progress leading to the current focus group as an aid to review material and options already considered. Reviews will also assist in checking the accuracy of the outcomes and degree of consensus reached. If a fresh opinion or cross-check to prior design decisions by other focus groups is being sought from an entirely new group of people, this review may be unnecessary or undesirable.

Don't keep your appreciation a secret! Acknowledge group members for the contribution they are making to the design and development of your HMD (even if you pay them).

Document the outcomes of each session, and distribute the document to the group members, if possible. This may also stimulate additional input at the next session.

11.2.4   Performance testing

After the scope-limiting factors surrounding testing are established (e.g., user, setting, and scenarios of use are defined, *and* the goal of the test is established) the performance-evaluation tasks must be defined. In comparison to the way users might do or already do their tasks using an HMD, the performance evaluation tasks are intended to represent only the key elements of those activities and to provide sufficient additional details to create the task setting. For example, to assess the maintainability of the HMD we might define the line replaceable units of the HMD to be the HMD display (CRT or LCD), optics train, electronics subassemblies, helmet attachments, and configuration. We would then define the scenarios for assessing maintainability as presented in Table 11.2.

TABLE 11.2   Scenarios for Assessing Maintainability

Maintenance scenario	Description
Calibration of optics	Align HMD optics.
Removing and replacing each LRU	Disassemble unit and reassemble using standard tools. Use bare hands. (Other protective ensemble is at the discretion of the vendor.)
Connector access	Remove and disconnect connectors. Use bare hands. (Other protective ensemble is at the discretion of the vendor.)
Gaining and closing accesses	Locate, open, and close each of the accesses provided for maintenance. Use bare hands. (Other protective ensemble is at the discretion of the vendor.)
Finding and identifying the functional use of test points	Verify that test equipment leads and/or probes have access to test points.
Documentation	Provide copies of all documentation
Troubleshooting	Vendor will demonstrate any troubleshooting fault-isolation procedures and provide any available documentation.

Notice that seven areas of maintenance activities are identified in Table 11.2. Only the review of documentation could be expected to be done off-line. All the remaining maintenance scenarios would need to be done in real time by evaluating a trained technician's performance. However, not all aspects of the HMD deserve the attention of performance testing. In the context of maintenance activities, only a subset of the defined maintenance scenarios may actually need to be conducted or would even be of interest. For example, the optics subsystem may be factory-sealed and therefore could logically be excluded from any performance demonstration. In fact, many aspects of the HMD can be evaluated on the basis of their attributes, as already discussed in Section 11.2.1.

Establishing the test focus. We might take photographs of the interior of the HMD electronics subsystem and count off-line the number of test points, document the function of each, and compare it to established maintainability standards for testing. It is important to consider limiting performance testing to the key elements and to test as many subsystems of the HMD with as few separate tests as possible.

This testing philosophy is holistic and systematic, as well as efficient. Efficiency is important given that time and money for testing are rarely abundant. Designing performance tasks draws on all the information collected to establish the user setting, scenarios of use, and, in particular, the goal of testing the HMD. The goal of testing the HMD identifies what components should be included as the most important aspects of the intensive performance-based task evaluation. For example, if we establish the design goal of evaluating a design modification for focus adjustment, then our performance task would involve the user in making various periodic adjustments under the important operational conditions (e.g., ambient light, image lighting and contrast, degree of see-through, static or moving imagery, and location of imagery in the FOV).

Often, designing performance-based tasks goes beyond all other information and requires additional interviews with SMEs or reviews of requirements to fully elaborate performance aspects of these tasks and measurement of performance-defined variables. The identification and taking of measures of performance (MOPs) are themselves extremely complex areas of performance-based task design that will only be introduced conceptually here. There are a number of excellent texts on HPT&E that discuss in depth the design of appropriate MOPs (see, e.g., Meister, 1985; O'Brien and Charlton, 1996). Performance testing follows the fundamental approach of HPT&E by considering how the tasks to be included in the performance evaluation contribute to an overall understanding of the HMD (e.g., HPT&E's holistic system-oriented approach). This is especially important to consider, since both software and hardware development are often required to implement even a rather modest performance task. It is also important for such development, given the associated costs and effort, to follow a systematic approach. To support this process, three basic steps are recommended in performance testing: pretest planning, performance test execution and post-test analyses/reporting. The basics of performance testing are illustrated in the hypothetical performance task design problem described next.

Suppose you were a member of a research firm that specializes in developing training programs for remotely controlled robotics applications, such as those used for undersea exploration. Suppose you were asked to determine if a given manufacturer's HMD meets your requirements for image update in wide FOVs that may require rapid head movements by the user to maintain location of the camera image in the center of the FOV. Given that undersea training for such image tracking would be costly, it would be an ideal application for an HMD-based portable simulation training environment that could even be used on-board the ship. In considering this problem, you recognize that an important HMD related issue is the *image transport delay time,* which is the time it takes an image sent from the computer image generator to be visible to the user on the HMD display. A second HMD-related issue is the *tracking delay* of head coupling related to HMD image positioning. A final HMD-related issue is maintaining *image quality* in imagery associated with high-speed movement in the HMD FOV. This is a concern when lowered image quality distorts an image's features or obscures its identity altogether. As is often the case in HPT&E, test time is precious and costly, involving expensive people as users and evaluators. Therefore we are asked to design a single task to address the HPT&E concerns of these three issues.

Pretest planning. In the pretest planning phase, the task is first developed conceptually from the requirements for transport delay and tracking. An object initially in view needs to change position periodically to require the HMD field of regard (FOR) to move within the FOV. Are the composition of the moving object and the viewing background crucial aspects of this task? They could be if the viewing background (e.g., water quality or ocean bottom) were murky or deep in sediment, resulting in poor visibility. Considering such issues would be examples of high-fidelity task concerns. In contrast, we may want to consider only the somewhat unrealistic situation of perfect viewing conditions, to establish a basic, lower-fidelity task as a starting point for our task development. Following the logic of a lower-fidelity task as a starting point, the key elements of the task are a tracking cursor and a moving image. The tracking cursor is boresighted to the head movements through a tracking system. The task begins with the image at some distance away from the tracking cursor. The object of the task is to acquire the image by placing the tracking cursor around it. The basic measure of performance is the time needed for the acquisition, which represents the combined delays of image-transport delay time and tracking delay time. A storyboard, similar to the one shown below in Fig. 11.1, is developed to illustrate the major actions involved in the task for a single trial. At the beginning of a trial an audio warning (a beep) is sounded, and the HMD display appears, as depicted in Fig. 11.1a. The boresighted cursor moves with head movements and is updated on the display. If the target image were not in view, the FOV could be moved by placing the cursor in a box on the perimeter of the view to "move" the display in that direction. Once the target image is in sight, placing the cursor over the image acquires it, as illustrated in Fig. 11.1b.

After the target is acquired, the beep sounds a second time, indicating that the target is again moving to a new location. We then repeat a second acquisition similar to the beginning of the trial. We could have ended the trial with acquisition and started each trial discretely, transitioning from the illustration in Fig. 11.1a to that in Fig. 11.1b (acquisition). However, by incorporating reacquisition as the next step

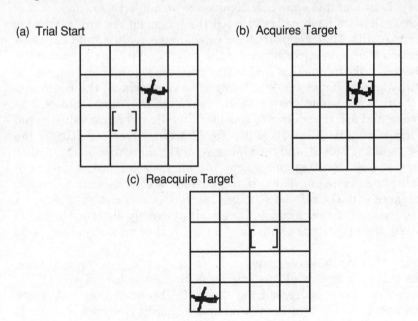

(a) Trial Start

(b) Acquires Target

(c) Reacquire Target

Figure 11.1. Transport delay tracking task.

in the task, we have made it a continuous-performance task. We now can see that the image-transport delay and the tracking delay will not be measured independently by this task but will be measured in the reacquisition time by the combined delays of these two factors. As an initial assessment task we determine that this promotes test efficiency by measuring these related effects but also prevents us from forming any conclusions on the individual contributions of these factors without additional testing. We decide to accept this constraint and reserve some test time for separate evaluation of image-transport delay time and tracking delay only if we notice objectionable time lags.

A second advantage of this continuous-performance task with reacquisition of the target is that the viewer is more likely to be observing the image closely. This addresses the third aspect of this task: assessment of image quality. To incorporate this final issue, image quality of high-speed movements, we can add a judgment by the viewer after observing a number of reacquisitions that may move in various directions with varying magnitude. There are a variety of methods that can be employed to assess image quality. Two basic approaches are assessment by subjective rating and assessment by magnitude estimation. For example, we could use a simple bipolar rating of image quality on a six-point scale varying from poor to very good. For each of the occasions on which we want to assess image quality, we could present the rating scale and ask for a current assessment of image quality. One criticism of this approach, common to all subjective ratings of this type, is that we have not referenced the judgments to any particular image quality. This means that the rated image quality could vary as a function of a changing criterion for the image quality on the part of the observer. A technique that avoids this criticism is to employ magnitude estimations of image quality. Using this approach, at the beginning of the task we would show a static display of the image that will be moving and ask the viewer to consider this the reference value of 100 in image quality. After 10 sequences of acquisition–reacquisition, the task would be halted, and the viewer would evaluate the image quality observed over the 10 sequences considering the reference as the image quality anchor point. Thus, if the current image appears degraded compared with the reference image, the viewer may choose to assign a magnitude of 90 to the current image. If a severely degraded image is viewed, the magnitude estimate of image quality may be judged as 30 or 40.

Now that we have designed a basic candidate performance task, does this task contribute to an overall understanding of the HMD? Overall, we have addressed the three HMD-related issues of image transport time, tracking delay, and image quality. However, in order to create a single task to address these three issues we have appropriately constrained both the image and background and the exact meaning of image transport delay time, tracking delay, and image quality. These constraints and the fidelity of the overall task will be important areas to document in our reporting and interpretation of the results of our performance evaluation.

Performance test execution. Conduct of test is important for several reasons. First, if comparisons are to be made across individual evaluators, consistency in the test conduct is crucial. Any differences in how the task is conducted may contaminate the evaluation results. A second reason is the typical hectic schedule testing requires, with the usual breakdowns in equipment and personnel before and during testing. A backup plan of action usually requires redundant measurement of key information in the performance task and a liberal time schedule to allow for breakdowns. In deciding on a general testing approach, it is also wise to include an additional day for makeup tests if possible. In our example, testing conditions such as lighting and other comfort factors such as temperature may be important, and if repeated evaluations of transport delay and image quality are requested, these comfort factors may reduce the fatigue of the evaluator and improve consistency. Data loss and testing interruptions are test conduct issues that must be considered by the HMD test evaluation team *before* concluding testing. It is always less of a threat to achievement of the testing goals to have as complete and consistent a data set as possible. Estimating values for missing data or employing special statistical procedures will reduce the overall power of the results and the interpretation of the data. An extreme case, such as eliminating an entire test activity (e.g., canceling maintainability evaluations), may mean being unable to meet the basic goal of the intended testing. In the confusion often surrounding testing, such a compromise can seem to be a good decision at the time if testing goals and the specific test activities that support those goals are not carefully compared.

Post-test analyses and reporting. Equally important to the test process is documenting the method of test. This has several distinct benefits: it communicates the test approach to others involved in the test process; it clearly summarizes the requirements for each test objective; it provides a summary of resource and data requirements accumulated from the resource and data requirements of the individual test objectives; and it defines a summary of the schedule of activities and indicates when various personnel, equipment, and material resources are required. When testing is completed there can be a staggering amount of data to reduce and analyze. Clearly defining data requirements when designing tasks and collecting the data with structured forms simplify the job. Computer-based data collection should be used when time and money permit, to ensure precision and consistency in the data. Real-time data reduction and summary are also advantages that computer-based data collection can offer. Such real-time collection can be used to troubleshoot any testing-related problems as well as to reduce the effort of post-test analyses. Employing ratings over open-ended questionnaires also can simplify the task of post-test analyses. Several guidelines should be considered as ways to enrich the collected data, including the following:

- Videotape or record test activities to document procedures and capture aspects of task performance that may not be directly measured—*record both what the evaluators see and their reaction to it.*

- Provide an observation room adjacent to the test area where testing can be viewed without disrupting it.

- Perform simple calculations of means and standard deviations to reveal where the HMD may need improvement. Graphs and charts of the summarized data are also invaluable in gaining initial impressions of the outcomes of testing and can assist in determining where more sophisticated statistical data analyses may be needed.

Alternatives to performance testing. Up to this point in our discussion of performance testing we have portrayed performance testing in its objective empirical-test-approach form. This form of performance testing is preferable whenever possible because it provides the most sensitive and controlled evaluation. However, it is not always an efficient use of the available test resources (e.g., time, money, and available users and evaluators) to conduct empirical performance testing. Actually, performance testing, in general terms, exists on a continuum ranging from highly objective and structured tasks, such as the example of a continuous tracking task described earlier in this section, to subjective and informal tasks, which include usability tests and usability inspection methods (Wiklund, 1991; Nielsen and Mack, 1994). Usability testing focuses on the user interface and users' reactions to it. Typically testing involves self-exploration as well as directed tasks with the HMD being tested. A self-exploration session with an HMD might be structured by establishing objectives such as the following:

- Collecting users' suggestions for improving the design
- Developing a baseline for completing a given usage scenario by measuring the time to completion and the number of errors committed

For instance, we could begin the self-exploration session with a candidate HMD for our undersea exploration by telling the user:

> Imagine you are using the HMD for that stated purpose. You should then don the HMD, assume it to be fully functional, and spend about 10 minutes exploring how the HMD works. Expect that in about 10 minutes the formal training session would start. Please talk aloud as you explore. Tell us what you are thinking and what decisions you are making. Tell us what aspects of the HMD please or disturb you. Ask for assistance only if it is absolutely necessary.

A post-exploration interview may be used to further elaborate the usability test. Questions such as the following should be asked:

- What do you think of the HMD so far?
- Was it clear how to operate the HMD and perform routine adjustments?
- What aspects of the HMD design made learning to use it easy or hard?

Directed-task sessions would be more structured, asking the user to complete a series of realistic task scenarios with the HMD focusing on typical tasks as well as those that would be expected to give the user trouble. Videotaping and post-session questionnaires can be employed as for the self-exploration usability test. *Usability inspection* is a generic term for a collection of methods for having evaluators or experienced users inspect or examine usability related aspects of the user interface. Some of the usability methods are not actually performance-based tests. However, those that rely on some form of scenario-driven walkthrough of the typical tasks in using the HMD would qualify as performance tests. Consider the maintenance example presented above in which we conducted a performance evaluation of selected activities for which we first received training and then participated in executing the actual maintenance activities. A much less time-intensive approach would involve rating a trained technician carrying out the maintenance activities on a series of rating dimensions such as those included in Table 11.3. All scales are six-point, bipolar-rating dimensions.

These ratings could be completed by one or more individuals within a very limited period of time as an efficient alternative to objective empirical performance testing.

11.3 Conclusions

The holistic and systematic testing approach that is the philosophy of HPT&E is intended to support the selection and development of the best

TABLE 11.3 **Response Alternatives for Subjective Assessments of Maintainability**

Rating dimension	Lower response alternative anchor	Upper response alternative anchor
Completing each maintainability scenario	Difficult adjustment (1)	Very easy adjustment (6)
Maintenance safety. Overall, how safe were the maintenance procedures to execute?	Unsafe (1)	Very Safe (6)
Maintenance information and documentation. Was the printed information and documentation adequate to support the maintenance activities demonstrated?	Inadequate (1)	Extremely adequate (6)
Handling and removal for replacement. Could the removals for replacement be executed easily?	Difficult (1)	Very easy (6)
Fasteners and special tools. To what extent were fasteners and special tools required to complete the maintenance scenarios?	Frequently (1)	Never [not needed] (6)
Accessibility. Overall, what is the accessibility to the components requiring maintenance?	Inaccessible (1)	Very accessible (6)

possible HMD for each application of intended users. After reading this chapter you should be prepared to participate in this valuable process to select the right evaluation team and make the important decisions to buy an off-the-shelf HMD, or to define and direct the development process to create a custom HMD. Often the reaction to any systematic approach is that it is too strict, confining, or limiting for practical, applied decision making. Rather than setting limits, this approach establishes clear and definable goals and conditions of satisfaction for reaching them early in the decision-making process. It is still the decision-maker's role to weigh the importance of formalizing the HPT&E process. Generally, the more formal the process, the greater the risks associated with the appropriateness of the decision outcome. Even formulating the problems of selecting and developing HMDs in this framework will be useful in writing clear statements and recommendations for purchasing HMDs. With the continued rapid advancements in the enabling technologies that underlie HMDs, no one can afford to waste money on an HMD that will not be useful, especially if it will be obsolete in several months or a year. Following the recommendations of this approach will ensure that your investment is well spent and enable your HMD users to achieve their application goals.

11.4   References

Benedict, C. P., & Gunderman, R. G. (1992). Helmet-mounted systems test and evaluation process. *Helmet Mounted Displays III, Proceedings of the SPIE, 1695*, 8–12.

Brown, R. C., Folds, D. J., Gerth, J. M., Kline, K., Losier, P., Mitta, D. A., Robinson, W. G., Sarathy, S., Weiler, D. S., Whaley, C. J., & Wightman, D. (1996, March). *Human factors evaluation of current helmet mounted display systems for U.S. Army training applications* (Final Report, Contract N61339-93-C-0134). Naval Air Warfare Center/Training Systems Division for U.S. Army Simulation, Training and Instrumentation Command (STRICOM).

Kelly, M. J. (Ed.). (1995). *Human factors handbook for advanced traffic management center design.* Atlanta: Georgia Institute of Technology.

Nielsen, J., & Mack, R. L. (Eds.). (1994). *Usability inspection methods.* New York: John Wiley & Sons.

Meister, D. (1985). *Behavioral analysis and measurement methods.* New York: John Wiley & Sons.

Newman, R. L., & Haworth, L. A. (1995). Helmet-mounted display requirements: Just another HUD or a different animal altogether? *Helmet- and Head-Mounted Displays and Symbology Design Requirements, Proceedings of the SPIE, 2218*, 226–237.

O'Brien, T. G., & Charlton, S. G. (1996). *Handbook of human factors testing and evaluation.* Mahwah, NJ: Lawrence Erlbaum.

Wiklund, M. (1991, July). Human factors design: Usability tests of medical products as prelude to clinical trial. *Medical Device and Diagnositic Industry*, 46–49.

12

Glossary of HMD Terms

Accommodation A number of oculomotor adjustments mediated by the autonomic nervous system that collectively influence the optical refraction of the crystalline lens of the eye. The range of accommodation can extend from greater than 10 diopters (the near point) to beyond optical infinity (the far point), with an equilibrium position, typically between 1 and 2 diopters, referred to as the **dark focus** of accommodation. The accommodative response is primarily driven by blur but is also interlinked with the vergence system. Accommodation is also affected by nonvisual factors such as emotion, stress, and learning.

Acuity The ability to see detail in high-contrast patterns. Acuity may be expressed in minutes of arc visual angle of a target's critical detail, or as referenced to a 1-minute critical detail at 20 feet. For example, 20/20 indicates that a 5-minute target with a 1-minute critical detail (acuity targets are typically five times the size of the critical detail) can be resolved at 20 feet, 20/40 that the target needs to be 2 minutes for successful resolution, 20/80 that the target needs to be 4 minutes, etc. Normal acuity is considered to be 20/20. Displays, including HMDs, are sometimes described as *having* an acuity of 20/40 or 20/60. This may mean that the detail of a 20/40 or 20/60 character can be seen on the display, but probably means that the angular subtense of a single display pixel or element corresponds to the critical detail of a 20/40 or 20/60 character (2 and 3 arc minutes, respectively).

Addressability The number of discrete horizontal and vertical elements or groups of color elements on the surface of a display or image source. For example, a VGA display has an addressability of 640 × 480 display elements or **pixels**. The utility of these display elements is dependent on factors such as luminance, contrast, and focus. See **Display format.**

Aliasing Creation of spurious low-frequency information when high-frequency information is undersampled, as by the photoreceptor mosaic of the human retina.

Anatomical landmarks Points on the human body that indicate the location of features such as pupils of the eyes or tip of the nose that can anatomically be considered the same (homologous) from person to person.

Anterior   Position of a landmark or body segment that is in front of a reference point.

Astigmatism   (1) Unequal refraction in different meridia of the eye, caused by the optical surfaces (most notably the cornea) not being exactly spherical. (2) Also an optical aberration in which the sagittal rays have a different focal length than the tangential rays.

Atmospheric haze   A distance cue in which objects appear hazy, less distinct, and bluer the farther away they are, because of the interaction of light with dust and moisture particles in the air.

Augmented reality   A display in which simulated imagery, graphics, or symbology is superimposed on a view of the real environment.

Beamsplitter   An optical component used to combine two or more optical paths. A beamsplitter located in front of the eye combines display imagery with a view of the outside world as in **augmented reality** applications. Sometimes referred to as a combiner.

Binocular HMD   An HMD with independent left and right image sources and optical relays. This configuration supports stereo imagery.

Binocular disparity   The separation of the eyes creates a slightly different view of the outside world for each eye. Retinal stimulation by near and far objects is disparate, and this binocular disparity is interpreted as stereo depth. Large amounts of binocular disparity can cause **diplopia**, visual suppression, and fatigue. Binocular disparity can be used in a **binocular HMD** to display stereo imagery.

Binocular fusion   The process by which small **binocular disparities** are fused in the visual system, resulting in single images.

Binocular overlap   The central display field of an HMD that is observable by both eyes.

Binocular rivalry   The alternation or suppression of a percept over time between one eye's view and the other eye's view when the two eyes view very different imagery.

Biocular HMD   An HMD where the left and right eyes view the same image source through left- and right-eye optical channels.

Biofeedback   Training technique that may be used to facilitate **brain-actuated control**. The user learns to self-regulate brain activity by observing real-time information concerning the activity of the particular brain response, component, or pattern that is to be used as the control signal.

Bitragion breadth   The horizontal breadth of the head is measured from the right **tragion** to the corresponding **tragion** of the left ear. This measurement is taken with a spreading caliper.

Brain-actuated control (BAC)   Technique in which brain's electrical activity is recorded and applied in real time as a direct control interface to operate physical devices and computers. May also be referred to as a direct brain interface or EEG-based control. See also **Brain interface, EEG-based control.**

Brightness   The attribute of a visual sensation by which a stimulus appears more or less intense or appears to emit more or less light. Although frequently used interchangeably with **luminance,** brightness is not a photometric standard and should not be used in conjunction with photometric units such as footlamberts.

Center of gravity (of the head)   The point about which the sum of moments of weights of all components on one side of a point is equal to the sum of moments of weights of all components on the other side.

Contrast   The ratio of the **luminance** of an object and background or peaks and valleys of a periodic image. Contrast can be used to indicate visibility or **image quality.**

Contrast sensitivity function (CSF)   A graph depicting a person's ability to see targets of various spatial frequencies. The spatial frequency of the test target is plotted on the horizontal axis, and the sensitivity (reciprocal of threshold **contrast**) is plotted on the vertical axis. Usually these are logarithmic axes.

Convergence   (1) A **vergence** eye movement whereby the eyes counter-rotate to fixate on a closer object. (2) Alignment of a **binocular** or **biocular** display that places the **vergence** point of the corresponding images at a distance less than optical infinity.

Convergent/divergent displays   Display configurations that determine whether the monocular regions of a partial-overlap binocular display are nasal or temporal. These configurations can be simulated by viewing a distant scene through a window (convergent) or by looking at a book held at arm's length (divergent).

Critical flicker fusion (CFF)   The lowest refresh rate of a display that will render the display flicker free. CFF is dependent on **luminance,** phosphor persistence, and display size.

Cybersickness   The analog to **simulator sickness** that specifically applies to HMDs and immersive visual environments.

Cyclofusion   See **rotational alignment.**

Dark adaptation   The increase in visual sensitivity that accompanies time in darkness following exposure to light.

Dark focus   The resting position of visual **accommodation** found in a dark environment or empty field. Values typically range from 1 to 3 **diopters** (1 to $\frac{1}{3}$ m).

Depth of focus   The physical or optical distance over which objects or images remain "in focus" or sharply imaged. The depth of focus is inversely proportional to **pupil** size.

Depth perception   Visual discrimination of absolute and relative distances using monocular (perspective, known size, **texture gradients, motion parallax**) and binocular (**convergence, stereopsis**) cues.

Dichoptic   The presentation of distinct images to each eye.

Diopter   A measure of optical distance, calculated as the reciprocal of distance in meters.

Diplopia   Double vision due to significant **binocular disparity** or misalignment.

Dipvergence   (1) An eye movement whereby the eyes counter-rotate in the vertical direction—typically in response to prism displacement, binocular displays, and other artificial stimulation. (2) Vertical misalignment of a binocular or biocular display that requires counter-rotation of the eyes to fuse the imagery.

Display format   The HMD display **addressability** (e.g., 640×480), aspect ratio (e.g., 4:3), or video standard (e.g., VGA).

Distal stimulus   A physical object (or event) in the environment that can be perceived.

Distortion   (1) The change in size, distance, or shape of objects due to factors such as optical curvature and refraction and perturbations in the atmosphere. (2) An optical aberration in which the true height of an off-axis image point differs from the paraxial position.

Divergence   (1) A **vergence** eye movement where the eyes counter-rotate to fixate a distant object. (2) Misalignment of a **binocular** or **biocular** display that places the **vergence** point of the corresponding images beyond optical infinity.

Dorsal   The back half of the head or body that has been divided into front and back halves by a plane.

EEG self-regulation   Voluntary modulation of ongoing brain electrical activity. This usually requires training and is used in **brain-actuated control** systems as a basis for control signal system input.

EEG-based control   See **Brain-actuated control**.

Egocentric perspective   View of the world from the observer's point of view or perspective.

Electroencephalogram (EEG)   Graphic recording of electrical activity from the cerebral cortex, observed as voltage potentials recorded from the scalp surface.

Electromyogram (EMG)   The recording and study of the electrical properties of the skeletal muscles.

Endogenous brain response   A component, pattern, or response within the EEG that corresponds predictably to specific perceptual or cognitive operations or to the preparation for body movement. May be adapted to serve as an EEG-based control signal with or without the use of operant conditioning or biofeedback.

Electro-oculogram (EOG)   The recording and study of electrical changes that occur during **eye movements.**

Engineering anthropometry   The study of human body measurement as it applies to the design and performance of equipment systems.

Environmental stressors   Stressors in the environment (e.g., heat, cold, unwanted auditory noise, or disability glare) that can affect performance.

Exit pupil   The image of an aperture stop in an optical system in image space. In a **pupil forming system,** the area through which all light rays pass.

Exocentric perspective   View of the world from a bird's or God's eye view.

Exogenous brain response   A component, pattern, or response within the **EEG** that corresponds strictly to a specific external stimulus. Use as an EEG-based control signal requires stimulus presentation and usually also will require biofeedback or some other form of training to enable learned voluntary self-regulation.

Eye dominance   The eye normally used for sighting (looking through a microscope or telescope, aiming a pool cue or rifle, etc.).

Eye movements (vergence)   Smooth, disjunctive movements where the eyes counter-rotate, including **convergence** and **divergence. Vergence** is synkinetically associated with **accommodation** and pupillary constriction, as well as with **binocular fusion.**

Eye movements (vestibular)   Eye movements that counteract head tilt and rotation in an attempt to stabilize the image of the outside world.

Eye movements (saccadic)   High-velocity, ballistic eye movements that are used in searching and reading.

Eye relief   The distance between the last surface in an optical system and the **exit pupil.**

Far point   The greatest optical distance to which the eyes can diverge and/or focus, which is frequently beyond optical infinity. The far point may be either the **accommodative** response of the observer or the optical distance of a target at the onset of noticeable blur.

Face breadth   The straight line distance between the right and left **zygion** landmarks. This measurement is taken with a spreading caliper.

Face length   The vertical distance between **menton** and **sellion.** This measurement is taken with a sliding caliper.

Feature envelopes   Anthropometric design envelopes that describe the spatial location and orientation of features (e.g., pupils) with respect to a well-defined, easily duplicated coordinate system.

Field curvature   If no **astigmatism** is present in an optical system, the tangential and saggital rays fall on the curved Petzval surface. Field curvature is inward for a positive lens and outward for a negative lens.

Field of view (FOV)   The angular extent of a display or aperture, usually expressed in degrees of visual angle and given in terms of a diameter, a diagonal, or in horizontal and vertical dimensions. The corresponding clinical term is visual field and involves a mapping of the perimeter of visibility of the eyes.

Fixation point   A point in space at which the eyes are pointing or directed, frequently referred to as the **line of sight (LOS).** Small errors in the direction of the eyes at this point are measured as the fixation disparity. Although it is frequently assumed that the eyes are focused (or **accommodated**) to the distance of this point, this is not necessarily the case.

Flicker   See **critical flicker fusion (CFF).**

Focus   The adjustment of a lens (including the crystalline lens of the eye) to attain good image quality.

Foreshortening   A visual cue to distance in which the surface extending into space appears more compressed at distances farther from the observer. For example, ties in a railroad track appear more closely spaced as a function of distance from the observer.

Fovea   A central area of the retina that supports detail and color vision and provides the best visual **acuity.**

Frame rate   The number of unique video frames presented per second—typically by a computer display. A high frame rate supports the appearance of smooth and continuous dynamic imagery. The frame rate is limited by the display refresh rate but is otherwise independent.

Frankfurt plane   A standard plane for orientation of the head. The Frankfurt plane is established by a line passing through the right **tragion** (the front of the ear) and the lowest point of the **infraorbitale.**

Ghost image   An observable (and generally undesirable) secondary reflection from an optical surface.

Glabella   The **anterior** point on the frontal bone midway between the brow ridges. The investigator stands on the right side and locates the landmark by inspection.

Head anatomical origin   The point in the head defined by the intersection of the three anatomical axes defined as follows: the y-axis positive direction being a vector from the right **tragion** to the left **tragion**; the x-axis positive direction being a vector normal to the y axis and extending forward from the y axis to the **infraorbitale** and then translated to the midline on the face; and the z-axis positive direction being a vector normal to the intersection of the x axis and the y axis and extending up through the top of the head.

Head breadth   The maximum horizontal breadth of the head taken above the ears (usually slightly above and behind the ears). This measurement is taken from the rear of the subject with the spreading calipers.

Head circumference   The maximum circumference of the head above the browridge or **glabella.** This measurement is taken with a tape measure.

Head length   The distance from **glabella** to the most **posterior** point of the back of the head. This measurement is taken with a spreading caliper.

Head movement   Movement of the head that causes a relative displacement between the environment and the eyes and head, usually accompanied by **eye movements.** Head movements compensate for body motion or are used in the acquisition of a peripheral target. In acquiring a peripheral target, the sequence is typically an eye movement, followed by a head movement, followed by an **eye movement** in the opposite direction. This compensates for the head movement and results in the eyes pointing straight ahead (relative to the head) and at the target.

Head tracker   A device that mounts on the head and tracks up to six types of movement (x, y, z, pitch, roll, and yaw). Typically used to stabilize a visual image in the presence of head motion.

Heterophoria   Horizontal, vertical, or rotational deviations of the eyes kept latent by **binocular fusion**.

HMD   A diverse family of viewing systems where one or more displays and sets of optics are attached to the head (*head-mounted display*) or helmet (*helmet-mounted display*).

Hue   The attribute of color perception denoted by color names such as blue, green, yellow, red, and purple. The psychological correlate of dominant wavelength.

Hybrid systems   Term used to refer to alternative control systems which employ brain electrical acitivity (**EEG**) in combination with other bioelectric signals such as forehead muscle activity (**EMG**) or eye movements (**EOG**). See **Brain-actuated control**.

Illuminance   The density of the luminous flux incident on a surface; the quantity of visually effective light with which a surface is illuminated. Typical units are footcandles (fc) and lux (lx) or lumens per square meter (lm/m^2). One lx or lm/m^2 = 10.76 fc. External lighting requirements for surface-mounted controls and displays should be expressed in units of illuminance.

Image quality   An objective measurement or subjective rating of the quality of a display system.

Immersion   The sensation of being in a synthetic world or at a remote location. Immersion is associated with **WFOV HMDs** with **head tracking**.

Inferior   A position landmark that is located below a particular reference point or away from the top of the head.

Infraorbitale   The lowest point on the anterior border of the lower ridge of the bony eye socket.

Interaxial   The distance between the optical axes of symmetry in a stereoscopic imaging system. This can be measured between the lenses of camera system or calculated between the centers of projection in a computer-generated display.

Interposition   A monocular depth cue based on occlusion of a distant object by a closer object.

Interpupillary distance (IPD)   The distance between the centers of the pupils of the eyes when the eyes are parallel. This measurement is taken with a pupillometer. Also referred to as interocular distance.

Instrument myopia   The tendency to overaccommodate when viewing through an optical instrument, due to a return to the intermediate resting position of **accommodation** found in darkness (the **dark focus**). The instrument's focus is evidently adjusted to correspond to the observer's resting position. This occurs even when the viewed scene is rich in detail and provides a good stimulus to **accommodation**.

Keystone   A trapezoidal image **distortion,** as when a rectilinear image is projected at an angle onto a plane.

Lateral   The position or motion of a body segment that is relatively farther away from the midline of the head or body segment.

Linear perspective   The convergence of lines that makes a two-dimensional representation of a scene appear to be three-dimensional. An example would be the edges of a straight road that appear to converge near the horizon.

Line of sight (LOS)   A line joining the center of the pupil to the object of regard. The line connecting the point of fixation and center of the entrance pupil of the fixating eye.

Luminance   The amount of visually effective light emitted by an extended source. Typically expressed in footlamberts (fL) or candelas per square meter (cd/m^2).

- One fL = 3.43 cd/m^2
- One cd/m^2 = 0.292 fL

Although often used interchangeably with **brightness,** luminance is a photometric standard, whereas **brightness** is a perceptual judgment.

Luning   The crescent-shaped "shadows" observed with a partially-overlapped binocular display that lie to the outside of the binocular region. **Binocular rivalry** is probably involved in this phenomenon in that the luning shadows alternate over time. The magnitude of the luning effect may be affected by the binocular-display configuration—**convergent** or **divergent.**

Magnification difference   The relative size of the left and right images of a binocular display. A small magnification difference can be tolerated by the observer.

Magnocellular pathway   Visual pathway of the human visual system that conveys information about motion. Phylogenetically, this pathway is more primitive than the **parvocellular pathways.**

Mass-moment of inertia   In an HMD, the resistance to movement or rotation of the head while wearing head-mounted equipment. This is the result of the distribution of the mass on the head.

Menton   The lowest point on the mandible (bottom of the chin) in the midsagittal plane. The menton is located by palpation.

Miniature display   An electronic display small enough to be mounted on the head. Typical diagonal dmensions are 2–4 cm.

Modulation transfer function (MTF)   The frequency response of an optical system defined as the ratio of the input to output modulation.

Monochrome   An image or display that emits a single hue but that may vary in intensity and saturation. Single-phosphor **miniature displays** used in many **HMDs** are typically green and are used to display symbology or sensor imagery.

Monocular HMD   An HMD that only provides imagery to one eye using a single optical train and image source.

Motion parallax   A potential source of monocular depth information based on differences in relative motion between images of objects located at different distances from an observer who is moving his or her head.

Mu rhythm   Endogenous sensorimotor rhythm (8–12 Hz frequency range) normally associated with physical movement. Can be used as a control signal in a **brain-actuated control** interface.

Near point   The closest point that the eyes can see clearly without **diplopia,** a function of both the near limits of **accommodation** and **convergence**.

Near response   The response to optics and apertures near the eye—resulting in over-**accommodation**. See **Instrument myopia.**

Non-pupil-forming   Referring to an optical system analogous to a simple magnifier without an external **exit pupil**. This design is typically shorter than a **pupil-forming** configuration.

Operant conditioning   Involves the use of reinforcement and feedback to produce a trained response. Can be used to train **EEG self-regulation.**

Orthoscopy   The characteristic of a display system that maintains exact correspondence with human visual perceptions of size, perspective, and depth.

Panum's (fusional) area   Objects appear single over a range of horizontal and vertical **binocular disparities.** This range is referred to as Panum's Area.

Parallax   A linear measurement, parallel to the baseline of the eyes, of the distance between homologous points of a **stereoscopic** display.

Parvocellular pathway   Pathway of the human visual system that conveys information about fine spatial detail and color. Phylogenetically, this is a more recent pathway than the **magnocellular pathway.**

Phoria (heterophoria)   The alignment of the eyes in the absence of **binocular fusion,** when each eye sees a different or dissociated image. Primarily used as a clinical evaluation of binocular imbalance.

Photopic   **Luminance** levels that allow the discrimination of color vision because the cone photoreceptors operate at these levels—usually daylight levels of illumination.

Pixel   The smallest individually addressable unit of an image that can be rendered or displayed.

Posterior   The position of a landmark, head, or body segment that is behind or in back of a referenced point.

Proximal stimulus   Information about a distal stimulus that reaches the receptors, such as a visual image on the retina or sound at the ears.

Pupil (and pupil diameter)   The eye aperture bounded by the iris. The diameter of the pupil, which depends on both the amount of light entering the eye and on the state of the autonomic nervous system (e.g., stress and emotion), may constrict to less than 2 mm and dilate to 7 mm.

Pupil-forming   Referring to an optical system analogous to a compound microscope in which an internal stop is imaged in space. This design is typically longer and heavier than a **non-pupil-forming** design but gives the designer greater flexibility to locate components on the head.

Pupillary axis   A line joining the center of curvature of the cornea to the center of the entrance pupil.

Reflection coefficient   Percentage of a beam of light that is reflected from a optical surface.

Relative size   A potential monocular cue to distance. An object farther away will have a smaller proximal (retinal) image size than that given object viewed from a closer distance.

Resolution   A measure of display or **image quality** involving the number of displayed or resolvable elements per unit visual angle. The resolution of an electronic display may be expressed as spot size, **pixel** size, line width, trace width, number of television lines, and **MTF.**

Rotational difference or alignment   Relative rotation of the left and right image of a **binocular** display. Small amounts of rotation can be tolerated by the observer due to sensory fusion and torquing of the eyes.

Scan data   An electronic image of the surface of an object such as a helmet system or the human head.

Scotopic   **Luminance** levels that do not allow the discrimination of color because only the rod photorecetors operate at these levels—usually dim levels of illumination as at nighttime.

See-through HMD   An HMD that uses **beamsplitters** to optically combine synthetic imagery with the real world, or that electronically combines synthetic with video imagery. See **Augmented reality.**

Sellion   The point of the deepest depression of the nasal bones at the top of the nose. This is not palpated or marked on the subject but is located by visual inspection during the computerized landmarking procedure when the data set is turned to a right profile.

Simulator sickness   Disorientation, dizziness, and nausea resulting from simulators that use visually-induced motion—possibly resulting from a mismatch between the visual and vestibular inputs. May also be referred to as **cybersickness.**

Single vision   The ability to discount small amounts of **binocular disparity** and see an object as single rather than double. See also **diplopia.**

Size constancy   The tendency for an object's perceived size to remain constant despite changes in the size of the retinal image of that object as viewing distance varies.

Size-distance invariance   The proximal (retinal) image of an object decreases as a function of the distance of that object from the nodal point of the eye.

Spatial resolution   The spatial resolution of a display is often specified in terms of the number of pixels in the horizontal and vertical direction (i.e., the distal stimulus). However, this spatial resolution could be given in terms of seconds or minutes of visual angle (i.e., the proximal stimulus), which partially depends on the height and width of the display as well as on the distance of the observer from the display.

Steady-state visual evoked response (SSVER) Exogenous brain response to visual stimulation (e.g., sinusoidally modulated light) delivered at a rate that produces successive response overlap. Can be used as a control signal in a **brain-actuated control** interface.

Stereogram A pair of pictures that produce **stereopsis** when viewed through an appropriate stereoscopic display device.

Stereopsis Binocular depth perception which results from binocular parallax (each eye sees a slightly different view of close objects) created by the horizontal separation of the two eyes (**interpupillary distance**).

Stereoscopy The technology of artificially inducing **steropsis** using optical, photographic, or electronic means.

Superior The position landmark that is located above a particular reference or closer to the top of the head.

Telepresence Enabling a operator to feel present at remote location. Telepresence be facilitated by an **HMD** with a **WFOV** and **head tracking**. See also **Immersion.**

Temporal resolution Capability of the physical display to temporally modulate the image.

Temporal sensitivity Ability of observer to detect temporal change (e.g., of a temporally modulated **contrast** of a visual sinewave grating). The lower the **contrast** at which temporal modulation can be detected, the more sensitive the observer is at that rate of modulation (usually measured in Hertz).

Texture gradient A form of perspective in which the density of a surface's texture increases with distance, providing information about the slant of the surface. The rate at which a light (or a sound) varies over time—usually measured in cycles per second (Hertz).

Three-dimensional (3D) surface scanning Optically acquiring geometry of the surface of an object such as the human head in electronic form.

Tiling Displaying a large image using adjoining discrete displays. In a **binocular HMD,** a central display can be binocular and adjoining display tiles can be monocular.

Tragion The superior point on the junction of the cartilaginous flap of the ear (forward of the ear canal) with the head.

Transmission coefficient Percentage of a beam of light that is transmitted through an optical surface.

Transmission efficiency In a optical system, the percentage of light emitted by the image source that is observed at the eye.

Update rate Rate at which an image can be changed or updated, usually expressed in Hertz (or cycles per second). The update rate is limited by, but is not the same as, the **frame rate** of the display.

Virtual environment   A synthetically created environment that usually consists of multimodal inputs (e.g., visual, auditory, and haptics) and that is interactive.

Ventral   Referring to the front half of the head or body that has been divided into front and back halves by a plane.

Vergence   Coordinated turning inward (**convergence**) or turning outward (**divergence**) of the eyes to view near or far objects.

Vestibulo-ocular reflex (VOR)   The VOR acts to stabilize the visual world by generating compensatory **eye movements** in response to head motion.

Visual acuity   A measure of the smallest visual detail that can be seen in standardized high-**contrast** characters.

Visual angle   The angle subtended at the eye (pupil plane or first nodal point) by an object. The magnitude of the angle depends on both the physical extent of the visual object and the distance of that object from the nodal point of the eye's lens.

Visual fatigue   Also referred to as eyestrain and discomfort. An ill-defined group of symptoms relating to eyestrain and discomfort of the eyes.

Wide field of view (WFOV)   A display **field of view** of sufficient extent to support sensations of **immersion** or **telepresence**.

Zygion   The lateral point on the zygomatic arch. It is located by palpation and by using a spreading caliper.

Index

ABOUT THE AUTHORS

James E. Melzer is a Senior Project Engineer at Kaiser Electro-Optics. He has 14 years of experience in optical and displays engineering, and is an expert in display design for head-mounted systems, aviation life-support, and user interface. Mr. Melzer has also been awarded three patents in head-mounted display design. Kirk Moffitt, Ph.D., is a Human Factors Specialist in Visual Displays, also at Kaiser. He has extensive background in software development involving hypertext, prototyping and the user interface, data analysis, and process control. Dr. Moffitt holds a patent for a method to improve image quality in binocular HMDs.